Endocrine Disruption in Invertebrates: Endocrinology, Testing, and Assessment

Other publications by the Society of Environmental Toxicology and Chemistry (SETAC):

Reproductive and Developmental Effects of Contaminants in Oviparous Vertebrates
Di Giulio and Tillitt, editors
1999

Multiple Stressors in Ecological Risk and Impact Assessment
Foran and Ferenc, editors
1999

Linkage of Effects to Tissue Residues: Development of a Comprehensive Database for Aquatic Organisms Exposed to Inorganic and Organic Chemicals
Jarvinen and Ankley
1999

Ecotoxicology and Risk Assessment for Wetlands
Lewis, Mayer, Powell, Nelson, Klaine, Henry, Dickson, editors
1999

Uncertainty Analysis in Ecological Risk Assessment
Warren-Hicks and Moore, editors
1998

Ecotoxicological Risk Assessment of the Chlorinated Organic Chemicals
Carey, Cook, Giesy, Hodson, Muir, Owens, Solomon, editors
1998

Sustainable Environmental Management
Barnthouse, Biddinger, Cooper, Fava, Gillett, Holland, Yosie, editors
1998

Ecological Risk Assessment Decision-Support System: A Conceptual Design
Reinert, Bartell, Biddinger, editors
1998

Principles and Processes for Evaluating Endocrine Disruption in Wildlife
Kendall, Dickerson, Giesy, Suk, editors
1998

For information about any SETAC publication, including SETAC's international journal *Environmental Toxicology and Chemistry*,
contact the SETAC Office nearest you:

1010 N. 12th Avenue
Pensacola, Florida, USA 32501-3367
T 850 469 1500 F 850 469 9778
E setac@setac.org

Avenue E. Mounier 83, Box 3
1200 Brussels, Belgium
T 32 2 772 72 81 F 32 2 770 53 83
E setac@ping.be

http://www.setac.org

Endocrine Disruption in Invertebrates: Endocrinology, Testing, and Assessment

Edited by

Peter L. deFur
Virginia Commonwealth University
Center for Environmental Studies, Richmond, VA, USA

Mark Crane
Royal Holloway
University of London, Egham, Surrey, UK

Christopher G. Ingersoll
U.S. Geological Survey
Columbia, MO, USA

Lisa J. Tattersfield
Zeneca Agrochemicals
Bracknell, Berkshire, UK

Proceedings of the Workshop on
Endocrine Disruption in Invertebrates: Endocrinology, Testing, and Assessment
12–15 December 1998
Noordwijkerhout, The Netherlands

SETAC Technical Publications Series

SETAC Liaison
Greg Schiefer
SETAC/SETAC Foundation

Current Coordinating Editor of SETAC Books
C. G. Ingersoll
U.S. Geological Survey

Publication sponsored by the Society of Environmental Toxicology
and Chemistry (SETAC) and the SETAC Foundation for Environmental Education

Cover by Michael Kenney Graphic Design and Advertising
Copyediting and typesetting by Wordsmiths Unlimited
Indexing by IRIS

Library of Congress Cataloging-in-Publication Data

Endocrine disruption in invertebrates : endocrinology, testing, and assessment / edited by Peter L. deFur ... [et al.].

p. cm. -- (SETAC technical publications series)

"Proceedings from the International SETAC Workshop on Endocrine Disruption in Invertebrates: Endocrinology, Testing, and Assessment, 12–15 December 1998, Noordwijkerhout, The Netherlands."

Includes bibliographical references.

ISBN 1-880611-27-9 (alk. paper)

1. Invertebrates--Effect of chemicals on--Congresses. 2. Endocrine toxicology--Congresses. I. deFur, Peter L. 1950- II. International SETAC Workshop on Endocrine Disruption in Invertebrates: Endocrinology, Testing, and Assessment (1998 : Noordwijkerhout, Netherlands) III. Series.

QL364.E48 1999
573.4'9512--dc21

99-045687
CIP

International Standard Book Number 1-880611-27-9
Printed in the United States of America
06 05 04 03 02 01 00 99 10 9 8 7 6 5 4 3 2 1

♾ The paper used in this publication meets the minimum requirements of the American National Standard for Information Sciences—Permanence of Paper for Printed Library Materials, ANSI Z39.48-1984.

Reference Listing: deFur PL, Crane M, Ingersoll CG, Tattersfield LJ, editors. 1999. Endocrine disruption in invertebrates: Endocrinology, testing, and assessment. Workshop on Endocrine Disruption in Invertebrates: Endocrinology, Testing, and Assessment; 1998 Dec 12–15; Noordwijkerhout, The Netherlands. Published by the Society of Environmental Toxicology and Chemistry (SETAC). 320 p.

The SETAC Technical Publications Series

The SETAC Technical Publications Series was established by the Society of Environmental Toxicology and Chemistry (SETAC) to provide in-depth reviews and critical appraisals on scientific subjects relevant to understanding the impacts of chemicals and technology on the environment. The series consists of books on topics reviewed and recommended by the SETAC Board of Directors and approved by the Publications Advisory Council for their importance, timeliness, and contribution to multidisciplinary approaches to solving environmental problems. The diversity and breadth of subjects covered in the series reflects the wide range of disciplines encompassed by environmental toxicology, environmental chemistry, and hazard and risk assessment. Despite this diversity, these volumes share the goal of presenting the reader with authoritative coverage of the literature. In addition, paradigms, methodologies and controversies, research needs, and new developments specific to the featured topics are presented in these volumes.

The SETAC Technical Publications are useful to environmental scientists in research, research management, chemical manufacturing, regulation, and education, as well as to students considering or preparing for careers in these areas. The series provides information for keeping abreast of recent developments in familiar subject areas and for rapid introduction to principles and approaches in new subject areas.

Contents

List of Figures

List of Tables

Acknowledgments

The EDIETA Steering and Organizing Committee wish to thank the contributors listed below, whose funding made this workshop possible, and all the participants without whose expertise and scientific knowledge this workshop would not have been possible. We give a well-deserved thank you to the Work Group Chairs and Reporters, as well as to the Plenary Chair and Reporter; these individuals did an excellent job under significant time constraints. We also wish to thank the SETAC Offices in Europe and North America for the logistical support to conduct the workshop. The participants appreciate the cooperation and assistance of colleagues who provided references, perspectives, and feedback, as well as the fine research that made this effort possible. Finally, we thank the 4 peer reviewers, Gary Ankley, Milton Fingerman, Michael Depledge, and John Walker, whose comments and suggestions improved this publication.

Contributors to the Workshop on Endocrine Disruption in Invertebrates: Endocrinology, Testing and Assessment (EDIETA):

CMA	Chemical Manufacturers Association, Arlington, VA, USA
ECETOC	European Centre for Ecotoxicology and Toxicology of Chemicals, Brussels, Belgium
EC DGXII	European Commission DGXII, Brussels, Belgium
ECPA	European Crop Protection Association, Brussels, Belgium
EMSG	Endocrine Modulators Steering Group, a CEFIC Task Force, Brussels, Belgium
Exxon	Exxon Chemical Company, Houston, TX, USA
MAFF	Ministry of Agriculture, Fisheries and Food, London, UK
OECD	Organization for Economic Cooperation and Development, Paris, France
RIKZ	Rijksinstituut voor Kust en Zee (National Institute for Coastal and Marine Management), Directorate General of Transport, Public Works and Water Management, The Hague, The Netherlands
UK DETR	United Kingdom Department of Environment, Transport and the Regions, London, UK
UK EA	UK Environment Agency, London, UK
USEPA	United States Environmental Protection Agency, Washington, DC, USA
USFWS	U.S. Fish and Wildlife Service, Washington, DC, USA
USGS	U.S. Geological Survey, Washington, DC, USA
VROM	Directorate General of Public Works and Water Management, Amsterdam, The Netherlands

Workshop Steering and Organizing Committee *

Nick DeOude *(ex officio)*
SETAC Europe
Brussels, Belgium

Kim Courjaret *(ex officio)*
SETAC Europe
Brussels, Belgium

Pamela M. Campbell[5]
Procter & Gamble Inc.
Toronto, Canada

Kathy Cameron
Department of the Environment, Transport and the Regions (DETR)
London, UK

Marie-Chantal Huet
Organization for Economic Cooperation and Development (OECD)
Paris, France

Peter Matthiessen[3]
Centre for Environment, Fisheries and Aquaculture Science (CEFAS)
Burnham-on-Crouch, UK

Mark Crane[4]
Royal Holloway
University of London
Egham, Surrey, UK

Peter L. deFur[2]
Virginia Commonwealth University
Richmond, VA, USA

Christopher G. Ingersoll[3]
U.S. Geological Survey (USGS)
Columbia, MO, USA

Gerald A. LeBlanc[3]
North Carolina State University
Raleigh, NC, USA

Greg Schiefer *(ex officio)*
SETAC North America
Pensacola, FL, USA

Rod Parrish *(ex officio)*
SETAC North America
Pensacola, FL, USA

Rosalind Rolland
Tufts University
Boston, MA, USA

A. Dick Vethaak
Rijksinstituut voor Kust en Zee (RIKZ)
Middleburg, The Netherlands

Ralph G. Stahl Jr.[1]
E.I. du Pont de Nemours and Company
Wilmington, DE, USA

Lisa J. Tattersfield[1]
Zeneca Agrochemicals
Berkshire, UK

[1] Steering / Organizing Committee co-chair
[2] Plenary chair
[3] Work group chairs
[4] Plenary reporter
[5] Group reporter

* These affiliations were current at the time of the workshop.

Workshop Participants and Contributing Authors*

Endocrinology Work Group

Richard P. Brown
Exxon Biomedical Science
East Millstone, NJ, USA

Pamela M. Campbell
Procter & Gamble Inc.
Ontario, Toronto, Canada

John Edwards
Pest Management Strategies
Sand Hutton, York, UK

Ernest S. Chang
Bodega Marine Laboratory
University of California
Bodega Bay, CA, USA

Joel R. Coats
Iowa State University
Department of Entymology
Ames, IA, USA

Peter L. deFur
Virginia Commonwealth University
Center for Environmental Studies
Richmond, VA, USA

Pieter den Besten
Rijksinstituut voor Integraal Zoetwaterbeheer en Afvalwaterbehandeling (RIZA)
Department of Chemistry and Ecotoxicology
Lelystad, The Netherlands

Tarlochan Dhadialla
Rohm & Haas Company
Spring House, PA, USA

Gerald A. LeBlanc
North Carolina State University
Department of Toxicology
Raleigh, NC, USA

Fumio Matsumura
University of California
Department of Environmental Toxicology
Davis, CA, USA

Lynn M. Riddiford
University of Washington
Department of Zoology
Seattle, WA, USA

Michael G. Simpson
University of Liverpool
School of Biological Sciences
Liverpool, UK

Terry W. Snell
Georgia Institute of Technology
School of Biology
Atlanta, GA, USA

Michael Thorndyke
Royal Holloway
University of London
School of Biological Sciences
Egham, Surrey, UK

Field Assessment Work Group

Toshihiro Horiguchi
National Institute for Environmemtal Studies
Regional Environmental Division
Ibaraki, Japan

Patricia Cameron
World Wildlife Fund (WWF), Germany
Marine and Coastal Division
Bremen, Germany

A. Dick Vethaak
Rijksinstituut voor Kust en Zee (RIKZ)
Department of Ecotoxicology
Middelburg, The Netherlands

Ian M. Davies
FRS Marine Laboratory
Aberdeen, UK

Zoe Billinghurst
University of Plymouth
Plymouth Environmental Research Centre (PERC)
Devon, Plymouth, UK

Peter Matthiessen
Centre for Environment, Fisheries and Aquaculture Science (CEFAS)
Burnham Laboratory
Burnham-on-Crouch, Essex, UK

David R. Mount
U.S. Environmental Protection Agency (USEPA)
Office of Research and Development (ORD)
Duluth, MN, USA

Jörg Oehlmann
IHI Zittau
Human- and Ecotoxicology Department
Zittau, Germany

Tom G. Pottinger
Institute of Freshwater Ecology
Windermere Laboratory
Far Sawrey, Ambleside, Cumbria, UK

Trefor Reynoldson
Environment Canada
Burlington, Canada

Helen Thompson
Central Science Laboratory (CSL)
Ministry of Agriculture, Fisheries and Food (MAFF)
Slough, Berks, UK

David W. Brassard
USEPA, USEPA Headquarters
Washington, DC, USA

Paul K. Sibley
University of Guelph
Centre for Toxicology
Guelph, Ontario, Canada

G. Thomas Chandler
University of South Carolina
Environmental Health Sciences
School of Public Health
Columbia, SC, USA

Ted DeWitt
USEPA, Coastal Ecology Branch
Newport, OR, USA

Stanley Dodson
University of Wisconsin
Department of Zoology
Madison, WI, USA

Andreas Gies
Umweltsbundesamt, Federal Environment Agency
Berlin, Germany

Christopher G. Ingersoll
U.S. Geological Survey (USGS) Biological Resources Division
Environmental & Contaminants Research
Columbia, MO, USA

Charles L. McKenney Jr.
USEPA, Gulf Ecology Division
Gulf Breeze, FL, USA

Eva Oberdörster
Clemson University
Environmental Toxicology Department
Pendleton, SC, USA

David Pascoe
Cardiff University
School of Biosciences
Cardiff, UK

Oliver Warwick
National Centre for Ecotoxicity and Hazardous Substances, Environment Agency
Evenlode House, Wallingford, Oxfordshire, UK

Ralph G. Stahl Jr.
E.I. du Pont de Nemours and Company
Corporate Remediation
Wilmington, DE, USA

Lisa J. Tattersfield
Shell Chemicals Ltd.
Shell Centre, London, UK

Laboratory Testing Work Group

Mark Crane
Royal Holloway
School of Biological Sciences
University of London
Egham, Surrey, UK

Marie-Chantal Huet
Organization for Economic Cooperation and Development (OECD)
Paris, France

Tom Hutchinson
Zeneca Ltd.
Brixham Environmental Lab, Freshwater Quarry
Brixham, Devon, UK

Donald J. Versteeg
Procter & Gamble Company
Environmental Science Department
Cincinnati, OH, USA

* These affiliations were current at the time of the workshop.

Preface

The planning for this workshop on invertebrate endocrine disruption began more than 2 years ago when environmental scientists around the world were engaged in a great discussion on endocrine disruption. All of the members of the steering committee and most, if not all, of the participants were so engaged in the summer of 1997. At that time, many of us in North America, Europe, Japan—and elsewhere, I am sure—were participants in workshops, conferences, and advisory committees on endocrine disruption in fish, birds, wildlife, and humans. The public discourse was quite heated and centered on human health conditions such as breast cancer and sperm count, and on the relationship between effects in wildlife and in humans. In the midst of this mixture of public and scientific debate, a number of scientists who later formed the Endocrine Disruption in Invertebrates: Endocrinology, Testing, and Assessment (EDIETA) Steering Committee stepped back from the discussion to evaluate the situation. We observed that all the ongoing debates and discussions omitted the majority of animals on earth—the invertebrates—and that this error should be corrected.

Quite independently, 2 groups formed in Europe and North America to begin planning a workshop to correct this huge omission. The 2 groups quickly learned of each other's activities, established a dialog, and reached common cause to carry out this workshop. Then began the arduous task of forming a steering committee whose membership spanned the Atlantic Ocean, blended cultures, and met the challenges of schedules and communications. Obviously the endeavor succeeded and did so with remarkable dispatch. The EDIETA Steering Committee Co-chairs, Lisa Tattersfield and Ralph Stahl, did a fabulous job of leading and coordinating this crew of scientists and deserve a great deal of credit for the success of this workshop.

All of us knew we had to work quickly to produce a volume that was both current and timely enough to inform the global initiatives on endocrine disruption. And we did. At this writing, we anticipate publication 10 months after the workshop! We also knew that we could not capture all of the relevant literature and that new research would come to light while we prepared this book. In fact, we include here some of our expectations of what scientists will find in the area of invertebrate endocrine disruption once the research effort increases.

Following our December 1998 workshop, important research results were, in fact, reported at international scientific meetings. I refer to the 2 meetings of which I am aware because I attended them. The first of these 2 meetings was the annual meeting of the Society for Integrative and Comparative Biology (SICB, formerly the American Society of Zoologists), held in January 1999 in Denver, CO. The abstracts appear in the *American Zoologist* (38[5]:1a–207a). The Division of Comparative Endocrinology papers include one on echinoderms (by K.M. Wasson, G.A Hines, and S.A Watts) showing a functional role for vertebrate reproductive hormones, and one by

X.L. Yu and D.L. Mykles reporting a new hormone that controls the timing of regeneration and molting in crustaceans. The second meeting, the Fifth International Congress of Comparative Physiology and Biochemistry, was held in August 1999 in Calgary, Alberta, Canada, while the editors read the first set of proofs for this book. The abstracts appear in *Comparative Biochemistry and Physiology* (August 1999, 124A[Suppl]:s1–s148). An excellent symposium presentation by N. Sherwood on evolutionary trends of gonadotropin-releasing hormone (GnRH) revealed that GnRh is found in 5 invertebrate phyla, including tunicates, mollusks, and coelenterates, and has a role in gonad function. I expect that presentations from other meetings are to be found in the literature.

These results provide further evidence of and insight into the evolutionary relationships among animal phyla. These relationships were also the subject of the poetry of an early zoologist, Walter Garstang, who wrote in the late 19th century on how ontogeny recapitulates phylogeny.

I leave you with his words:

"The Ancestry of Vertebrates"

Gill-slits, Tongue bars, Synapticulae,
Endostyle and Notochord: all these you will agree
Mark the Protochordate from the fishes in the sea,
And tell alike for them and us our lowly pedigree.

Thyroid, Thymus, Subnotochordal Rod:
These we share with Lampreys, the Dogfish and the Cod, —
Relics of the food-trap that served our early meals,
And of the Tongue-bars that multiplied the primal water-wheels.

Larval Forms (Oxford UK: Basil Blackwell, 1966)

Peter L. deFur
Virginia Commonwealth University
Richmond, VA, USA
September 1999

EXECUTIVE SUMMARY

Workshop on Endocrine Disruption in Invertebrates: Endocrinology, Testing, and Assessment (EDIETA)

Peter L. deFur, Mark Crane, Lisa J. Tattersfield, Christopher G. Ingersoll, Ralph G. Stahl Jr., Peter Matthiessen, Gerald A. LeBlanc

Scope and Objectives

The Workshop on Endocrine Disruption in Invertebrates: Endocrinology, Testing, and Assessment (EDIETA), held 12–15 December 1998 in Noordwijkerhout, The Netherlands, brought together more than 40 scientists from Europe, Japan, and North America. The objectives of the workshop were to

- increase the basic understanding of invertebrate endocrinology as it relates to endocrine disruption (ED),
- determine the applicability of invertebrates in testing for the ED mechanism, and
- determine the applicability of invertebrates as sentinels or monitors for potential environmental changes due to ED.

In order to address these objectives, the workshop participants were divided into 3 major work groups:

- endocrinology,
- laboratory testing, and,
- field assessment.

The workshop also identified data gaps and areas in which additional research is needed. This publication details workshop findings and suggests areas for future research.

Endocrine Disruption in Invertebrates: Endocrinology, Testing, and Assessment. Peter L. deFur et al., editors.

Introduction

Invertebrates account for roughly 95% of the known species of animals on our planet, yet our knowledge of their basic endocrinology is limited, and only recently have we begun to understand their value in alerting us to potential environmental ED. Among environmental toxicologists, there is widespread acknowledgment that invertebrates have been, and will continue to be, an important class of organisms for routine ecotoxicological testing and for monitoring environmental conditions. In this regard, they will be important in ecological risk assessment and in shaping environmental decisions in the future.

The U.S., Japan, Canada, Great Britain, and Germany are among the nations now embarking on programs to assess the phenomenon of ED in the environment. Several efforts recognize the need to include invertebrates. For these efforts to be meaningful, they must seek to include or develop

- detailed information on endocrinology,
- validated test methods that are available or might be adapted, and
- the best methods for assessing ED in exposed populations in the field.

Given the substantial level of effort already devoted to ED, and that this effort seems destined to increase worldwide, many agencies and scientists will find this publication to be a useful summary of the state of knowledge of ED in and concerning invertebrates.

Full details of the discussions, recommendations, and conclusions of the work groups addressing endocrinology, laboratory testing methods, and field assessment are given in Chapters 2, 3, and 4 respectively. Major conclusions are summarized below.

Endocrinology

A fundamental tenet of animal physiology is that the endocrine system functions to regulate endogenous biological processes in response to environmental or physiological stimuli. This basic endocrine strategy to regulate biological processes has been widely conserved across animal phyla. Nevertheless, the individual components intrinsic to the endocrine system have undergone significant evolutionary divergence, resulting in distinct differences in the endocrine systems of various taxa. Invertebrates represent more than 30 different phyla within the animal kingdom. Accordingly, components and strategies of the endocrine system of invertebrates are also quite diverse. Chapter 2 of this publication summarizes basic endocrinology of invertebrates for those phyla in which some aspects of the organism's endocrinology have been investigated.

Invertebrates use steroid hormones as well as peptide hormones and terpenoids, but by far the most common invertebrate hormones are the peptides. Invertebrates rely on peptides secreted by neurosecretory structures, cells and tissues of neurological origin. In general, many of the processes known to be under endocrine regulation in vertebrates (i.e., development, growth, maturation, reproduction, water, and ion balance) are also under endocrine regulation in invertebrates. In addition, processes unique to some taxa of invertebrates (i.e., molting, limb regeneration, diapause, pheromone production, pigmentation and color change, metamorphosis) are also under endocrine control.

Among the arthropods (insects, crustaceans, and some minor groups), ecdysone and related compounds, termed "ecdysteroids," are major endocrine regulators of molting, embryonic development, metamorphosis, reproduction, and pigmentation. Ecdysteroids have been reported in a few other invertebrate groups (i.e., annelids, mollusks); however, their function in these organisms has not been established.

In insects, juvenile hormone (JH) functions in concert with ecdysone to mediate different regulatory functions of ecdysone. Methyl farnesoate (MF) seems to function in crustaceans as JH does in insects. A wide variety of neuropeptides function as endocrine regulators in all invertebrate groups thus far evaluated. These neurohormones include molt stimulating or inhibiting hormones (arthropods), allotropins and allostatins (insects and others), diuretic and antidiuretic hormones (insects and others), cardioacceleratory peptides (mollusks, insects, and others), insulin-like peptides (mollusks), egg-laying hormones (mollusks), regeneration-stimulating hormones (annelids and crustaceans), and glycine-leucine tryptophan amides (GLWamides) or metamorphosis-stimulating hormones (coelenterates).

Neuropeptides may directly regulate an endocrine process or may modulate the production or secretion of the terminal hormone (i.e., ecdysone or JH). Androgenic hormone, a putative peptide product of the androgenic gland of some crustaceans, regulates sexual differentiation by stimulating the development of male primary and secondary sex characteristics and inhibiting the development of female sex characteristics.

Evidence suggests a role for vertebrate-type sex steroids in echinoderms and mollusks. Although these hormones have been detected in representatives of other invertebrate groups, their function remains equivocal.

The majority of the evidence for chemically induced ED in invertebrates stems from exposures to pesticides specifically designed to disrupt endocrine-regulated processes such as growth, metamorphosis, and molting. These chemicals act as ecdysone antagonists or JH analogs. Existing evidence indicates that the invertebrates are susceptible to the same modes of ED (i.e., hormone agonists and antagonists, modulation of endogenous hormone levels) as observed with vertebrates. The antifouling agent tributyltin (TBT) seems to function by altering normal androgenic mechanisms in gastropod snails that are afflicted by a condition known as "imposex" and described in Chapter 4.

Some chemicals act on invertebrate growth, reproduction, development, or molting, yet do not interact with hormonal processes, making clear identification of ED chemicals complex. Considering the diversity of endocrine processes and hormones in invertebrates, coupled with the current lack of knowledge of the endocrinology of many invertebrate phyla, the Endocrinology Work Group recommended against use of a single representative invertebrate for endocrine toxicity evaluations.

Laboratory Testing

The primary objective of the Laboratory Testing Work Group was to evaluate the extent to which methods for conducting toxicity tests with invertebrates in the laboratory can be used to assess endocrine-disrupting compounds (EDCs). None of the standard methods for conducting toxicity tests with invertebrates were designed specifically to evaluate effects associated with EDCs. However, many methods describe approaches that can be used to measure endpoints that may be responsive to effects of EDCs (e.g., development, growth, reproduction). As such, these methods are integrative of potential adverse effects, irrespective of mechanisms, and such tests represent a greater level of potential to determine ecological risk than does mechanistic evidence of endocrine activity *per se*. Toxicity testing alone, whether acute or chronic, cannot be used to identify the specific effects or identity of EDCs. To do this, it is necessary to measure changes in hormone levels, receptor binding, or some other marker or specific indicator of hormonal function.

Sufficient information is outlined in existing publications to summarize methods for conducting toxicity tests with representatives of cnidarians, turbellarians, mollusks, annelids, mites, millipedes, centipedes, crustaceans (amphipods, cladocerans, isopods, fairy shrimp, copepods, mysids, mud crabs, grass shrimp, lobsters, crayfish, barnacles), insects (midges, other aquatic insects, and terrestrial insects), echinoderms, and tunicates. We were unaware of information that could be used to adequately describe methods for conducting toxicity tests with sponges, jellyfish, anemones, bryozoans, leeches, spiders, sea stars, brittle stars, or sea cucumbers.

Additional endpoints that could be measured in laboratory tests and may provide information indicative of ED might include

1) sex ratio (of adults or offspring),
2) mating success,
3) production of ephippia (resting eggs in cladocerans),
4) egg hatching success,
5) viability of the offspring,
6) morphological abnormalities,
7) molting frequency and timing,
8) time to emergence or pupation,

9) metamorphic or larval developmental success, or
10) pigmentation.

A variety of surrogate organisms could be used with limited methods development to conduct partial or full life-cycle exposures in the laboratory (including transgenerational exposures). However, there is limited knowledge of the endocrinology controlling molting, growth, reproduction, or sexual determination for most of these species. Insects and crustaceans are the only groups of organisms to the best of our knowledge where there are surrogates available for life-cycle testing with a relatively well-understood endocrine system. An important research need is to determine whether surrogates of a particular group (e.g., *Chironomus* spp.) are representative of the range of responses in a larger group (e.g., insects).

There is a need to validate responses of a number of species and endpoints with a diverse set of reference compounds likely having an ED mode of action, or known EDCs (candidate reference compounds are recommended in Chapter 3). Based on the data obtained from these validation tests, the most useful of these methods should be standardized for routine assessment and screening of EDCs.

Field Assessment

The field assessment chapter addresses the following issues:

1) cases where ED has been identified as causing effects in either individuals, populations, or communities in the field;
2) practical methods for examining effects in the field at community, population, and individual levels; and
3) new areas or methods to be researched.

Data exist that link population-level effects in terrestrial arthropods to insect endocrine disrupters. This is unsurprising, as several pesticides have been developed specifically to act on insect endocrine systems, and effects on development in nontarget species have been identified. In marine environments, the only clear example of cause and effect is that of TBT in mollusks, although other examples exist by inference (e.g., in crustaceans) but have not been categorically demonstrated. In fresh water, many effects have been observed that could involve endocrine systems, (e.g., alterations in behavior, growth, and reproduction), but the mechanisms have not been demonstrated. In addition to the examples cited above, it is likely that ED is occurring more widely in invertebrates and that the absence of many examples is primarily because the appropriate fieldwork, coupled with basic endocrine function, has yet to be done.

Existing biological monitoring programs do not generally incorporate information on variables that are indicative of ED in invertebrates (nor do they generally provide diagnostic information). In freshwater, marine, and terrestrial ecosystems, monitoring programs are based almost entirely on the collection of structural data on populations and communities. Clearly there is a need to develop field monitoring programs and suitable biomarkers that can provide information on whether ED effects are occurring. Recommendations for research on suitable methods have been made in Chapter 4.

However, field assessments alone will never be able to confirm a mechanism as ED, the determination of which requires supportive experimental work on both endpoints and mechanisms. Nevertheless, the disadvantage of field studies with regard to their diagnostic capability is balanced by the advantage of being the true measure of ecological conditions.

Conclusion

The output from the EDIETA Workshop is a balanced attempt, based on current knowledge, to determine the potential for the use of invertebrates for the testing or field assessment and monitoring of ED. The current state of our knowledge is presented in this publication. It is recognized that there are substantial gaps in our knowledge, and further research is recommended to address some of these issues. The invertebrates will offer a wealth of knowledge in understanding comparative and ecological aspects of ED. For these reasons, we believe invertebrate systems need to be a high priority for research, screening and testing, and methods development. The conclusions and recommendations of the EDIETA Workshop provide input for the development of research and regulatory programs that should be considered by international and national bodies addressing ED in relation to invertebrates.

CHAPTER 1

Introduction to the Workshop on Endocrine Disruption in Invertebrates: Endocrinology, Testing, and Assessment

Ralph G. Stahl Jr., Lisa J. Tattersfield, Pamela M. Campbell, Toshihiro Horiguchi, Peter L. deFur, A. Dick Vethaak

Introduction

Scope

The introduction provides a synopsis of the genesis of the Workshop on Endocrine Disruption in Invertebrates: Endocrinology, Testing, and Assessment (EDIETA) and the rationale behind it. It details the workshop and work group purposes, as well structural and functional elements of the workshop and this book.

Purpose and Objectives of the Workshop

Invertebrates account for about 95% of the known species of animals on earth (Barnes 1980), yet knowledge of their basic endocrinology is limited, and only recently have scientists begun to understand their value in signaling potential environmental endocrine disruption (Alvarez and Ellis 1990; Crisp et al. 1997; LeBlanc and Bain 1997). With this in mind, the broad purposes of this workshop were to increase scientific understanding of the use of invertebrates in testing for endocrine disruption (ED), as sentinels for potential environmental changes stemming from exposure to endocrine-disrupting chemicals (EDCs), and to summarize and perhaps thus improve the understanding of invertebrate basic endocrinology.

The specific objectives for the workshop were to

1) review the current knowledge of invertebrate endocrine systems and the factors that can cause disruption of important endocrine processes in invertebrates, focusing on unique features and common elements (Chapter 2);

Endocrine Disruption in Invertebrates: Endocrinology, Testing, and Assessment. Peter L. deFur et al., editors.

2) review methods for conducting laboratory tests with invertebrates and identify their potential usefulness in detecting ED, suggest practical modifications to improve existing tests where necessary, and identify potential new tests (Chapter 3); and
3) review the use of invertebrates for environmental monitoring in aquatic and terrestrial systems, and determine whether species and endpoints currently being used are appropriate for the detection of ED (Chapter 4).

To address these 3 objectives, more than 40 scientists from Europe, Japan, and North America, representing the academic, business, nonprofit, and governmental sectors, were invited to participate in the workshop. The workshop was held 12–15 December, 1998, in Noordwijkerhout, The Netherlands. Participants were divided among 3 work groups:

1) endocrinology;
2) laboratory testing, and
3) field assessments.

A schematic of how the various work groups integrated across the ED issues is shown in Figure 1-1.

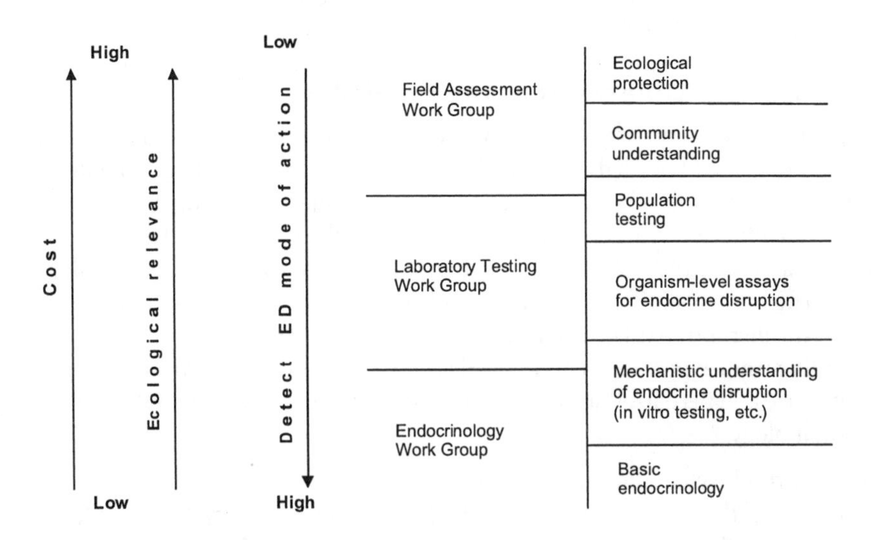

Figure 1-1 Integration of the Endocrinology, Laboratory Testing, and Field Assessment Work Groups at the EDIETA Workshop

Reasons for the Workshop

This workshop on invertebrates was held for diverse reasons:

1) Two previous workshops, one in North America (Ankley et al. 1998), the other in Europe (Holmes et al. 1997), focused on ED in vertebrate wildlife. Both workshops identified invertebrates as important organisms to address but left the task to others. The current workshop took up that task.
2) Invertebrates have been and will continue to be an important class of organisms for routine ecotoxicological testing (ECETOC 1996) and for monitoring environmental conditions (LeBlanc and Bain 1997). Thus, they will be important in ecological risk assessment and in shaping environmental decisions in the future.
3) The diversity and abundance of invertebrates far exceeds that of vertebrates; hence, focusing on invertebrates means addressing the largest and most widespread group of animals known.
4) There is a high level of uncertainty remaining in the understanding of ED in wildlife, particularly with respect to endocrine-regulated processes in the invertebrates (Colburn 1995; Kavlock and Ankley 1996; Campbell and Hutchinson 1998; Stahl and Clark 1998). This workshop was designed to reduce uncertainty with respect to ED in invertebrates.
5) In Europe, Japan, and North America, government and other regulatory groups have either begun, or plan to begin, environmental studies of major aquatic ecosystems for EDC, and scientists anticipate that invertebrates will be important in those efforts (Ankley et al. 1997; Crisp et al. 1997; Langer and Sang 1997; Tattersfield et al. 1997).
6) A number of legislative initiatives (i.e., U.S. Food Quality Protection Act of 1996; Amendments to the U.S. Safe Drinking Water Act of 1996) mandate the development of testing regimes and/or monitoring programs for ED in a wide variety of organisms, including wildlife. As noted above, invertebrates may become important components of these testing and monitoring programs.
7) Numerous invertebrates are highly valued by humans for commercial or recreational purposes, and it is important to understand the factors that might influence long-term population viability of these valued resources.
8) Many aquatic and terrestrial invertebrates serve important positions in the food web of vertebrate wildlife, and their loss could jeopardize the viability of a diverse array of organisms.

With these broad objectives, it was impossible to be comprehensive in all aspects of the workshop. The perspectives and experiences the participants brought to the workshop, and the ensuing scientific discussions, heavily influenced the recommendations and conclusions. Perhaps others will see areas highlighted by this effort that are in need of further study and will, in due course, take the initiative to address them.

Endocrine Disruption in Invertebrates

Hormonal regulation of biological functions is common to both vertebrates and invertebrates. While a basic endocrine strategy to regulate biological processes has been widely conserved, the individual components intrinsic to the endocrine system have undergone significant evolutionary divergence, resulting in distinct differences in the endocrine systems of various taxa (Chapter 2). Through evolution, species of invertebrates have developed a diversity of life histories and hence a multitude of unique approaches to growth, development, and reproduction. These approaches include processes of metamorphosis, diapause, and pupation that are life history traits not evident in most vertebrates, even though larval fish and amphibians undergo metamorphosis during maturation. The regulation of these processes by the neuroendocrine system in invertebrates is therefore considerably more diverse than that found in the vertebrates. Unfortunately, with a few exceptions that are not recent (Highnam and Hill 1977; Laufer and Downer 1989), the endocrine system of invertebrates has not been documented comprehensively.

Vertebrate-type hormones have been detected in a number of invertebrate phyla (Voogt et al. 1985; de Loof and de Clerck 1986; Hines et al. 1992) however, firm evidence of their role in the endocrine system of invertebrates is, for most phyla, still lacking (Pinder and Pottinger 1998). In general, invertebrate endocrine systems have not been as well investigated as those of vertebrates, nor have responses of invertebrate endocrine systems been as well studied with respect to ED (Fox 1992; Colburn et al. 1993; Baldwin et al. 1995). One reason for the lack of comprehensive study may be that the diversity in life histories of aquatic and terrestrial invertebrates presents a challenge to those scientists wishing to use these organisms in endocrine studies, or to survey feral populations for endocrine-type effects (Donker et al. 1993).

Many aquatic and terrestrial invertebrates have complicated life histories, display various forms of hermaphrodism (Gorbman and Davey 1991), and in some cases, display poorly resolved sexual dimorphism (Sellmer 1967; Eckelbarger 1974). Their reproductive cycle can be highly complex (Sastry 1968, 1970) and controlled by various environmental stimuli, including light intensity, temperature, desiccation, and diet (Ansell and Trevallion 1967; Largen 1967; Copeland and Bechtel 1974; Young 1978; Tessier et al. 1983). Determining whether a chemical can impact endocrine-driven functions in a feral population of invertebrates, or other wildlife for that matter, is difficult and may require a prolonged and in-depth understanding of their life history, morphology, and the influence of the local environmental conditions (Stahl and Clark 1998). There is evidence, however, that some invertebrates can respond to extremely low levels of behavior-modifying biochemical stimuli originating from nearby predators or similar organisms, and this behavior/response may be partly controlled by the endocrine system (Hidu 1969; Dodson and Hanazato 1995).

One of the challenges to scientists working on ED is determining whether a chemical is indeed an endocrine disrupter. Assays will need to distinguish chemicals that interact with the endocrine system to provoke some adverse response (i.e., ED) from environmental conditions (including chemicals) that affect or inhibit hormonally regulated physiological processes via some other mechanism. Such events can be misinterpreted as ED if mechanistic links to effects on the underlying hormonal processes are not made. This type of situation underscores 2 points: the need to know the underlying mode of action of the putative endocrine-disrupting chemical in order to identify it as an ED, and that adverse effects on reproduction and development can also occur via mechanisms distinct from hormonal mediation.

The most well-known invertebrate endocrine system is that of the insects, owing to their economic, ecological, and agricultural importance. Among aquatic invertebrates, one of the best-studied endocrine systems is that of crustaceans, particularly the decapods. Many species are important commercially and recreationally, which may have stimulated many of the studies.

Identification of the hormones controlling insect growth and development has permitted the synthesis of compounds designed to mimic, block, or otherwise interact with normal hormone systems (i.e., ED) of the targeted insect. Juvenile hormone has been the basis of some insecticides that attempt to mimic the hormone and disrupt normal maturation of the insect (Slama 1971; Gorbman and Davey 1991; Jepson 1993; Lagadic et al. 1994; Nimmo and McEwen 1994).

Some natural compounds have estrogenic or other hormonal properties. Given the diversity of physiological and morphological changes that occur during an insect's lifetime, it can be difficult to distinguish endocrine-type effects caused by chemicals from those caused by naturally occurring materials such as phytoestrogens. There are, however, examples when endocrine-type effects caused by chemicals have been distinguished from those caused by non-anthropogenic materials. Tributyltin (TBT)-induced imposex and intersex in gastropod mollusks provides some of the strongest evidence for the occurrence of ED in the field (Gibbs et al. 1988). In addition, as discussed above, chemicals have been purposefully synthesized to disrupt the endocrine system of a number of insects in order to reduce risk of disease or destruction of food crops.

The examples above illustrate that invertebrates are indeed susceptible to ED, and attention must be paid to developing methodology to both test chemicals for ED in invertebrates in the laboratory (Chapter 3) and to monitor effects in the field (Chapter 4). This need has been recognized by a number of national and international regulatory authorities, and plenary presentations, given at the workshop, on current perspectives in this arena are summarized below.

Endocrine Disruption in Invertebrates and Regulatory Perspectives from Europe, Japan, and North America

Opening plenary remarks

The recent interest in the science and policy of chemicals that alter hormone systems in animals (endocrine disrupters) began in the early 1990s after reports of reproductive and developmental problems in wildlife were linked to previous chemical exposures (Colborn and Clement 1992). Laboratory and field evidence indicated that certain chemicals (notably polychlorinated biphenyls [PCBs], and dichlorodiphenyltrichloroethane [DDT], among others) altered normal reproduction and fetal or egg development in fish, birds, and wild mink (Gilbertson and Fox 1977; Rolland et al. 1997; see also Kendall et al. 1998). At about the same time, researchers studying the effects of certain chemicals, especially diethylstilbestrol (DES) (e.g., McLachlan et al. 1992), on human health, realized the 2 groups of scientists were investigating similar questions in different animal systems. The resulting scientific interest generated a number of symposia, conferences, and research papers on the subject (see Kavlock et al. 1996; Ankley et al. 1998; Kendall et al. 1998; Rolland et al. 1998).

The resulting publications demonstrate that the interest in ED focused on vertebrates, with no small attention to human health effects. In all the discussion of scientific evidence, case studies, and research planning, little if any mention was made of the invertebrates. Despite the number and variety of invertebrate animals, and the importance of such animals as insects and crustaceans, the focus of even SETAC's efforts was on vertebrates. In Kendall et al. (1998), only a few brief paragraphs in the entire volume raise examples from the invertebrate literature.

More important, much of the discussion seemed to reflect a rather startling (at least disappointing) level of knowledge and awareness of invertebrate endocrinology and even general biology (see, e.g., Endocrine Disrupter Screening and Testing Advisory Committee [EDSTAC] 1998). This lack of knowledge and awareness is even more disconcerting given that the insect endocrine system is quite well understood and has been for some time. And there are several excellent examples of ED in invertebrates taken from both the general environmental toxicology literature and from the development of insecticides that intentionally affect insect hormones to focus on target species.

Europe

In a number of European countries (e.g., Germany, the UK, The Netherlands), a significant pattern of regulatory and research activity is emerging. A number of regulatory agencies in the various countries are discussing plans for implementing ED-assessment programs in water bodies, some of which are likely to utilize fish and invertebrates as target species. In the UK, for example, a program of research (which is a joint initiative of the Department of Environment, Transport and the Regions [DETR]; the Ministry of Agriculture, Fisheries and Food [MAFF]; the Environment Agency; the Scotland and Northern Ireland Forum for Environmental Research [SNIFFER]; and the European Chemical Industry Council) has been initiated to investigate whether there is evidence of changes in the reproductive health of marine life and, if so, to seek and identify possible causes and potential impacts on populations. This program, Endocrine Disruption in the Marine Environment (EDMAR), will focus on fish and invertebrates, in particular marine crustaceans.

The European Community Program of Research on Environmental Hormones and Endocrine Disrupters (COMPREHEND) will assess the extent of ED in wildlife on a Europe-wide basis. It will focus on assessment of a range of domestic and industrial effluents for estrogenic activity, including chemical analysis, impact studies on fish and the applicability of a 3-generation life-cycle test using the harpacticoid copepod *Nitrocera spinipes.*

In The Netherlands, a large-scale 3-year baseline study entitled "National Investigation into Estrogenic Compounds in the Aquatic Environment" (LOES) is currently being carried out. The project will assess, among others, the estrogenic and reproductive impact of chemicals on marine and freshwater fish and on mussels.

Japan

In Japan, the Japanese Environment Agency (JEA) has been actively studying the impact of TBT on several species of gastropods inhabiting coastal waters. These data have shown a clear correlation between the use of TBT and an increased incidence of imposex in these organisms. In response, the JEA has reassigned a substantial amount of its funds to conduct further studies of ED in invertebrates and other wildlife. The content of these research programs is currently in development.

Canada

In Canada, various government departments have been actively assessing the endocrine disrupter issue, resulting in regulatory changes and the allocation of government research funds to further understand this issue.

The Canadian Environmental Protection Act (Bill C32, CEPA) is currently being revised and, at the time of this writing (April 1999), was due to be reported to the House of Commons by the Standing Committee on Environment and Sustainable Development. This updated Act contains a definition of endocrine disruption in the research section (Section 3) of the bill, and directs the Canadian government to conduct research in this area. Prior to this, the Canadian government had already recognized the emerging endocrine issue and had established working groups to assess the issue and the appropriate government response. One of these groups was the "4NR" and consisted of the 4 government agencies dealing with natural resource issues.

Environment Canada has conducted monitoring programs to assess the estrogenicity of sewage effluents in Canada, and is currently focusing its efforts on assessing the impact of hormone inputs to soil and water systems from agricultural practices, e.g., hog farm manure and runoff. In addition, both Health Canada and Environment Canada are active participants in the Organization for Economic Cooperation and Development (OECD) Endocrine Disrupter Testing and Assessment Working Group, which is focused on developing a screening and testing strategy for EDCs.

The Toxic Substances Research Initiative (TSRI), which was recently launched in Canada, is designed to enhance Canadian environmental and health science capacity by providing $40 million for research for the fiscal years 1998–2002. This initiative, administered by Health Canada and Environment Canada, provides research funds for 5 priority research areas, one of which is to "Accelerate research activities designed to establish an adequate understanding of the implications of endocrine-disrupting chemicals."

Interestingly, a recent public survey found that about 25% of the Canadian public had some awareness of the endocrine disrupter issue. Hence it appears that in Canada, as in the U.S., the endocrine disrupter issue will continue to attract significant scientific, regulatory and public interest for some time.

United States

In the U.S., the U.S. Environmental Protection Agency (USEPA) and other federal agencies have been putting together plans for both internal and extramural research (Ankley et al. 1997; Crisp et al. 1997). A major endeavor is not only to understand the mechanisms of ED in a wide variety of mammalian and wildlife species, and to augment existing testing schemes to account for the ED endpoint, but also to develop simple, rapid, inexpensive in vitro screening tests that can be used to test a large number of chemicals. There is evidence both from a regulatory and legislative standpoint that the issue of ED in humans, mammals, and wildlife will remain an important element of the public debate in the U.S. for some time (Daston et al. 1997).

Terminology and Summary of Major Findings

Terminology

The workshop was not intended to develop yet another definition of ED, even one specific to invertebrates, nor was it intended to review and attempt to revise existing definitions. Limited time made it impossible to engage in a discussion of the many nuances and meanings related to ED. It was recognized, however, that the definition of ED was important and would be helpful in focusing discussions among participants. The participants generally agreed that an ED is a substance that interacts with the endocrine system and alters some normal biological processes; the definition from the workshop in Weybridge UK (Holmes et al. 1997) was used as a practical working definition:

"An endocrine disrupter is an exogenous substance that causes adverse health effects in an intact organism, or its progeny, secondary to changes in endocrine function. A potential endocrine disrupter is a substance that possesses properties that might be expected to lead to ED in an intact organism."

It was noted that some disruptive effects caused by a chemical may mimic those due to ED but are not a result of direct endocrine interference. A case in point among the invertebrates is that of the chitin synthesis inhibitors, such as diflubenzuron, which inhibit chitin synthesis and therefore prevent normal cuticle production in insects and crustaceans. This is discussed more fully in Chapter 2 and summarized in Chapter 5, but the basic point of the discussion was that chemicals need to interact with the hormone system, not simply alter a process that is under normal endocrine control.

Endocrinology Work Group *(Gerald LeBlanc, chair; Pamela Campbell, reporter)*

The Endocrinology Work Group focused on defining endocrine processes in aquatic and terrestrial invertebrates. Particular emphasis was directed at mechanisms involved in development, maturation, and reproduction, but mechanisms responsible for key invertebrate behavior were also considered.

The primary objectives of the work group were to

1) assess what was known of invertebrate endocrine systems and the physiological processes they control, and include a summary of the systems or processes for which there may currently be very limited information;
2) identify important invertebrate endocrine systems and mechanisms that are known to respond to or are likely susceptible to EDCs, both natural and anthropogenic; and
3) identify important invertebrate groups and prioritize the particular endocrine systems and processes of those groups that require focus in consideration of testing or surveying effects of ED.

Arguably, the Endocrinology Work Group had the most difficult challenge. During the planning stage, many of the workshop steering committee felt that invertebrate endocrinology was the area most in need of addressing, and the workshop was designed to alleviate some of the uncertainties in this area. This is particularly important because in order to use species of invertebrates in testing, and in understanding field effects resulting from exposure to EDCs, the underlying role of the endocrine system in reproduction and development, behavior, gender plasticity, and a host of other key biological processes should be well understood. Without this basic information, potential conclusions drawn from laboratory testing or field studies could be highly inappropriate. Some examples of the importance of the basic underlying endocrine function arose in the workshop discussions (e.g., chitin synthetase inhibitors and parasitic castration).

Laboratory Testing Work Group *(Chris Ingersoll, chair; Tom Hutchinson, reporter)*

The goal of the Laboratory Testing Work Group was to focus primarily on laboratory testing methods for aquatic invertebrates, with consideration of terrestrial invertebrates as appropriate.

The primary objectives of this group were to

1) assess the extent to which currently available testing methods with invertebrates are useful for the detection and assessment of endocrine disrupters;
2) identify the strengths and weaknesses of current tests with invertebrates and their potential applicability in ecological risk assessment;
3) suggest potential improvements to existing tests and, where appropriate, recommend the development of new tests; and
4) identify suitable model species and recommend suitable endpoints to be analyzed.

The Laboratory Testing Work Group was challenged with sorting substantial existing information on test methods, given the considerable invertebrate ecotoxicology database. There has been widespread use of invertebrates as surrogate and indicator species in environmental toxicology, over many years, and particularly in aquatic environments (Gannon and Stemberger 1978; Miller et al. 1985; Brinkhurst 1993; Buikema and Voshell 1993). Similarly, there is a long history of using aquatic invertebrates in biomonitoring (Tyler-Schroeder 1979; Nimmo and Hamaker 1982; Hopkin 1993; Rosenberg and Resh 1993), and assessing waste sites (Miller et al. 1985). Because of this long usage in ecotoxicology, there are a significant number of standardized laboratory tests using invertebrates (Chapter 3).

Recently, biomarkers of exposure and effects have been developed for some invertebrates, for use in ecological risk assessment (Depledge and Fossi 1994; Lagadic et al. 1994). One obvious problem is whether there are biomarkers specific for ED and, if so, understanding their link to potential effects in the organism or the population. This question was raised several times during the workshop and was the subject of substantial interaction and discussion among all 3 work groups.

Field Assessment Work Group *(Peter Matthiessen, chair; Trefor Reynoldson, reporter)*

The scope of the Field Assessment Work Group was directed at addressing real-world exposures of invertebrates to potential EDCs in the environment, both natural and anthropogenic, with particular emphasis on the aquatic and terrestrial compartments and effects at the population level.

The primary objectives of this group were to

1) develop an approach for the detection of effects, assignment of causality, sources, and sinks of the potential disrupting compounds, and the biological and chemical tools that will need to be developed or exploited for these means;
2) assess the usefulness of current approaches to environmental monitoring of invertebrates for determining exposure to and effects of endocrine disrupters;
3) understand the underlying variability in invertebrate populations, life histories, and environmental conditions, so that a baseline can be developed to detect population changes due to endocrine disrupters;
4) recommend suitable biomarkers and endpoints of exposure and effects for detecting EDCs in invertebrates;
5) determine possible triggers or thresholds for field monitoring strategies and identification of causality of effects noted in the field; and
6) develop a framework for monitoring in terms of what background, exposure, and effects information is required and how it should be collected.

The Field Assessment Work Group focused on the use of invertebrates as sentinels, particularly where individual or population level impacts may have been observed. This group was also charged with developing an understanding of what factors might improve scientists' ability to decipher cause and effect relationships in field studies. As is appropriate in these types of workshops, the participants relied on a significant amount of discussion with the other work groups.

Book Structure

In this book, the scientific understanding of invertebrate endocrinology, the use of invertebrates in testing, and the use of invertebrates as sentinels of environmental change are summarized and interrelated. By bringing these 3 aspects of the subject together in a single workshop, the participants endeavored to improve the understanding of ED in both aquatic and terrestrial invertebrates. Key findings of the workshop are presented in the Executive Summary. Detailed results from the Endocrinology, Laboratory Testing, and Field Work Groups are found in Chapters 2, 3, and 4, respectively. Chapter 5, Conclusions, provides a summation of the workshop and areas where additional research is needed.

References

Adiyodi KG, Adiyodi RG. 1970. Endocrine control of reproduction in decapod Crustacea. *Biol Rev* 45:121–165.

Alvarez MMS, Ellis DV. 1990. Widespread neogastropod imposex in the Northeast Pacific: Implications for TBT contamination surveys. *Mar Pollut Bull* 21: 244–247.

Ankley GT, Johnson RD, Detenbeck NE, Bradbury SP. 1997. Development of a research strategy for assessing the ecological risk of endocrine disrupters. *Rev Toxicol* 1: 231–267.

Ankley G, Mihaich E, Stahl R, Tillitt D, Colborn T, McMaster S, Miller R, Bantle J, Campbell P, Denslow N, Dickerson R, Folmar L, Fry M, Giesy J, Gray LE, Guiney P, Hutchinson T, Kennedy S, Kramer V, LeBlanc G, Mayes M, Nimrod A, Patino R, Peterson R, Purdy R, Ringer R, Thomas P, Touart L, Van der Kraak G, Zacharewski T. 1998. Overview of a workshop on screening methods for detecting potential (anti-) estrogenic/androgenic chemicals in wildlife. *Environ Toxicol Chem* 17: 68–87.

Ansell AD, Trevallion A. 1967. Studies on *Tellina tenuis* Da Costa I. Seasonal growth and biochemical cycle. *J Experim Mar Biol Ecol* 1: 220–235.

Baldwin WS, Milam DL, LeBlanc GA. 1995. Physiological and biochemical perturbations in *Daphnia magna* following exposure to the model environmental estrogen diethylstilbestrol. *Environ Toxicol Chem* 14: 945–952.

Barnes RD. 1980. Invertebrate zoology. Philadelphia PA: W.B. Saunders.

Brinkhurst RO. 1993. Future directions in freshwater biomonitoring using benthic macroinvertebrates. In: Rosenberg DM, Resh VH, editors. Freshwater biomonitoring and benthic macroinvertebrates. New York NY: Chapman & Hall. p 442–460.

Buikema AL, Voshell JR. 1993. Toxicity studies using freshwater benthic macroinvertebrates. In: Rosenberg DM, Resh VH, editors. Freshwater biomonitoring and benthic macroinvertebrates. New York NY: Chapman & Hall. p 344–398.

Campbell PM, Hutchinson TH. 1998. Wildlife and endocrine disrupters: Requirements for hazard identification. *Environ Toxicol Chem* 17: 127–135.

Colborn T. 1995. Statement from the work session on environmentally induced alterations in development: A focus on wildlife. *Environ Health Perspectives* 103: (Supplement 4) 3–5.

Colburn T, Clement C, editor. 1992. Chemically induced alterations in sexual and functional development: The wildlife/human connection. Volume XXI, Advances in modern environmental toxicology. Princeton NJ: Princeton Scientific.

Colborn T, vom Saal FS, Soto AM. 1993. Developmental effects of endocrine-disrupting chemicals in wildlife and humans. *Environ Health Perspectives* 101: 378–384.

Copeland BJ, Bechtel TJ. 1974. Some environmental limits of six gulf coast estuarine organisms. *Contribut Mar Sci* 18: 169–204.

Crisp TM, Clegg ED, Cooper RL, Anderson DG, Baetcke KP, Hoffmann JL, Morrow MS, Rodier DJ, Schaeffer JE, Touart LW, Zeeman MG, Patel YM, Wood WP. 1997. Special report on environmental ED: an effects assessment and analysis. Washington DC: USEPA. EPA/630/R-96/012.

Damkaer DM, Dey DB, Heron GA. 1981. Dose/dose-rate responses of shrimp larvae to UV-B radiation. *Oecologia* 48: 178–182.

Daston GP, Gooch JW, Breslin WJ, Shuey DL, Nikiforov AI, Fico TA, Gorsuch JW. 1997. Environmental estrogens and reproductive health: A discussion of the human and environmental data. *Reprod Toxicol Rev* 11: 465–481.

de Loof A, de Clerck. 1986. Vertebrate-type steroids in arthropods: Identifications, concentrations, and possible functions. In: Porchet M, Andries J-C, Dhainaut A. Advances in invertebrates reproduction. Amsterdam: Elsevier. p 117–123.

Depledge MH, Fossi MC. 1994. The role of biomarkers in environmental assessment: Invertebrates. *Ecotoxicol* 3: 161–172.

Dodson SI, Hanazato T. 1995. Commentary on effects of anthropogenic and natural organic chemicals on development, swimming behavior, and reproduction of *Daphnia*, an important member of aquatic ecosystems. *Environ Health Perspect* 103: (Supplement 4) 7–12.

Donker MH, van Capelleveen HE, van Straalen NM. 1993. Metal contamination affects size-structure and life-history dynamics in isopod field populations. In: Dallinger R, Rainbow PS, editors. Ecotoxicology of metals in invertebrates. Boca Raton FL: Lewis. p 383–400.

European Centre for Ecotoxicology and Toxicology of Chemicals (ECETOC). 1996. Environmental oestrogens. A compendium of test methods. Brussels, Belgium: ECETOC Document No. 33.

Eckelbarger KJ. 1974. Population biology and larval development of the terebellid polychaete *Nicoleas zostericola*. *Mar Biol* 27: 101–113.

[EDSTAC] Endocrine Disrupter Screening and Testing Advisory Committee report to EPA. 1998., Washington DC: U.S. Environmental Protection Agency USEPA, OPPTS.

Ellis DV, Pattisina LA. 1990. Widespread neogastropod imposex: A biological indicator of global TBT contamination. *Mar Pollut Bull* 21: 248–253.

Fox GA. 1992. Epidemiological and pathobiological evidence of contaminant-induced alterations in sexual development in free-living wildlife. In: Colburn T, Clement C, editors. Chemically-induced alterations in sexual and functional development: The wildlife/human connection. Volume XXI, Advances in modern environmental toxicology. Princeton NJ: Princeton Scientific. p 147–158.

Gannon JE, Stemberger RS. 1978. Zooplankton (especially crustaceans and rotifers) as indicators of water quality. *Transactions American Microscopic Society* 97: 16–35.

Gibbs PE, Pascoe PL, Burt GR. 1988. Sex change in the female dog-whelk, *Nucella lapillus*, induced by tributyltin from antifouling paints. *J Mar Biol Ass UK* 68: 715–731.

Gilbertson M, Fox GA. 1977. Pollutant associated embryonic mortality of Great Lakes herring gulls. *Environ Pollut* 12: 211–216.

Goldberg ED. 1980. The international mussel watch. Washington DC: National Academy of Sciences.

Gorbman A, Davey K. 1991. Endocrines. In: Prosser CL, editor. Comparative animal physiology. New York: Wiley-Liss. p 693–754.

Hidu H. 1969. Gregarious setting in the American oyster *Crassostrea virginica* Gmelin. *Chesapeake Science* 10: 85–92.

Highnam KC, Hill L. 1977. The comparative endocrinology of invertebrates. New York: American Elsevier. 270 p.

Hines GA, Watts SA, Sower SA, Walker CW. 1992. Sex steroid levels in the testes, ovaries, and pyloric caeca during gametogenesis in the sea star *Asterias vulgaris*. *Gen Comp Endocrin* 87: 451–460.

Hobbs HH. 1991. Decapoda. In: Thorp JH, Covich AP, editors. Ecology and classification of North American freshwater invertebrates. San Diego CA: Academic Pr. p 823–858.

Holmes P, Harrison P, Bergman A, Brandt I, Brouwer B, Keiding N, Randall G, Sharpe R, Skakkebaek N, Ashby J, Barlow S, Dickerson R, Humfrey C, Smith LM. 1997. European workshop on the impact of endocrine disrupters on human health and wildlife. Proceedings of a workshop; 2–4 December 1996. Weybridge UK: MRC Institute for Environment and Health. Report No. EUR 17549.

Hopkin SP. 1993. In situ biological monitoring of pollution in terrestrial and aquatic ecosystems. In: Calow P, editor. Volume 1, Handbook of ecotoxicology. Oxford UK: Blackwell Scientific. p 397–427.

Jepson PC. 1993. Insects, spiders and mites. In: Calow P, editor. Volume 1, Handbook of ecotoxicology. Oxford UK: Blackwell Scientific. p 299–325.

Kavlock RJ, Ankley GT. 1996. A perspective on the risk assessment process for endocrine-disruptive effects on wildlife and human health. *Risk Analysis* 16: 731–739.

Kavlock RJ, Daston GP, DeRosa C, Fenner-Crisp P, Gray LE, Kaatari S, Lucier G, Luster M, Mac MJ, Maczka C, Miller R, Moore J, Rolland R, Scott G, Sheehan DM, Sinks T, Tilson HA. 1996. Research needs for the risk assessment of health and environmental effects of endocrine disrupters: A report of the USEPA sponsored workshop. *Envirol Health Perspect* 104 (Suppl 4): 715–740.

Kendall R, Dickerson R, Giesy J, Suk W. 1998. editors. Principles and processes for evaluating ED in wildlife. Proceedings from a workshop; 6 March 1996; Kiawah Island SC. Pensacola FL: Society of Environmental Toxicology and Chemistry (SETAC). 515 p.

Lagadic L, Caquet T, Ramade F. 1994. The role of biomarkers in environmental assessment (5) Invertebrate populations and communities. *Ecotoxicol* 3: 193–208.

Langer J, Sang S. 1997. ED: Cause for concern, cause for action. *Canadian Chem News* 49: 16–17.

Largen MJ. 1967. The influence of water temperature upon the life of the dog-whelk *Thais lapillus* (Gastropoda: Prosobranchia). *Journal Animal Ecology* 36: 207–214.

Laufer H, Downer RGH. 1989. Endocrinology of selected invertebrate types. New York: Alan R. Liss.

LeBlanc GA, Bain LJ. 1997. Chronic toxicity of environmental contaminants: Sentinels and biomarkers. *Environ Health Perspect* 105: 65–80.

Matthiessen P, Gibbs PE. 1998. Critical appraisal of the evidence for tributyltin-mediated ED in mollusks. *Environ Toxicol Chem* 17: 37–43.

McLachlan JA, Newbold RR, Teng CT, Korach KS. 1992. Environmental estrogens: Orphan receptors and genetic imprinting. In: Colburn T, Clement C, editors. Chemically-induced alterations in sexual and functional development: the wildlife/human connection. Volume XXI, Advances in modern environmental toxicology. Princeton NJ: Princeton Scientific. P 107–112.

Miller WE, Peterson SA, Greene JC, Callahan CA. 1985. Comparative toxicology of laboratory organisms for assessing hazardous waste sites. *J Environ Qual* 14: 569–574.

Moore CG, Stevenson JM. 1991. The occurrence of intersexuality in harpacticoid copepods and its relationship with pollution. *Mar Pollut Bull* 22: 72–74.

[NRC] National Research Council. 1991. Animals as sentinels of environmental health hazards. Washington DC: National Academy Pr.

Nimmo DR, Hamaker TL. 1982. Mysids in toxicity testing - a review. *Hydrobiologia* 93: 171–178.

Nimmo DR, McEwen LC. 1994. Pesticides. In: Calow P, editor. Volume 2, Handbook of ecotoxicology. Oxford UK: Blackwell Scientific. p 155–203.

Pinder LCV, Pottinger TG. 1998. Endocrine function in aquatic invertebrates and evidence for disruption by environmental pollutants. Institute of Freshwater Ecology, Centre for Ecology and Hydrology, Natural Environmental Research Council.

Raffaelli D. 1982. An assessment of the potential of major meiofauna groups for monitoring organic pollution. *Mar Environ Res* 7: 151–164.

Rolland RM, Gilbertson M, Petersen RE, editors. 1998. Chemically induced alterations in functional development and reproduction of fishes. Proceedings of a conference and the Wingspread Conference Center; 21–23 July 1995; Racine WI, Pensacola FL: Society of Environmental Toxicology and Chemistry (SETAC). 224 p.

Rosenberg DM, Resh VH. 1993. Freshwater biomonitoring and benthic macroinvertebrates. New York NY: Chapman & Hall.

Sastry AN. 1968. The relationship among food, temperature, and gonad development of the bay scallop *Aequipecten irradians* Lamarck. *Physiolog Zool* 41: 44–53.

Sastry AN. 1970. Reproductive physiological variation in latitudinally separated populations of the bay scallop, *Aequipecten irriadians* Lamarck. *Biolog Bull* 138: 56–65.

Sellmer GP. 1967. Functional morphology and ecological life history of the gem clam, *Gemma gemma* (Eulamellibrachia:Veneridae). *Malacologia* 5: 137–223.

Slama K. 1971. Review of juvenile hormone analogs. *Ann Rev Biochem* 40: 1079–1102.

Smith BS. 1981a. Male characteristics on female mud snails caused by antifouling bottom paints. *J Appl Toxicol* 1: 22–25.

Smith BS. 1981b. Tributyltin compounds induce male characteristics on female mud snails *Nassarius obsoletus* = *Ilyanassa obsoleta*. *J Appl Toxicol* 1: 141–144.

Smith PJ, McVeagh M. 1991. Widespread organotin pollution in New Zealand coastal waters as indicated by imposex in dogwhelks. *Mar Pollut Bull* 22: 409–413.

Stahl RG Jr. 1997. Can mammalian and non-mammalian sentinel species data be used in evaluating potential human health impacts? Invited debate. *Hum Ecolog Risk Assess* 3: 329–335.

Stahl RG Jr., Clark JR. 1998. Uncertainties in the risk assessment of endocrine modulating substances in wildlife. In: Kendall R, Dickerson R, Giesy J, Suk W, editors. Principles and processes for evaluating ED in wildlife, Pensacola, FL: Society of Environmental Toxicology and Chemistry (SETAC). p 431–448.

Tagatz ME. 1968. Biology of the blue crab, *Callinectes sapidus* Rathbun, in the St. Johns River, Florida. *Fish Bull* 67: 17–33.

Tattersfield L, Matthiessen P, Campbell P, Grandy N, Lange R., editors. 1997. SETAC-Europe / OECD / EC expert workshop on endocrine modulators and wildlife: Assessment and testing. EMWAT. Brussels, Belgium: SETAC-Europe.

Tessier AJ, Henry LL, Goulden CE. 1983. Starvation in Daphnia: Energy reserves and reproductive allocation. *Limnol Oceanog* 28: 667–676.

Thorp JH, Covich AP. 1991. Ecology and classification of North American freshwater invertebrates. San Diego CA: Academic Pr.

Tyler-Shroeder DB. 1979. Use of the grass shrimp (*Palaemonetes pugio*) in a life-cycle toxicity test. Philadelphia PA: American Society for Testing and Materials (ASTM). STP No. 667. p 159–170.

Voogt PA, Broertjes JJS, Oudejans RCHM. 1985. Vitellogenesis in sea star: Physiological and metabolic implications. *Comp Biochem Physiol* 80A: 141–147.

Young JPW. 1978. Sexual swarms in *Daphnia magna*, a cyclic parthenogen. *Freshwater Biol* 8: 279–281.

CHAPTER 2

The Endocrinology of Invertebrates

Gerald A. LeBlanc, Pamela M. Campbell, Pieter den Besten, Richard P. Brown, Ernest S. Chang, Joel R. Coats, Peter L. deFur, Tarlochan Dhadialla, John Edwards, Lynn M. Riddiford, Michael G. Simpson, Terry W. Snell, Michael Thorndyke, Fumio Matsumura

Introduction

Scope

This chapter provides basic information on invertebrate physiology and endocrinology and gives environmental toxicologists, administrators, and policy makers some understanding of invertebrate endocrinology. In addition, the authors offer their insight into endpoints that could be used to monitor for endocrine disruption (ED) in both laboratory toxicity evaluations (Chapter 3) and field monitoring (Chapter 4). Finally, this attempt to highlight and summarize important aspects of invertebrate endocrinology should serve as a scientific foundation to judge the utility of representative or surrogate species for assessing ED in invertebrates. The basic endocrinology of invertebrates, as currently understood, is described for those invertebrate phyla for which information is available. The information is organized by phyla, beginning with arthropods, progressing to the juncture of the evolutionary lineages, protostome and deuterostome, then covering some representatives of the deuterostome invertebrates. Finally, the endocrinology of cnidarians is presented. The cnidarians represent a taxon phylogenetically located before the divergence of the protostomes and deuterostomes. Each section consists of an overview of

1) the basic endocrinology of the organisms,
2) the extent of knowledge regarding hormonal mechanisms of action, and evidence of ED.

Coverage of some taxa is quite limited, owing to the paucity of information on both the endocrinology of these organisms and their susceptibility to ED.

Endocrine Disruption in Invertebrates: Endocrinology, Testing, and Assessment. Peter L. deFur et al., editors.
 ISBN 1-880611-27-9

Endocrinology overview

The use of hormones to regulate biological processes is a strategy common to both vertebrate and invertebrate animals. Endocrine systems of animals consist of widely scattered, ductless glands that synthesize and release chemical mediators, hormones, into the body. These hormones exert their regulatory effects at low concentrations at various target sites within the body. Hormones comprise a variety of molecular structures, including steroid, terpenoid, and peptide. Endocrine control strategies and basic hormonal mechanisms to regulate biological processes have been widely conserved among animal phyla. Nevertheless, individual components of endocrine systems have undergone significant evolutionary divergence, resulting in distinct differences among various invertebrate taxa.

Typically, endocrine systems function through cascades that are initiated by environmental or physiological signals and culminate with the action of the terminal hormone on the targeted organ. Such cascades have been best described for mammals that can serve to exemplify the general processes. For example, sexual maturation in male mammals includes the development of accessory sex organs and the development of secondary sex characteristics; both are under the regulatory control of testicular androgens. The hormonal cascade for sexual maturation originates at the hypothalamus and proceeds along the hypothalamic–pituitary–gonadal axis (Figure 2-1). In response to the signal for sexual maturation, the hypothalamus secretes the peptide hormone, gonadotropin-releasing hormone. This hormone travels from the hypothalamus to the pituitary gland via the hypophysial portal blood vessel and stimulates the pituitary to secrete peptide hormones, gonadotropins. The gonadotropins travel via the circulatory system to the testis, where they stimulate the secretion of the steroidal hormone testosterone. Testosterone then travels throughout the body via the circulatory system. Testosterone may function directly as the terminal regulating hormone, or it may be converted to the potent androgen dihydrotestosterone. The terminal androgen then orchestrates sexual maturation.

Two basic principles of hormone action have emerged:

1) the responsive target cells for a hormone express specific, high-affinity protein receptors for the hormone; and
2) hormone–receptor binding forms a complex that mediates biochemical activity in the target tissue.

The activity of all hormones is mediated by specific hormone receptors. The peptide and catecholamine hormones (i.e., epinephrine, glucagon, insulin) typically bind to receptors located on the surface of the cell, stimulating a series of intracellular reactions. The intracellular reactions constitute a signal transduction pathway that ultimately results in the regulation of the transcription of specific genes.

Steroid and thyroid hormones enter the cell where they bind to intracellular receptors. These hormone-receptor complexes interact with the cell's DNA to

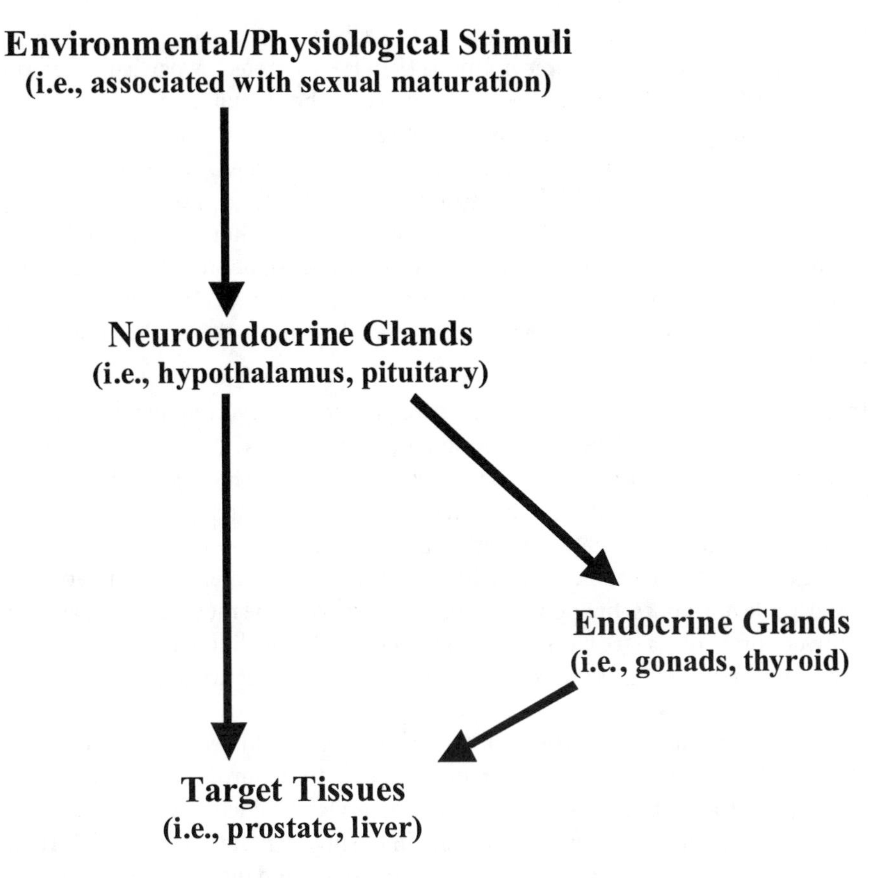

Figure 2-1 Typical neuroendocrine cascade with example components

regulate the transcription of responsive genes. Steroid hormones may also regulate gene transcription through cell surface receptors in association with steroid hormone-binding proteins located in the blood. Here, steroid hormones can regulate gene transcription using signal transduction pathways analogous to those used by peptide hormones. The action of hormones through cell surface receptors and signal transduction pathways may allow for the rapid response of the cell to the presence of the hormone and the rapid cessation of the response in the absence of the hormone. The actions of hormones through nuclear receptors provides for more prolonged response of the cell following initial stimulation by the hormone.

The use of hormone cascades to regulate physiological processes is a successful strategy that has been appreciably conserved during evolution. Cascades provide a communication link between the nervous system and the endocrine system. Cascades also provide for multiple sites at which hormone action can be regulated. As we learn more about invertebrate endocrinology, the use of endocrine cascades is also becoming evident, though alternative strategies also exist. For example, the neuropeptide prothoracicotropic hormone (PTTH) is secreted by particular cells in the central nervous system of insects in response to the stimuli to molt. PTTH stimulates the prothoracic glands to synthesize and secrete a steroid hormone, ecdysone. The fat body and a few other tissues then convert ecdysone to 20-hydroxyecdysone (20E), which initiates the appropriate action at the target cells. However, the precise action of 20E is controlled by another hormone, juvenile hormone (JH), a sesquiterpenoid. The production and secretion of JH is under the control of a distinct neuroendocrine cascade. Thus, 2 cascades function in concert to regulate a variety of endocrine processes mediated by the 2 terminal hormones, 20E and JH.

Invertebrates encompass more than thirty phyla and have evolved many different approaches to growth, development, and reproduction. It is not surprising that regulation of these processes by the neuroendocrine system is considerably more diverse in these animals than in the vertebrates, which comprise only a single phylum. Despite the diversity in invertebrate endocrinology, some basic generalities can be made. As noted above, invertebrate endocrine systems use steroid, terpenoid, and peptide hormones, but peptide hormones are by far the most common among all invertebrate phyla (see Gorbman and Davey 1991). The differences between steroid hormones, such as ecdysone and vertebrate reproductive hormones, and peptides are not at all trivial. The 2 types of compounds have vastly different physical and chemical properties, particularly solubility and resistance to degradation. In addition, the biological differences between the 2 hormone types are equally as great. Typically, steroids are secreted from true glands, ductless organs that synthesize and secrete chemical mediators. In the invertebrates, however, internal secretory structures are most often neuronal in origin (and function), and are called "neurosecretory cells" or "organs." Neurosecretory structures have cell bodies typical of nerve cells, except for the synthesis of secretory products, but the axons end in bulbous formations from which hormones are secreted. Neurosecretory structures embody a close link between the nervous and endocrine systems and are found in some form in all the invertebrate phyla examined thus far. Close connections between nervous and endocrine tissue are found in all animals but are most prominent in the invertebrates.

Endocrine disruption

Current concern over chemically induced ED stems from observations that some chemicals have the ability to bind to hormone receptors and modulate the activity of these receptors (McLachlan 1993; Kelce et al. 1995). Such receptor-binding activity of xenobiotics has been largely documented with steroid hormone receptors in vertebrate systems. Some xenobiotics have been shown to bind to steroid hormone receptors and to function like the hormone for that receptor (Arnold et al. 1996). That is, binding of the xenobiotic to the receptor results in the regulation of transcription of responsive genes. Such xenobiotics function as hormone agonists.

Xenobiotics can also bind to a hormone receptor without stimulating the activity of the receptor. As long as the xenobiotic is occupying the receptor, the endogenous hormone for that receptor cannot bind to the receptor. Thus, activity of the receptor is competitively inhibited. Chemicals that bind to a hormone receptor and inhibit the activity of that receptor function as hormone antagonists. The compound *o,p'*-DDT is an example of an environmental chemical that functions as a hormone (estrogen) agonist in vertebrates (Arnold et al. 1996). The DDT metabolite *p,p'*-DDE has been shown to function as a hormone (androgen) antagonist in vertebrates (Kelce et al. 1995). Xenobiotics can also act as endocrine disrupters without affecting hormone binding, by modulating endogenous hormone levels. For example, the fungicide ketoconazole can lower serum testosterone levels by inhibiting testosterone synthesis (Rajfer et al. 1986). As is discussed below, various components of the endocrine system of invertebrates are known or are likely to be susceptible to similar modes of disruption.

Overview of the invertebrates

Evolutionary biologists generally agree that the more than 30 phyla of animals diverged into 2 discrete lineages, the protostomes and the deuterostomes, during the early evolution of the animal kingdom (Figure 2-2). Less concordance exists among scientists with respect to the specific characteristics that differentiate these 2 phylogenetic groups. Historically, organisms were assigned to either of these 2 categories based upon whether, during gastrulation, the pore extending from the inner embryo to the external environment (the blastopore) developed into a mouth (protostomes) or an anus (deuterostomes). However, even this dichotomy appears not to be absolute. For purposes of this book, the important concept is that most of the major phyla of invertebrates (protostomes) diverged from the lineage including the vertebrates (deuterostomes) during early evolution (Figure 2-2). Accordingly, significant divergences in endocrine strategies would be expected between the vertebrates and most major groups of invertebrates; lesser divergence might be expected with a relatively few phyla. Furthermore, the high degree of evolutionary divergence among the individual invertebrate phyla would also allow for significant divergence in endocrine strategies used by the different phyla. As exemplified in this chapter, this is indeed the case.

This major bifurcation in the animal lineage also seems to mark a major divergence in endocrine strategy. Significant evidence exists to indicate that invertebrate deuterostomes use vertebrate-type sex steroids (androgens, estrogens, progestogens) as terminal hormones to regulate reproduction along extended neuroendocrine cascades. Existing data suggest that protostomes make limited use of vertebrate-type sex steroids for reproductive maturation and function. Rather, the lower protostomes use neuropeptides to regulate these processes; the insects and crustaceans use a neuroendocrine cascade with the ecdysteroid and terpenoid hormones as terminal hormones. The ecdysteroids are apparently not part of the deuterostome arsenal of hormones, and the deuterostomes make use of terpenoids (e.g., retinoic

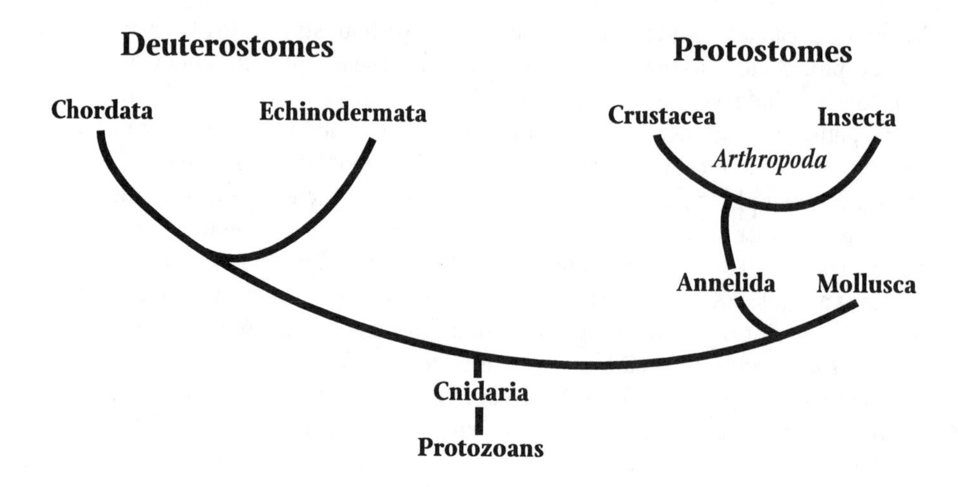

Figure 2-2 Phylogeny of major invertebrate groups discussed in this chapter

acid) primarily in patterning early development. Thus, the argument could be made that the endocrinology of deuterostome invertebrates such as sea stars and sea urchins is more homologous to that of the vertebrates in comparison with the protostome invertebrates.

Notwithstanding some gross similarities between the endocrinology of invertebrates and vertebrates, various aspects of invertebrate endocrinology extend beyond the realm of those paradigms established with vertebrates. Accordingly, precedents established with vertebrates may have limited utility when the endocrine toxicity of a chemical to invertebrates is evaluated. Further, endocrine toxicity to a representative of one invertebrate phylum has limited relevance to other phyla. A comprehensive understanding of invertebrate endocrinology and toxicology is required for the sagacious evaluation of the effects of environmental chemicals on this diverse assemblage of organisms.

Insecta

The insects represent the most successful group of invertebrates, as judged by the number of existing species, the anatomical and physiological diversity within the taxa, and the multiplicity of habitats in which they thrive. The insects, along with the crustaceans and some minor taxa, comprise the arthropods. The endocrinology of the insects has been described to a much greater extent than that of any other invertebrate group because of the silkworm industry and the need to control insect pests. Basic research sought to identify features of insect physiology that could be exploited for use in the development of insecticidal agents to confer greater specificity to chemical agents.

Endocrinology

The insect endocrine system is comprised of neurosecretory cells found primarily in the central nervous system, 3 endocrine glands (the corpora allata, the prothoracic glands, and the epitracheal glands), and the gonads (Figure 2-3, based upon a lepidopteran model). Morphologically, the secretory structures are compactly arranged; the brain, the corpora cardiaca, the neurohemal organ for many of the neurosecretory cells of the brain, the corpora allata, and the prothoracic glands are all in the anterior head and thorax of the insect. Intrinsic neurosecretory cells lie within the corpora cardiaca and in other ganglia of the ventral nervous system. In general, these neurosecretory cells, which release neuropeptides into the hemolymph, are responsible for coordinating growth, molting, and reproduction in response to changes in the external environment (photoperiod, temperature, food, etc.). These secretions also regulate processes such as metabolic homeostasis, water balance, and color change. The major insect hormones, their sources, functions, and chemical type are presented in Table 2-1.

Hormones involved in growth, molting, and reproduction will be considered in more detail below since most is known about endocrine mediation in these processes, which are also the endpoints primarily affected by endocrine disrupters. Some of the major neuropeptides involved in the control of metabolic and osmoregulatory homeostasis, sex pheromone release, and diapause are also listed in Table 2-1. The list is not meant to be exhaustive; many additional neuropeptides important in reproduction and homeostasis have been identified in various insect species. Most of these molecules appear to be unique to insects and related arthropods (e.g., crustaceans), and the majority seem not to occur or have hormonal function in deuterostomes (including vertebrates). By contrast, several vertebrate hormones have been detected in insects (e.g., bombyxin, which is highly similar to insulin, prostaglandins, etc.), although their function in insects is unclear.

Development

Molting

Insect growth and development is punctuated by a series of molts in which a new, larger cuticle (exoskeleton) is produced and followed by the shedding of the old cuticle (Nijhout 1994). Hardening of the new cuticle follows the molt and any molt-associated growth. The process of molting is triggered by the release of PTTH from neurosecretory cells in the brain (Figure 2-3). PTTH acts on the prothoracic glands to cause the synthesis and secretion of the molting hormone, ecdysone. Ecdysone and related compounds, the ecdysteroids, are derived from dietary cholesterol or

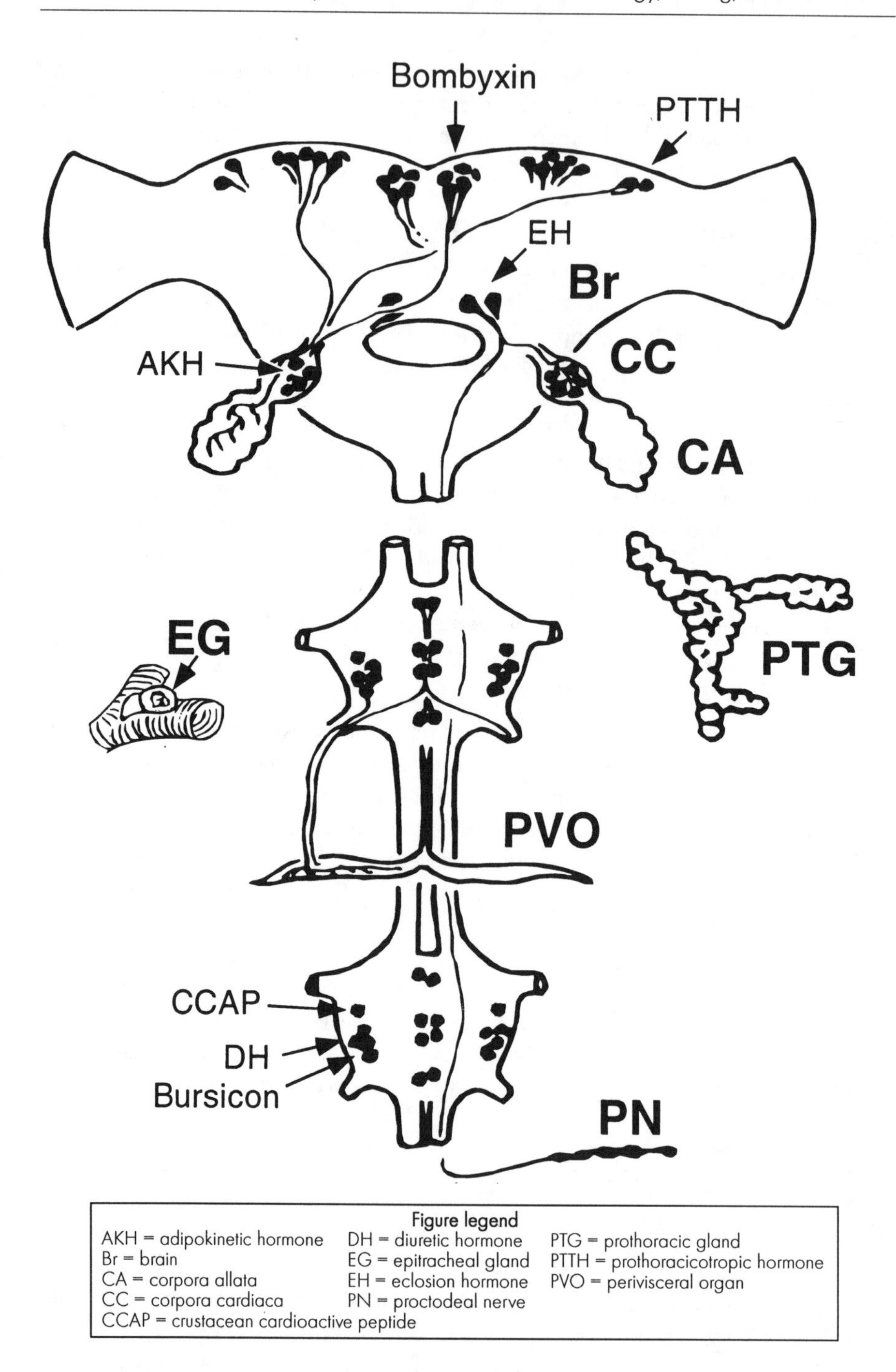

Figure 2-3 Neuroendocrine system of insects based on a lepidopteran model with some of the neuropeptide-producing cells indicated

Table 2-1 Some of the known insect hormones, their sources, functions, and chemical type[1]

Hormone	Source	Function	Chemical type
Ecdysteroids	Prothoracic gland	Initiation of molting and egg maturation	Steroid
Prothoracicotropic hormone (PTTH)	Brain	Control of ecdysteroid production	Peptide
Bombyxin	Brain	Ecdysteroid production, carbohydrate metabolism	Peptide (insulin-like)
Ecdysis-triggering hormone (ETH)	Epitracheal gland	Ecdysis behavior	Peptide
Eclosion hormone (EH)	Brain	Ecdysis behavior	Peptide
Crustacean cardioactive peptide (CCAP)	Ventral nerve ganglia	Ecdysis behavior, heart rate modulation	Peptide
Bursicon	Brain and ventral nerve ganglia	Tanning of cuticle	Peptide
Juvenile hormone (JH)	Corpora allata	Regulation of metamorphosis and reproduction	Terpenoid
Allatostatin	Brain	Inhibition of JH production	Peptide
Allatotropin	Brain	Stimulation of JH production	Peptide
Adipokinetic hormone (AKH)	Corpora cardiaca	Fat (energy) metabolism	Peptide
Hypertrehalosemic hormone	Corpora cardiaca	Control of sugar metabolism	Peptide
Diuretic hormone (DH)	Brain, corpora cardiaca, ventral ganglia	Water balance and urine production	Peptide
FMRFamides	Central nervous system	Neuromodulators, myosuppression	Peptide
Diapause hormone	Brain, subesophageal ganglion	Initiation of diapause in embryos	Peptide
Pheromone biosynthesis activating neuropeptide	Subesophageal ganglion	Initiates pheromone production	Peptide

[1] Not intended to be a comprehensive list

plant steroids, since insects and other arthropods cannot synthesize cholesterol. The secreted ecdysteroid is then transported by hemolymph (insects and many other invertebrates have open circulatory systems and a hemocoel) to the fat body and other peripheral tissues that convert ecdysone to 20E (Figure 2-4). Both ecdysone and 20E are important in the initiation of the molt. Ecdysone is involved in events of cell proliferation (Champlin and Truman 1998), and 20E is involved in events of differentiation such as the production of the new cuticle (Figure 2-5). However, in vitro studies have shown that low doses of 20E can mimic the actions of ecdysone on most tissues (Riddiford 1985; Champlin and Truman 1998). Production of the new cuticle does not begin until the ecdysteroid titer starts to decline. As the

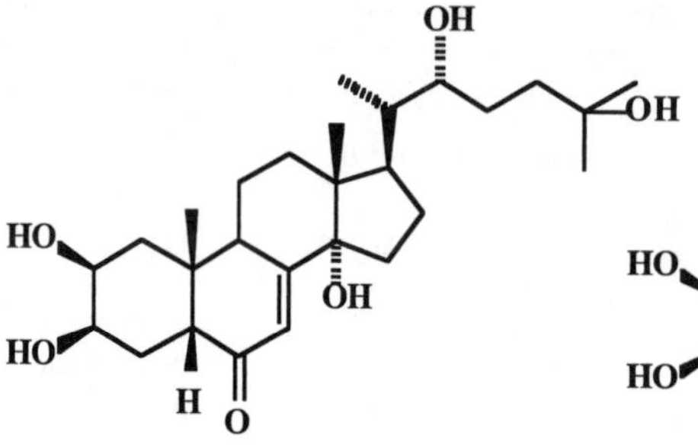

Ecdysone

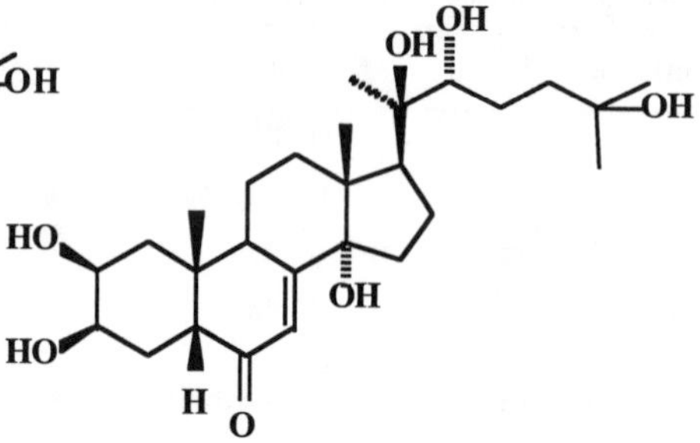

20-Hydroxyecdysone

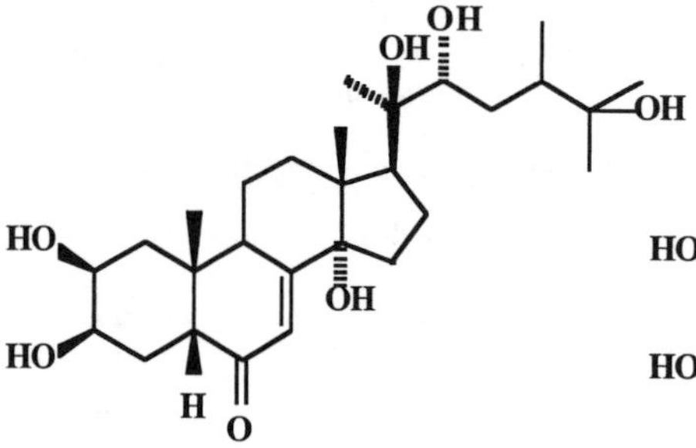

Makisterone A

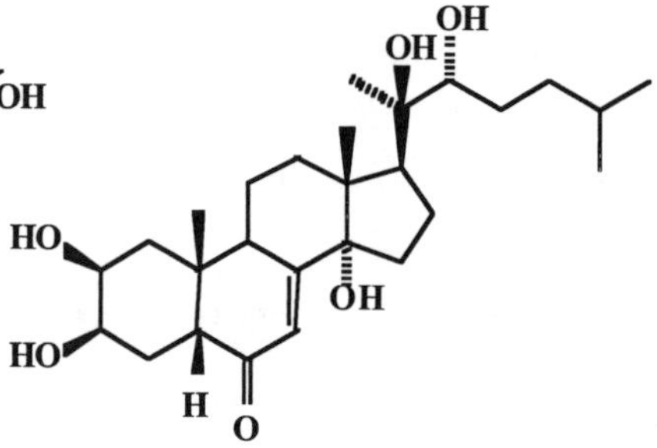

Ponasterone A

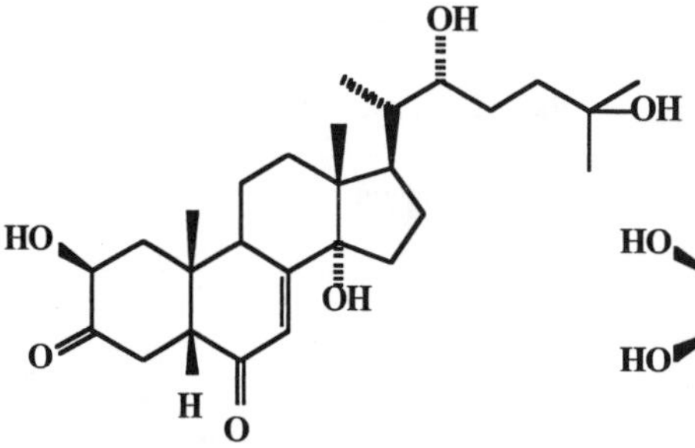

3-Dehydroecdysone

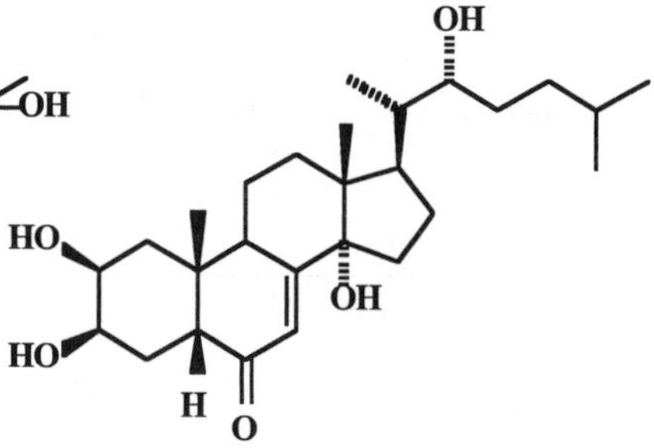

25-Deoxyecdysone

Figure 2-4 Chemical structures of the major ecdysteroids in insects and crustaceans

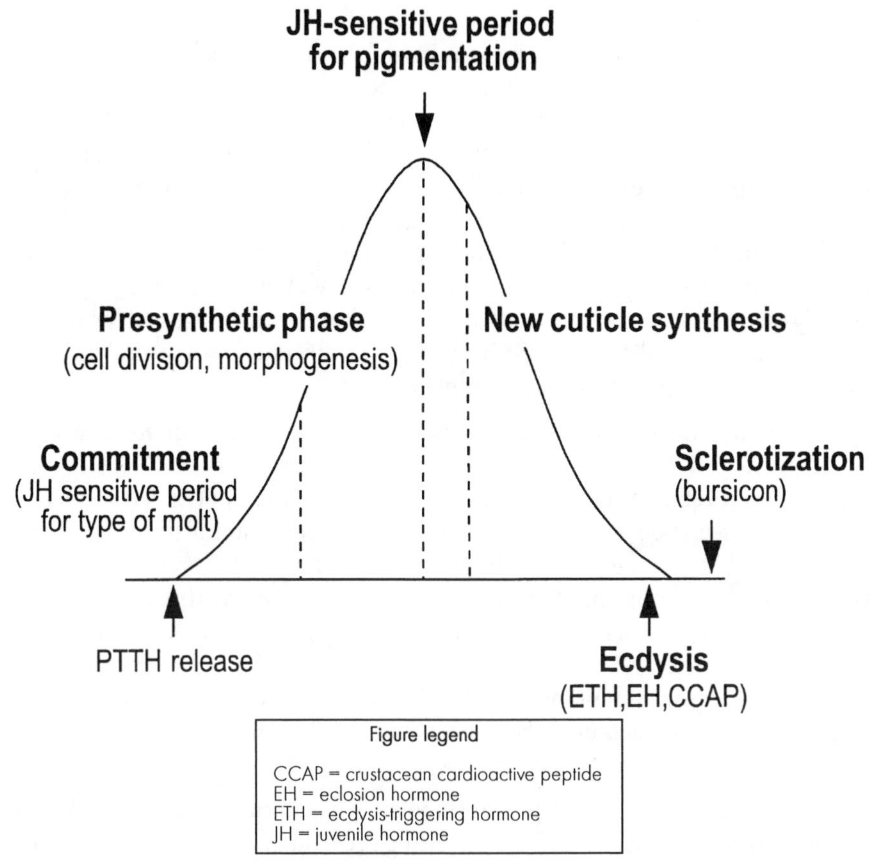

Figure 2-5 Diagrammatic ecdysteroid titer causing a molt, indicating various phases of molt and juvenile hormone-sensitive periods for preventing metamorphosis (commitment) and for regulation of pigmentation

ecdysteroid titer returns to basal levels, a series of neuropeptides are released causing the shedding of the old cuticle or ecdysis (Figure 2-5). Due to some as yet unknown signal, ecdysis-triggering hormone (ETH) is released from the epitracheal glands beneath the spiracles (Figure 2-3). This causes pre-ecdysial behavior and the release of eclosion hormone (EH) from the brain, both within the ventral nervous system and into the hemolymph (Zitnam et al. 1996; Ewer et al. 1997; Gammie and Truman 1999). EH then acts within the nervous system to trigger the release of crustacean cardioactive peptide (CCAP), which in turn inhibits the pre-ecdysis behavior and initiates ecdysis. In the hemolymph, EH acts on the epitracheal glands to ensure complete release of ETH and also on the dermal glands that release the waterproofing cement layer, which is spread over the cuticle during ecdysis. After ecdysis the new cuticle may expand, harden (sclerotization), and sometimes darken in response to the neuropeptide bursicon.

Metamorphosis

Two basic strategies of metamorphosis are used by insects: hemimetaboly and holometaboly. The lower insects (dragonflies, locusts, cockroaches, true bugs, and termites) are hemimetabolous and have incomplete metamorphosis. The larvae (usually called "nymphs") are morphologically very similar to the adult but lack functional reproductive organs and fully formed wings. The final nymphal molt directly gives rise to the adult stage with wings and genitalia. The higher insects (beetles, butterflies and moths, bees and ants, mosquitoes and flies) are holometabolous and undergo complete metamorphosis. Holometabolous larvae are morphologically quite distinct from the adult and first molt to a pupal stage with external characteristics resembling the adult. They subsequently undergo a second molt in which transformation to an adult is completed.

The type of molt (juvenile or adult) is determined by the presence or absence of the sesquiterpenoid JH (Figure 2-6) that is secreted by the corpora allata (CA) (Figure 2-3) (Riddiford 1994). Neurosecretory hormones from the brain regulate the synthesis and secretion of JH from the CA with allatotropin stimulating and allatostatin inhibiting the process. During larval life stages, JH is present during the feeding period and at the onset of the molt cycle. The results are that the initial rise of ecdysteroid titer occurs in the presence of JH, and the insect molts to another larval stage (Figure 2-5). During the final larval instar, the absence of JH at the time of ecdysone release ensures that metamorphosis will occur. The critical period for the anti-metamorphic action of JH occurs at the outset of the ecdysteroid rise (Figure 2-5).

With both holometabolous and hemimetabolous metamorphosis strategies, JH titer declines early in the final larval instar, allowing premetamorphic development of imaginal-specific structures. When the ecdysteroid titer rises during the subsequent molt, the absence of JH causes the adult molt in the hemimetabolous insects. By contrast, JH reappears during this molt in the holometabolous insects, and pupae are formed. The subsequent molt of the pupa to the adult occurs in response to an ecdysteroid rise in the absence of JH.

In most insects, the number of larval stages depends on various intrinsic and extrinsic factors. The presence of JH during the final larval stage produces excess (supernumerary) larval molting, and the absence of JH from earlier larval instars can cause premature transformation to adulthood (precocious metamorphosis) (Riddiford 1994). Additionally, exposure of a pupa to exogenous JH prevents development to the adult so that the organism undergoes a second pupal molt. In contrast, in the higher flies such as *Drosophila melanogaster*, the number of larval instars is fixed and most of the adult, except for the nervous system and Malpighian tubules, is formed from imaginal discs and other imaginal precursor cells, which proliferate during larval life (Riddiford 1993). Exogenous JH can only prevent metamorphosis of the nervous system (Restifo and Wilson 1998) and of the adult

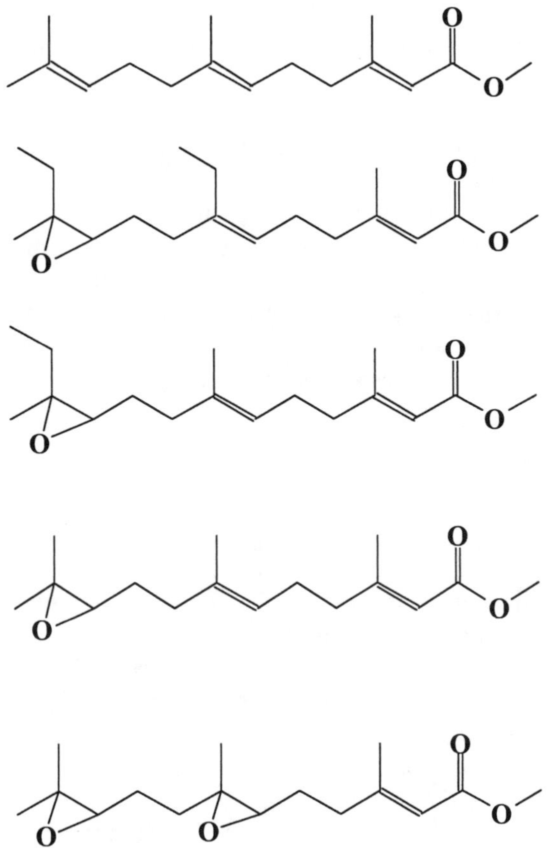

methyl farnesoate

JH-I

JH-II

JH-III

JH-III bis epoxide

Figure 2-6 Chemical structures of naturally occurring juvenile hormones and the methyl farnesoate of crustaceans

abdomen (Postlethwait 1974; Riddiford and Ashburner 1990) when applied during the final instar or at pupariation.

Embryonic development

Embryonic development in insects also is regulated by these same hormones, with ecdysteroids causing one or more molts within the egg, and JH being present during the final molt that results in the first instar nymphal or larval cuticle (Sbrenna 1991; Riddiford 1994). Exposure to JH during the JH-free stages can disrupt later embryonic development (Staal 1975; Riddiford 1994).

Pigmentation

In some insects, the presence or absence of JH at the time of the peak of the ecdysteroid titer during the molt cycle determines the subsequent larval pigmentation (Figure 2-5; Nijhout and Wheeler 1982). In the black mutant strain of the tobacco hornworm, *Manduca sexta,* CA do not secrete sufficient JH at this time during the molt, and the new cuticle is darkly pigmented (melanized) (Safranek and Riddiford 1975). Exposure to exogenous JH, at the time of head capsule slippage and the peak of the ecdysteroid titer, prevents this melanization in a dose-responsive manner and forms the basis of the black larval bioassay for JH (Fain and Riddiford 1975). Exposure of a wild type *Manduca* to substances that cause a precocious decline of JH during the molt cycle or a JH antagonist would cause melanization of the new cuticle and therefore could be used as a bioassay for disruption of JH-mediated processes.

Reproduction

Female

In the adult female, the CA produces JH, which regulates egg maturation (Wyatt and Davey 1996). In many insects, this JH stimulates the fat body to produce vitellogenin. JH increases the protein-synthesizing machinery within the cell and activates vitellogenin gene transcription. The vitellogenin is secreted from the fat body into the hemolymph, transported to the gonad, and taken up into the oocyte by specific receptors on the oocyte membrane. In some insects such as the mosquito (Raikhel and Lea 1985), the formation of this vitellogenin-receptor complex is under the control of JH. Also, the follicle cells must elongate, forming intercellular spaces through which the vitellogenin must pass in order to access the oocyte membrane. In locusts and the bloodsucking bug *Rhodnius prolixus,* this cell shape change is mediated by JH (Abu-Hakima and Davey 1977; Wyatt and Davey 1996). In other insects such as *Drosophila,* JH is also necessary for vitellogenin uptake, presumably by a similar mechanism (Riddiford 1993). In some insects, such as cockroaches that use accessory gland proteins to form egg cases, JH may regulate this process (reviewed in Wyatt and Davey 1996).

The follicle cells of the ovary synthesize various ecdysteroids, most of which are incorporated into the eggs as conjugates to be used for embryonic molts before the prothoracic glands become functional (Lagueux et al. 1979; Sbrenna 1991). In Diptera, ecdysone may be secreted into the hemolymph and induce vitellogenin synthesis in the fat body. The best known example is in the mosquito *Aedes aegypti,* where a blood meal causes the release of ovarian ecdysteroidogenic hormone (OEH), previously known as egg development neurosecretory hormone [EDNH]) from the brain, which stimulates ecdysone release from the ovary (Adams 1999). This ecdysone is converted to 20E that induces vitellogenin gene transcription in the fat body, but only if the fat body has previously been exposed to JH. In *Drosophila,* both the fat body and the ovarian follicle cells make vitellogenin, and both JH and 20E are necessary for the coordination of yolk production and deposition (Riddiford 1993).

In most insects, the female produces a sex pheromone that attracts the male. In the Lepidoptera, the synthesis of this pheromone is under the control of a neuropeptide, pheromone biosynthesis-activating neuropeptide (PBAN), which is released only at a certain time of day (Raina 1993). In some cockroaches and in houseflies, pheromone biosynthesis is regulated by JH and by ecdysteroids respectively (Blomquist et al. 1993).

Male

In the male, spermatogenesis is generally under control of the hormones that control metamorphosis and thus begins when ecdysteroids act in the absence of JH (Happ 1992). Ecdysteroids are also produced by testis of some insects, and they may play a role in local stimulation of growth factors necessary for genital duct development (reviewed in Adams 1997). The male accessory glands produce materials involved in spermatophore production, sperm activation, and stimulation of female oviposition behavior. The function of the accessory gland is often, but not always, under the control of JH (Happ 1992; Wyatt and Davey 1996; Wolfner 1997).

Physiological regulation

Many neurohormones are important in regulating metabolism, water and ion balance, and other physiological processes (see review by Gade et al. 1997 and articles in Coast 1998). For example, the adipokinetic hormone (AKH) causes mobilization of fatty acids from fat stores in locusts and is essential for long-term flight (Goldsworthy 1983). In locusts, the initial mobilization of fatty acids for the first few minutes of flight is stimulated by octopamine, which is released from the corpora cardiaca (CC) and acts as an "insect adrenaline" (Orchard et al. 1993). In other insects such as cockroaches and flies, carbohydrates are used as fuel for flight, and AKH-like hormones are necessary to mobilize the trehalose from the fat body stores (Gade et al. 1997). Bombyxin, a member of the insulin family, was first found during the isolation of PTTH of the silkworm *Bombyx mori;* it acted as a PTTH (causing synthesis and secretion of ecdysone) in *Samia cynthia* pupae but not in *Bombyx* pupae (Ishizaki and Suzuki 1992). Bombyxin is secreted in all life stages (Saegusa et al. 1992), but whether it is involved in carbohydrate metabolism or in the regulation of growth is still uncertain.

The diuretic hormones of the bloodsucking insects are released after a blood meal and act on the Malpighian tubules to cause rapid excretion of excess fluid (Phillips 1983; Coast 1998). A Malpighian tubule assay for compounds that cause this effect was developed (Ramsay 1954) and could be adapted to many different kinds of insects. In other insects, antidiuretic hormones and neuropeptides regulate ionic balance, depending on the adaptations needed for their survival (Gade et al. 1997).

Many other identified neuropeptides have either stimulatory or inhibitory action on various muscles, yet their precise target of action is still unknown (Gade et al. 1997; Duve et al. 1998; Nassel et al. 1998). The FMRFamides (phenylalanine-methionine-arginine-phenylalanine-amide compounds), first found in mollusks as cardiostimulators (Price and Greenberg 1977), are also found in the central nervous systems of insects, and one FMRFamide subfamily is found to have myosuppressive activity (Orchard and Lange 1998).

Diapause

Insects in temperate clines may overwinter in a state of relative quiescence and extremely low metabolism, called "diapause" and characterized by minimal oxygen consumption. In the more tropical areas, a similar diapause state may enable survival during the dry season. Diapause may occur during any life stage, although for a particular insect, it usually occurs only in a single life stage. Diapause may also occur during multiple life stages (Denlinger 1985). Diapause is typically hormonally controlled although not all mechanisms have yet been elucidated.

In the silkworm *Bombyx mori*, females reared under long-day conditions will release diapause hormone, a neuropeptide from the same gene that encodes PBAN (Yamashita 1996). This neuropeptide enables antifreeze proteins, pigment precursors and other metabolites to be deposited into the egg, resulting in diapause at the germ band stage of embryo development (Yamashita and Suzuki 1991). Other insects, such as the gypsy moth, diapause as fully developed embryos. The mechanism responsible for this late embryonic-stage diapause is unknown. Larval diapause may occur in any instar but often occurs just before metamorphosis (Chippendale 1983). In the case of the southwestern corn borer larva, JH levels remain intermediate during diapause, and a series of stationary molts ensue (Yin and Chippendale 1979). Pupal diapause results from the absence of PTTH release. Chilling and/or long photoperiods are necessary to reactivate those neurosecretory cells (Denlinger 1985). Adult-stage diapause commonly occurs due to the lack of JH release and may be triggered by either photoperiod or the lack of palatable food (De Wilde 1983). In female insects such as the Colorado potato beetle, no eggs are produced during diapause; in males, the accessory glands may stop functioning.

Mechanism of action of the insect hormones

Ecdysteroids

Ecdysone was the first steroid hormone shown to have an action directly on DNA. In 1960, Clever and Karlson showed that ecdysone applied to the salivary glands of the midge, *Chironomus tentans*, caused specific "puffs" (later shown to be sites of mRNA synthesis) within 15 minutes, with a second series of "puffs" appearing in 3 to 4 hours (Clever and Karlson 1960). This latter series was dependent on protein synthesis (Clever 1964). Similar studies in *Drosophila* led to a model for ecdysteroid action (Ashburner et al. 1974) in which the biologically active ecdysteroid, 20E, coupled to the ecdysone receptor (EcR) differentially regulates several classes of target genes. Transcription of "early" genes is activated by the 20E–EcR complex, whereas transcription of "late" genes is repressed by the 20E–EcR complex. Early genes have been shown to code for tissue-specific structural proteins and transcription factors that are found widely throughout most tissue types. In addition, transcription factors induced by the 20E–EcR complex appear with a delayed time course and may require the early transcription factors for expression. Thus, there is a 20E-induced interactive transcription cascade that directs the subsequent pattern of the "late" structural gene expression in specific tissues (Thummel 1996).

The *Drosophila* EcR gene is a member of the nuclear hormone receptor family that encodes 3 protein isoforms: EcR-A, EcR-B1, and EcR-B2. The three isoforms differ only in their N-terminal A/B domains and are thought to be important in transactivation by protein–protein interactions with other factors (Talbot et al. 1993). Similar isoforms have been identified in Lepidoptera (Jindra et al. 1996; Kamimura et al. 1997) and beetles (*Tenebrio*) (Mouillet et al. 1997). In all these taxa, the expression of a particular isoform in a tissue during molting and metamorphosis is correlated with a particular cellular activity such as proliferation or cuticle production. Isoform-specific expression can also be associated with a particular cell fate such as extremely high levels of EcR-A in *Drosophila* larval neurons that are destined to die at the end of metamorphosis (see also Robinow et al. 1993; Truman et al. 1994). Studies involving EcR-B mutants in *Drosophila* suggest that 2 EcR-B isoforms are involved in distinct aspects of adult neuronal development (Schubiger et al. 1998). EcR must form a heterodimer with another member of the nuclear hormone superfamily, Ultraspiracle (USP), in order to bind either ecdysteroid or DNA (Yao et al. 1992; Yao et al. 1993; Cherbas and Cherbas 1996). In *Drosophila,* only one isoform of USP has been identified (Henrich et al. 1994); however, 2 isoforms of USP have been identified in both the mosquito *Aedes* (Kapitskaya et al. 1996) and the lepidopteran *Manduca* (Jindra et al. 1997). Thus, the transcription factor cascade activated by the same rise in ecdysteroid titer can differ between cells and tissues, depending on which isoforms are involved in the EcR/USP heterodimeric complex.

USP is a homolog of the vertebrate retinoid X receptor (RXR) and can form heterodimers with the RXR partners, the retinoic acid receptor (RAR), the thyroid hormone receptor (TR), and the vitamin D receptor (Yao et al. 1992; Mangelsdorf and Evans 1995). The vertebrate orphan receptors, RIP14 (Seol et al. 1995) and the farnesol X receptor (FXR) (Forman et al. 1995), are homologs of EcR based on similarity of their DNA binding domains. These receptors can form heterodimers with RXR that are capable of binding to the DNA of the heat shock protein (hsp) 27 ecdysone response element (EcRE) normally bound by the EcR/USP heterodimer to activate hsp27 transcription. Activation of transcription from this EcRE by the FXR–RXR heterodimer occurs with either farnesol or high concentrations (15 to 50 mM) of JH III, but not with methoprene, a JH analog (Figure 2-7). By contrast, methoprene and its acid as well as hydroprene acid but not JH III can activate transcription from an RXR response element (Harmon et al. 1995). In this case, the acids of the JH analogs were shown to compete for binding of 9-*cis*-retinoic acid, the normal ligand for the RXR homodimer.

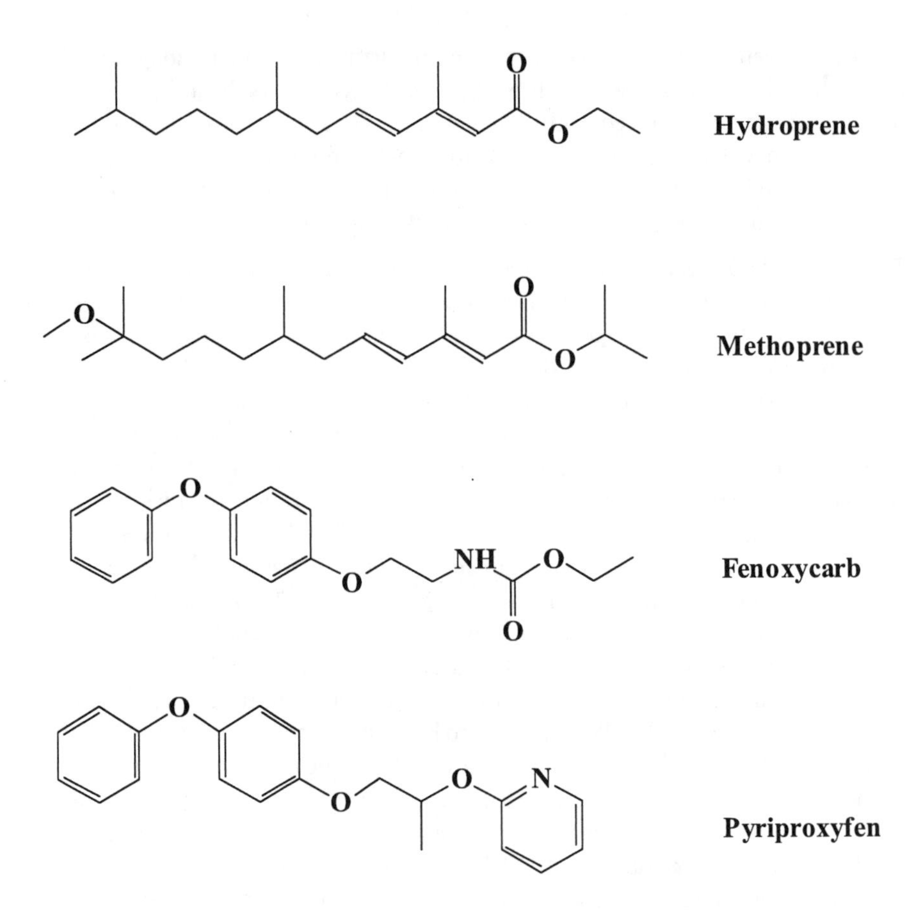

Figure 2-7 Chemical structures of some biologically active insect juvenile hormone agonists

Recently, the cDNA for the EcRs and the RXRs of arachnid ticks (Guo, Harmon, Jin et al. 1998; Guo, Harmon, Laudet et al. 1998) and the crustacean fiddler crab (Chung et al. 1998) have been isolated and sequenced. These EcRs are quite similar to those of the insects. However, the RXRs are more similar to the vertebrate RXRs in the ligand-binding domain (~70% identity) than are the USPs of the higher insects Lepidoptera and Diptera (~48% identity). The ligand for the vertebrate RXR homodimer, 9-*cis*-retinoic acid, does not allow the tick RXR to activate transcription from the RXR response element. Whether this lack of activity is due to its lack of binding of the ligand or to the differences in its C-terminal AF2 activation domain (Guo, Harmon, Laudet et al. 1998) is unknown.

Juvenile hormone

A receptor for JH has not yet been identified. However, this hormone appears to elicit pleiotropic actions on target cells (Riddiford 1994; Jones 1995; Wyatt and Davey 1996). JH appears to act intracellularly to modulate ecdysteroid action and stimulate vitellogenin synthesis. Jones and Sharp (1997) showed that JH III binds with low affinity ($>10^{-7}$ M) to USP produced in yeast and causes USP oligomer formation in a yeast 2-hybrid system. They suggest that USP is the JH receptor, but no further reports on its role in an insect system are yet available. The methoprene-tolerant gene in *Drosophila* conveys about a 10-fold increase in resistance to methoprene and JH III (Wilson and Fabian 1986) and has been found to encode a protein that has about 35 to 40% homology to the dioxin aryl hydrocarbon receptor (AhR) in the bHLH-PAS gene domains (Ashok et al. 1998). Its role in the cell is not yet understood.

The stimulation of vitellogenin uptake by follicle cells is mediated by the action of JH through a membrane receptor that has been partially characterized in *Rhodnius* and *Locusta* (reviewed in Wyatt and Davey 1996). JH acts through this receptor to cause a rapid increase in the ion transport protein sodium/potassium ATPase activity, leading to water loss and cell shrinkage. This action appears to be mediated by the protein kinase C pathway.

Allatostatins and allatotropins: endogenous regulators of JH production

In some (possibly all) insects, the activity of the CA appears to be under the control of regulatory neuropeptides that either stimulate (allatotropins) or inhibit (allatostatins) the production of JH. The allatotropin from *Manduca sexta* (Kataoka et al. 1989) was the first to be characterized and interestingly stimulates the adult, but not the larval, CA. The allatotropin gene is expressed in the larval nervous system as well as in that of the pupa and adult (Taylor et al. 1996). Recently, this allatotropin has been found to reversibly inhibit larval midgut ion transport (Lee et al. 1998), but whether this is a normal physiological role is unknown. Allatostatins were first characterized from the brains of the cockroach *Diploptera punctata* (Woodhead et al. 1989), and the gene encoding 13 allatostatin-type peptides was later isolated (Donly et al. 1993). Recently, similar genes encoding a variable number of related peptides have been found in several other cockroach species where some of the peptides are functional inhibitors of JH synthesis, both in vitro and in vivo (Stay et al. 1994; Weaver et al. 1994, 1998). Similar peptides (or genes putatively encoding them) have been discovered in dipterans, orthopterans, hymenopterans, and lepidopterans, although the functional significance of these molecules in these taxa is questionable (see Duve et al. 1998; Weaver et al. 1998).

Another peptide, unrelated to the cockroach peptides, was purified from brains of the lepidopteran *M. sexta* and strongly inhibits in vitro JH biosynthesis by CA of both fifth-stage larvae and adult females (Kramer et al. 1991). Evidence for the existence of an identical peptide has been found in the noctuid moths *Pseudaletia unipuncta* (Jansons et al. 1996) and *Lacanobia oleracea* (Audsley et al. 1999), but its function is unknown.

The occurrence of other allatostatins, allatostatin-related peptides, or allatostatin-like immunoreactivity has been reported in several other invertebrates, but their function is still unclear. Cockroach allatostatin-like peptide immunoreactivity is found in the nervous system of the crab *Cancer borealis*, and four of the cockroach allatostatins have actions on the rhythmic movements of the crab stomach (Skiebe and Schneider 1994). Immunoreactivity to the *Diploptera* allatostatins is also found in the nervous systems of freshwater snails *Bulinas globosus* and *Stagnicola elocles* (Rudolph and Stay 1997) and of *Hydra oligactis* (Hydrozoa), *Moniezia expansa* (Cestoda), *Schistosoma mansoni* (Trematoda), *Artioposthia triangulata* (Turbellaria), *Ascaris suum* (Nematoda), *Lumbricus terrestis* (Oligochaeta), *Limoc pseudoflavus* (Gastropoda), and *Eledone cirrhosa* (Cephalopoda) (Smart et al. 1994). The distribution of allatostatin-like immunoreactivity within the nervous system of a range of helminths has led to speculation that they may have neurotransmitter/neuromodulation functions with a role in locomotion, feeding, reproduction and sensory perception (Smart et al. 1995).

Neuropeptides

Neuropeptides act on membrane receptors and therefore stimulate various types of second messengers within the cell. Hormones such as PTTH (Gilbert et al. 1996), AKH (Van Marrewijk and Van der Horst 1998), diuretic hormone (Coast 1998), and bursicon (Reynolds 1985; Gade et al. 1997) are thought to act through cyclic adenosine monophosphate (cAMP) and therefore activate the protein kinase A pathway. Calcium entry is also important in AKH and PTTH action. The diuretic hormone receptor has been identified and found to be in the calcitonin–secretin receptor family (Reagan 1994). Eclosion hormone activates cyclic guanosine monophosphate (cGMP) in its target cells (Ewer et al. 1994), although the details of that pathway are not yet understood.

Endocrine disruption

Ecdysteroid-mediated processes

Effects of the insect ecdysteroids can be interrupted by inhibition of synthesis and/or release from the prothoracic glands, interference with the peripheral metabolism of 20E, or by competition at its receptor level. The interaction of a compound at the ecdysone receptor can be either agonistic or antagonistic in its activity. If agonistic, its persistence can cause effects attributed to hyperecdysonism (persistent 20E). If antagonistic, the effects will mimic those seen when 20E levels are low or absent. It should be noted that an ecdysteroid agonist/antagonist in one order of insects does not automatically mean that the same compound will have similar disruptive effects in other organisms, which may have highly similar but not identical receptors.

Ecdysteroid agonists
Although a number of phytoecdysteroids have been extracted from plants and some of them found to be more active than 20E in laboratory assays, their use as insecticides has been limited due to their structural complexity for synthesis and relative metabolic instability in the target organisms (Watkinson and Clark 1973). Recently, a nonsteroidal agonist of 20E, tebufenozide, has been discovered and commercialized as an insecticide. Tebufenozide is extremely selective in its toxicity to lepidopteran larval pests and its virtual inactivity on species from other orders of arthropods, especially beneficial arthropods (Dhadialla et al. 1998). Tebufenozide has very low mammalian toxicity and has favorable ecotoxicological properties. It is nonmutagenic in a variety of standard in vitro and in vivo laboratory assays. It causes neither adverse developmental effects after oral exposure nor reproductive toxicity in multigeneration dietary studies. Tebufenozide is also completely non-oncogenic in chronic feeding studies in both mice and rats.

The mode of action of tebufenozide is very well understood. It manifests its effects through interaction with the ecdysone receptor complex. Paradoxically, even though almost all insects and arthropods use 20E as their molting hormone, tebufenozide is selectively toxic to species from only one order of insects (Lepidoptera). The lepidopteran selective toxicity of tebufenozide correlates directly with its very high binding affinity to the lepidopteran ecdysone receptor complex as compared to receptors from members of other orders of insects (Dhadialla et al. 1998). Hence, even though there is a high degree of amino acid identity and similarity among the ligand-binding domains of EcRs of insects from different orders (see pages 37 and 38 and Kothapalli et al. 1995), the little dissimilarity or lack of identity is sufficient to provide significant binding-affinity differences for ecdysone agonists like tebufenozide. Yet ecdysteroids like 20E, and especially ponasterone A, bind to the different EcRs with very similar affinities (Dhadialla et al. 1998). In other words, it seems that while the insect molting hormone tolerates slight differences in the amino acid sequences of ligand-binding domains of different insect EcRs, the same EcRs discriminate between nonsteroidal ecdysone agonists through their relative binding affinities. By extrapolation, one would expect that the vertebrate steroid nuclear receptors (homodimeric or heterodimeric functional partners), which have only 17 to 26% amino acid identity with *Drosophila* EcR in their ligand-binding domains (Baker 1997), would have only slight or no affinity for nonsteroidal ecdysone agonists like tebufenozide. An additional mechanism for selective toxicity of tebufenozide to lepidopteran insects versus dipteran insects may be due to its active exclusion from dipteran cells (Sundaram et al. 1998).

Symptomology
Larval lepidopteran pests intoxicated with tebufenozide stop feeding within 4 to 16 hours. During this time, the molting process is initiated, and by 24-hour intoxicated larvae prematurely slip their head capsule. However, the larvae are unable to complete the molt and subsequently die of starvation, blood loss due to hemor-

rhage, or predation. The inability of the intoxicated larvae to successfully molt is due to sustained effective titers of tebufenozide in the hemolymph (equivalent to hyperecdysonism), which in turn prevents the release of eclosion hormone that normally occurs upon the decline of 20E and thus the completion of a successful molt. As an endpoint marker, in insect larvae susceptible to tebufenozide or another 20E agonist, feeding will be inhibited, and the insect will form a double cuticle and, in the case of lepidopteran larvae, a slipped head capsule. Molting attempts in intoxicated larvae will be lethal. Tebufenozide or other agonists of 20E will initiate a premature lethal molt also in the final larval stage, resulting in a larval–pupal intermediate or a deformed pupa, depending on the JH titer at the time of contact.

Although 20E agonists initiate a molt attempt, its completion depends on the release of the ecdysis behavior initiated by the ETH-EH-CCAP cascade. Therefore, any chemical agent that interferes with the release or action of any of these peptides will result in the inability of the target insect species to successfully shed the old cuticle.

Ecdysteroid antagonists

So far, compounds with antagonistic molting hormone activity have not been found. However, such compounds would prevent occurrence of normal 20E-initiated molts. In addition, processes like spermatogenesis and egg development in some insects could be disrupted. Therefore, the potential effects of such compounds would be reduced fecundity or fertility of the affected insect.

Interference with ecdysone synthesis

Once again, there are not too many examples of compounds that inhibit the synthesis and/or production of ecdysone. Azadirachtin, a tetranotriterpenoid plant limnoid, is a potent insect antifeedant and has growth inhibitory properties (reviewed by Mordue et al. 1993). While the exact site and the molecular basis of its action are not understood, substantial evidence suggests that one of its target sites may be the CC, the release site for PTTH and other neuropeptides. It is known that azadirachtin does not affect the synthesis or release of ecdysone from the prothoracic glands and that it does not bind to the ecdysone receptor complex. Insects intoxicated with azadirachtin have been observed to have dramatically reduced titers of 20E or delayed ecdysteroid peaks as compared to control treated insects. In some insects, effects on JH titers have also been implicated due to the occurrence of supernumerary larval molts following treatment with azadirachtin. Dihydroazadirachtin, a derivative of azadirachtin with similar potency, has also been shown to bind to membrane proteins from the testis of the desert locust *Schistocerca gregaria* (Nisbet et al. 1995). This could also be the basis of reproductive effects observed in some insects (reviewed by Mordue et al. 1993).

Symptomology

A variety of morphological effects of azadirachtin have been observed in several insect species. Azadirachtin causes growth inhibition, malformation of developmental stages, and mortality in intoxicated larvae in a dose-dependent manner. Increased doses of azadirachtin in larval stages result in

1) adults with reduced longevity and fecundity;
2) wingless adults and adults with unsclerotized, deformed wing lobes;
3) lethal nymphal or larval molts;
4) severe pupal deformities; and
5) over-aged larvae with extended larval life.

Juvenile hormone-mediated processes

Juvenile hormone analogs

With the discovery of the chemical structure of JH in 1967 (Roller et al. 1967), attempts were made to produce synthetic analogs for use as "third generation" insecticides (Williams 1967). Ideally, such chemicals would specifically disrupt insect development and/or reproduction without impinging on physiological processes in other animals, where JH is not present and has no known function. Also, insects were thought unlikely to become resistant to their own hormones (Williams 1956). However, resistance to the effects of these chemicals has been shown to occur (Wilson and Fabian 1986; Dhadialla et al. 1998).

In general, the physiological effects of the naturally occurring JHs are mimicked by the effects of their synthetic analogs, and much of what we know about the physiological and morphological effects of excess JH derives from experiments with these juvenile hormone analogs (JHAs). Figure 2-7 shows the structures of some of the synthetic analogs that have been developed as insect control agents. These molecules have been commercialized and are therefore likely to enter the environment. For this reason, JHAs have potential to act as endocrine disrupters in nontarget insect species and in other invertebrates.

Developmental effects

Under normal circumstances, JH is present in insects at relatively high levels, beginning at about 60% of embryonic development until the beginning of the final larval instar (see above; Riddiford 1994). Therefore, exposure to JHAs only disrupts development when it is present during early embryonic development (either by contamination of gravid females or by exposure of newly deposited eggs) or during the final larval (nymphal) stage. The affected embryos do not hatch and show various morphogenetic anomalies, whereas application of excess hormone to final stage larvae and nymphs disrupts metamorphosis. With high doses (or more active compounds), supernumerary larval molting occurs in most insect orders (except the higher Diptera). These "giant" larvae continue to feed and may eventually develop into large adults when the JHA is removed, although the reproductive capacity of such insects is often compromised. Exposure to lower levels usually allows some aspects of metamorphosis to proceed. Among the hemimetabolous species (e.g., cockroaches or true bugs), this results in severely deformed adults (typically "twisted" wings), which are invariably sterile due to subtle and varied morphological defects (Staal 1975). Holometabolous larvae exposed to low doses of JHA may molt successfully to the pupal stage, but adult emergence is prevented, and permanent pupae, pupal–adult intermediates, or "adultoids" are formed. Similar effects can be produced by exposure of newly formed pupae to JHA. The pupal–adult intermediates rarely survive, and death often occurs as a result of desiccation due to damage to the cuticle (presumably caused by an unsuccessful attempt to complete adult eclosion). The adultoids are short-lived and reproductively sterile. Such graded deformities have been exploited in standardized assays for JH or JHA activity (e.g., the *Tenebrio* assay [Gilbert and Schneiderman 1960]).

Reproductive effects

In many adult insects, JH regulates vitellogenin production in the fat body and/or uptake by the ovaries (see p 36). In general, exposure of adult insects to JHA has little effect on survival or longevity, although it may affect fecundity if given to gravid females (see p 36). However, in some groups (e.g., ants), exposure of queens to JHA results in ovarian atrophy and sterility (Edwards 1975). The mechanisms inducing such sterility are not known. In other insects, particularly those in which regular cycles of oocyte production occur (e.g., cockroaches), exposure of adults to JHA increases the rate of ootheca production (Weaver and Edwards 1990). Furthermore, because of the relationship between JH and pheromone biosynthesis in some insect taxa, exposure to JHA in adult insects could conceivably stimulate pheromone production and disrupt normal mating behavior. Similarly, because some insect behaviors are known to be under endocrine control (e.g., the JH-mediated transformation of nurse honeybees into forager bees [Robinson 1987]), exposure of these adult insects to JHA could derange normal behavioral patterns.

Effects on polymorphism

In some social insects (e.g., ants and termites), polymorphism is under the control of JH, and exposure of certain species to JHA can result in a shifting of the sub-caste bias towards "soldiers" (Nijhout and Wheeler 1982; Noirot and Bordereau 1991). In other insects (e.g., locusts and butterflies), JH is an important determinant of phase polymorphism (Nijhout and Wheeler 1982; Pener 1991; Pener and Yerushalmi 1998). In these insects, exposure to JHA can alter phenotypic characters such as color or seasonal morph determination as in the black *Manduca* larva described above.

Aphids have complex life cycles that alternate between sexual and parthenogenetic forms and between winged (alate) and apterous forms, all under hormonal control (Hardie and Lees 1985). Several JH-sensitive periods for these alternations could be disrupted by JHA (Mittler 1991).

Anti-juvenile hormone agents

Because JHAs produce the majority of their effects only at the end of larval life after the major feeding damage by most insect pests occurs, the concept of anti-juvenile hormones (AJHs) as another insect control strategy was born. The idea was to cause precocious metamorphosis, thereby deleting the final, most destructive larval stage.

Here, the major classes of AJH are described, and their mode of action (where known) detailed. The diversity of chemical type and the variety of their actions indicate that the endocrine-disrupting potential of such molecules is substantial, despite the fact that currently none are available commercially. Most of these AJHs are not true JH antagonists in the sense that they do not prevent its action at the target tissue.

Compounds producing cytotoxic effects on the corpora allata

Bowers et al. (1976) reported that extracts from the plant *Ageratum houstonianum* contained naturally occurring, biologically active principles that induced premature metamorphosis and adult sterility in the hemipteran *Oncopeltus fasciatus*. The active molecules were identified as mono- or di-methoxy substituted 2,2-dimethyl chromenes (precocenes, Figure 2-8).

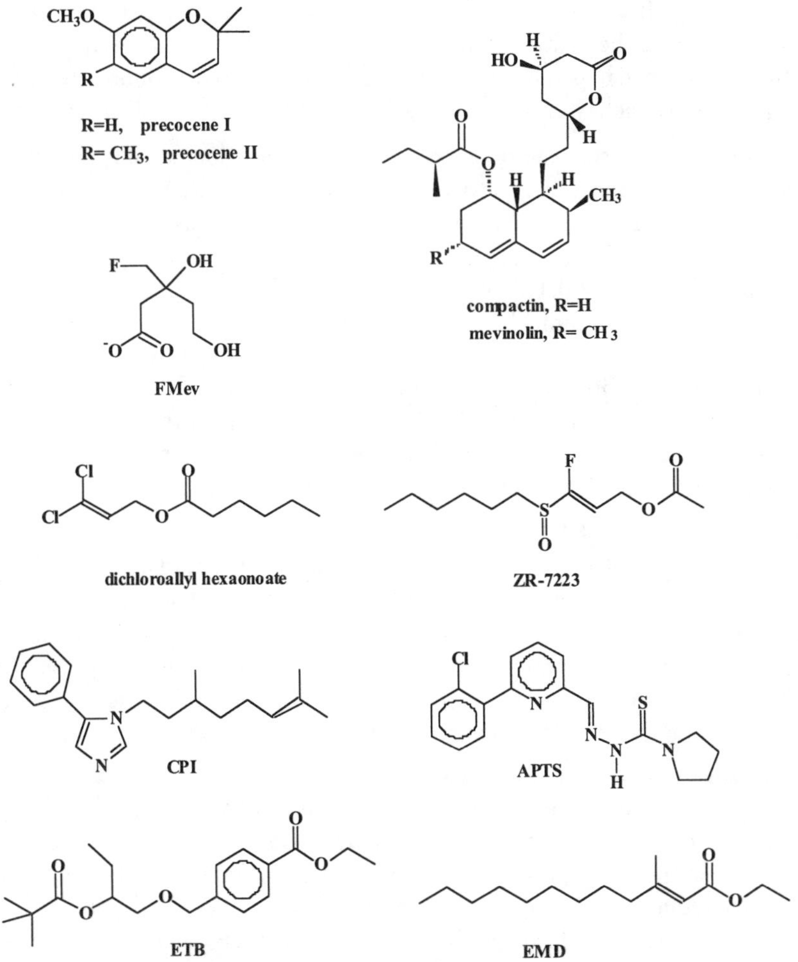

R=H, precocene I
R= CH_3, precocene II

compactin, R=H
mevinolin, R= CH_3

FMev

dichloroallyl hexaonoate

ZR-7223

CPI

APTS

ETB

EMD

Figure 2-8 Chemical structures of some molecules known to act as anti-juvenile hormone agents

These chromenes are selectively metabolized by the cells of the CA to highly unstable epoxide alkylating agents, which then destroy these cells (Pratt et al. 1980). Despite the fact that the precocenes were effective AJH agents, their activity appears to be limited to certain hemipteran and orthopteran insects and to some ticks (Staal 1986).

Synthetic inhibitors of juvenile hormone biosynthesis

Another AJH strategy was to inhibit key steps in the biosynthetic pathway leading to the production of juvenile hormone from acetate and other simple precursors through the mevalonate pathway. The fungal metabolite compactin (Figure 2-8), which inhibits the enzyme HMG-CoA reductase (Monger et al. 1982), was shown to be a highly potent inhibitor of JH biosynthesis in vitro (EC50 ~ 2×10^{-9} M) (Monger et al. 1982; Edwards and Price 1983). However, compactin was not sufficiently active in vivo for practical development against insect pests (Monger et al. 1982; Edwards and Price 1983; Hiruma et al. 1983). A variety of synthetic compactin analogs have been developed by the pharmaceutical industry as effective medicines for use in the treatment of human hypercholesterolemia. Some of these compounds are substantially more active than compactin and could potentially result in ED of insects.

Another inhibitor of cholesterol biosynthesis (fluoromevalonolactone [Fmev]; Figure 2-8) showed convincing in vivo AJH activity in *Manduca sexta* larvae (Quistad et al. 1981), the first in a holometabolous species. FMev was also shown to be mildly effective in delaying the JH-regulated production of oothecae in *Periplaneta americana* (Edwards et al. 1985). Other antimetabolites of intermediates of JH biosynthesis in Lepidoptera are 3,3-dichloroallyl hexanoate with weak AJH activity (Figure 2-8; Quistad et al. 1985) and a modified allylic ester ZR-7223 (Figure 2-8; Henrick 1991), with strong AJH activity. Foliar application of a 0.25% spray of the latter material caused premature development in 100% of second stadium of the beet armyworm *Spodoptera exigua* placed on treated foliage (Henrick 1991). Clearly, molecules of this type have real potential to act as endocrine disrupters in insects.

Kuwano et al. (1983, 1984) discovered AJH activity on *Bombyx mori* in a series of 1-citronellyl-5-substituted imidazoles (e.g., citronellyl phenyl imidazole ([CPI], Figure 2-8). These compounds may act by inhibition of the cytochrome P-450, which is necessary for epoxidation of methyl farnesoate and its homologs to JHs (Kuwano et al. 1983). However, these compounds have not been shown to be active in vivo in any species of insect other than the silkworm (Staal 1986), and so their status as potential endocrine disrupters remains unclear.

Miscellaneous anti-juvenile hormone agents

A variety of other molecules are known to modulate JH levels in insects, although their precise modes of action are not fully understood. For example, early studies indicated that the application of the JHA hydroprene to intact insects was highly effective in reducing JH biosynthesis (Tobe and Stay 1979) and endogenous JH levels (Edwards and Price 1983). Further studies showed that these significant feedback-mediated reductions in endogenous JH levels were apparent only with highly active

JH compounds (Edwards 1987), in which case the exogenous JHA compensated for the reduction in endogenous hormone levels.

Likewise, Staal (1986) reported the discovery of 2 compounds, ethyl 4-[2-(*tert*-butylcarbonyloxy) butoxybenzoate (ETB) and ethyl 3-methyl 2-dodecenoate (EMD) (Figure 2-8) which have mixed JH agonist and antagonist activity. Both of these compounds have relatively low JH activity in a wide variety of species, and both have AJH effects that are manifest at only a few intermediate doses in the sensitive species; at higher doses, the intrinsic JH agonistic effects override the antagonistic effects. Presumably, these compounds exert this limited AJH effect by suppressing endogenous JH titers through a stronger effect on those receptors that cause the feedback inhibition versus those receptors responsible for the morphogenetic effects. It has also been suggested that EMD acts by blocking the JH receptor (Staal et al. 1981).

A number of thiosemicarbazones, especially 6-(o-chlorophenyl) pyridyl-2-carboxyaldehyde pyrrolidinothiosemicarbazone (APTS; Figure 2-8), are active as AJHs on certain Lepidoptera, producing a variety of effects including precocious pupae (Barton et al. 1989). Curiously, these substances were only active when fed in the diet and not upon topical application. Their AJH effects could be "rescued" by treatment of larvae with the juvenoid fenoxycarb. Again, the mode of action of these compounds remains unclear. For a recent review of molecules acting as AJH agents, see Schooley and Edwards 1996.

Allatostatins and allatotropins as possible endocrine disrupters

Because the allatotropins and allatostatins regulate JH production (see p 41), there is interest in utilizing them to control insect development (Weaver et al. 1998). For instance, if allatostatic peptides could be introduced into insects in sufficient quantities, they could lower endogenous JH levels and perhaps cause precocious metamorphosis or prevent egg maturation. The allatostatins and allatotropins are peptides that are likely to be relatively unstable in the environment, unlikely to penetrate the insect cuticle, and probably inactivated following ingestion. Yet the recent development of more stable mimics (Garside et al. 1997; Nachman et al. 1997) warrants further exploration of their potential to disrupt nontarget endocrine-mediated activity in invertebrates.

JHA effects on nontarget invertebrates

Because several JHA have been used practically as insect pest control agents, there are already a few reports of endocrine-disrupting effects of such chemicals in nontarget insects. For example, fenoxycarb treatment in orchards has been shown to have deleterious effects on the development of honeybee broods (De Ruijter and Vandersteen 1987). There have also been numerous reports of the potential deleterious action of field applications of JHA on other beneficial insects such as insect predators and parasitoids. Similarly, some of the JHAs have been reported to cause detrimental effects (although not necessarily resulting from the disruption of endocrine processes) in several other invertebrate groups, including effects in cladocerans (Hosmer et al. 1998), polychaete annelids (Biggers and Laufer 1992), shrimp (Ahl 1986), and barnacles (Tighe-Ford 1977).

Juvenile hormone effects on vertebrates

JH is not known to occur in vertebrates, and JHAs are generally regarded as being highly specific insect control agents. However, recent reports have indicated that certain photodegradation products of one JHA (methoprene) may be a factor contributing to observed abnormal embryonic development in frogs (Ankley et al. 1998; La Clair et al. 1998). This developmental toxicity is presumed to be due to interactions of the JHA derivatives with either the RAR or the RXR. Thus, interactions between invertebrate hormone analogs (albeit degradation products) may impinge on developmental processes in some vertebrates.

Biomarkers and bioassays for detecting ED

Ecdysteroid disrupters

The two bioassays described below can be used to screen for potential endocrine-disrupting properties of environmental chemicals but would not serve as markers of ED in the field.

Calliphora pupariation assay

Traditionally, the *Calliphora* pupariation assay (Fraenkel 1935) has been used to detect compounds that stimulate insect molting. In this assay, mid-third instar *Calliphora* larvae are ligated in the middle and allowed to undergo pupariation. Only the anterior part containing the brain and prothoracic glands (Figure 2-4) undergoes pupariation, since the posterior part of the larva lacks 20E. The molt-inducing activity of a xenobiotic can be tested by briefly immersing the posterior end of a ligated larva into its solution in a solvent such as acetone or methanol. The posterior end of ligated larva then will pupariate if the xenobiotic has molt-inducing activity.

In order to screen for 20E antagonists and for compounds interfering with ecdysone synthesis, the anterior end could be injected with the xenobiotic and an inhibition of its pupariation assessed. If this were found, then one would have to use the ecdysone receptor assay to determine whether the compound was acting at the target site. If not, then it likely would be interfering with either PTTH or ecdysone release.

Competitive radioligand binding assay for agonists of 20E activity

The binding affinity of a xenobiotic to EcR/USP can be tested using radiolabeled ponasterone A (Cherbas et al. 1988) and either protein extracts from 20E-responsive tissues or insect cells, or proteins produced in vitro from cloned cDNAs encoding EcR and USP (Kapitskaya et al. 1996). The ability of a xenobiotic to bind EcR/USP is determined by using it as a competitor of tritiated ponasterone A binding. Data from competitive displacement of the radioligand with a wide range of concentrations of the test compound can be used to calculate its IC50 value.

Juvenile hormone disrupters

The main effects of substances acting as JH agonists in insects would be disruption of normal embryogenesis and, at the end of larval development, interference with the process of metamorphosis and the inhibition of normal adult eclosion. Frequently, adult insects that manage to survive these morphogenetic effects of JHA exhibit marked reduction in their ability to reproduce. Various morphological and phenotypic (e.g., color) parameters may be indicative of such effects. In some insects, JH agonism may reduce mating, induce partial or complete sterility, affect sex ratio, and induce or terminate diapause. Substances acting as AJH agents will induce premature metamorphosis of larval stages and suppress reproduction. Many of these morphogenetic effects caused by compounds interfering with the JH-based endocrine system in insects are easily recognized. Yet it should be remembered that similar effects might be induced if insects are exposed to certain physical, environmental, or sublethal chemical stress during critical stages of development.

Nonendocrine-disrupting effects on development

Chapter 1 noted the working definition the workshop members agreed to use, pointing out that ED actions involve some functional interaction with the hormone system. Endocrine systems are sufficiently complex that such interactions can and do occur at various points in the system, especially when hormonal cascades are complex and extensive. Furthermore, the physiological systems of many invertebrates are sufficiently interdependent that reproduction, growth, or development may be impaired by ED mechanisms or some other, nonhormone-mediated mechanism. This latter situation occurs in the chitin synthesis inhibitors, used as insecticides against certain insects.

The benzylphenyl ureas (Ishaaya 1992) represent a class of chitin synthesis inhibitors that interfere with an endocrine-regulated process in arthropods (molting) through a nonendocrine mechanism. Chitin synthesis inhibitors (CSIs) are selective insecticides that act on insects of various orders by inhibiting one of the steps of chitin synthesis. This synthesis usually takes place at or during the time of molting from one developmental stage to the other when the synthesis of new cuticle requires massive chitin synthesis. Thus, inhibiting chitin synthesis prevents normal cuticle formation and causes death during the molt, albeit not via an endocrine pathway. Molting in arthropods is an endocrine-regulated process that involves a combination of molecular, biochemical, physiological, and behavioral actions. Altering any one of the many processes in the molt may disrupt the outcome, even if ED is not involved. Thus, while molting of insects is under the regulatory control of ecdysteroids, the ability of a chemical to interfere with molting may not reflect ED. The inhibition of molting may be due to effects at nonendocrine sites, as demonstrated with CSI.

Crustacea

Most of what is known about crustacean endocrinology has been obtained through the observation of the larger decapods (crabs, lobsters, crayfish, and shrimp). Some other groups of crustaceans, however, have proven useful as model systems for the elucidation of some hormonal processes (most notable, the androgenic hormone of amphipods). Although there are some examples of laboratory-based experiments demonstrating ED in crustaceans, the conclusions are often equivocal. Field-based data are generally lacking. Crustaceans may serve as significant indicator species of ED due to their economic importance (aquaculture and fisheries), ecological significance (as keystone species in many food webs), and extensive use as model invertebrates in laboratory toxicity evaluations.

Molting

Ecdysteroids

The class of steroid molting hormones, the ecdysteroids, has been introduced above in the section on insects. Shortly after elucidating the chemical structure of insect ecdysone (Figure 2-4), the crustacean molting hormone was isolated and characterized in the lobster *Jasus lalandei* (Hampshire and Horn 1966; Horn et al. 1966). Although the crustacean hormone was originally called *crustecdysone* or *ecdysterone*, it was demonstrated that molting hormones from both insects and crustaceans were identical (20E).

The primary secreted product of insect prothoracic glands was determined to be ecdysone through the use of organ culture experiments (Chang and O'Connor 1977; Keller and Schmid 1979). The homologue of the prothoracic gland in crustaceans is termed the "Y-organ." Organ culture experiments with crustacean tissues, similar to those conducted with insects, resulted in the characterization of ecdysone as the primary secretory product of the Y-organ. In both insects and crustaceans, secreted ecdysone is rapidly hydroxylated to 20E by several tissues.

More recent experiments, however, indicate further complexities to this paradigm. Other reports have shown that the Y-organs of some crustaceans secrete other ecdysteroids. In some crabs, 3-dehydroecdysone and 25-deoxyecdysone (Figure 2-4) are the products of the Y-organs (Lachaise et al. 1989; Spaziani et al. 1989). The roles of these different ecdysteroids remain to be determined.

Neuropeptide

While ecdysone secretion by the prothoracic glands of insects is stimulated by a neuropeptide (PTTH; see section on insects), ecdysone secretion by the Y-organ of crustaceans is under the regulatory control of a neuropeptide inhibitor. Following the removal of both stalked eyes of the fiddler crab *Uca pugilator*, the experimental animals displayed a shortening of the interval of their molt cycle (Zeleny 1905).

This observation led to the postulation that an endocrine factor was present in the eyestalks that normally serves as a molt-inhibiting hormone (MIH) (Figure 2-9).

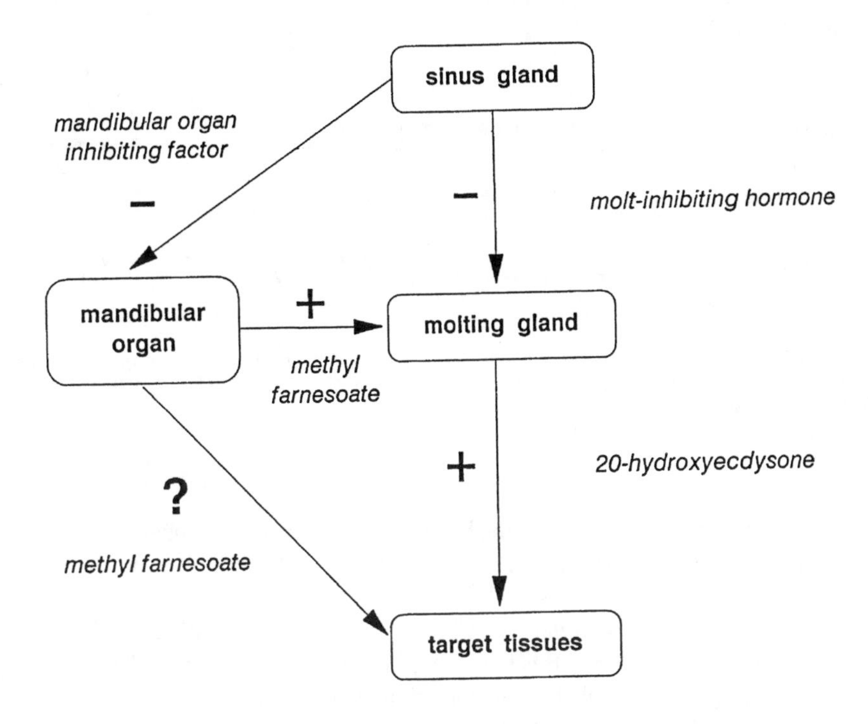

Figure 2-9 Schematic of the endocrine control of molting in a decapod crustacean

Microscopic examinations of the eyestalk neural tissue revealed a neurohemal (storage) organ. This organ is called the sinus gland and serves as a storage site for neurosecretory products. It consists of the enlarged endings of a group of neurosecretory neurons collectively called the "X-organ." The observed shortening of the molt interval following eyestalk ablation is likely a result of a rapid elevation in the concentration of circulating ecdysteroids due to the disinhibition of the Y-organ. Evidence for the ability of the X-organ–sinus gland to inhibit ecdysteroid secretion by the Y-organ has been presented (Soumoff and O'Connor 1982; Mattson and Spaziani 1985). These authors showed that in vitro secretion of ecdysone by crab Y-organs could be inhibited when the organs were cultured with conditioned medium that had previously been incubated with explanted sinus glands.

MIH is a member of an arthropod-specific neuropeptide family. In the lobster *Homarus americanus*, MIH is a 72 amino acid peptide (Chang et al. 1990; Tensen et al. 1991; Chang 1997). It is of interest that only a few changes in some of the amino acid residues result in an alteration of the peptide's biological activity. In the lobster, a related peptide, crustacean hyperglycemic hormone B (CHH-B) has no MIH biological activity but does have considerable hyperglycemic activity (Chang et al. 1990). There is a 96% identity between MIH (=CHH-A) and CHH-B (Figure 2-10; Tensen et al. 1991).

	1										11										21
Hoa-CHH-A/MIH	pE	V	F	D	Q	A	C	K	G	V	Y	D	R	N	L	F	K	K	L	D	R	V	C	E	D
Hoa-CHH-B	pE	V	F	D	Q	A	C	K	G	V	Y	D	R	N	L	F	K	K	L	N	R	V	C	E	D

					31										41						
Hoa-CHH-A/MIH	C	Y	N	L	Y	R	K	P	F	V	A	T	T	C	R	E	N	C	Y	S	N	W	V	F	R
Hoa-CHH-B	C	Y	N	L	Y	R	K	P	F	I	V	T	T	C	R	E	N	C	Y	S	N	R	V	F	R

	51										61										71	
Hoa-CHH-A/MIH	Q	C	L	D	D	L	L	L	S	D	V	I	D	E	Y	V	S	N	V	Q	M	V_{NH2}
Hoa-CHH-B	Q	C	L	D	D	L	L	M	I	D	V	I	D	E	Y	V	S	N	V	Q	M	V_{NH2}

Figure 2-10 Amino acid sequences of lobster (*H. americanus*) crustacean hyperglycemic hormone A (Hoa-CHH-A), which is the same as molt-inhibiting hormone (Hoa-MIH), and Hoa-CHH-B (Chang et al. 1990; Tensen et al. 1991; Chang 1997)

Another member of this neuropeptide hormone family is the vitellogenesis-inhibiting hormone (VIH) from *H. americanus* (Soyez et al. 1991). When injected into shrimp, lobster VIH inhibits vitellogenesis. It is unknown whether this peptide is active in lobsters. It also is unknown whether this peptide has either CHH or MIH activities. Another member of this family has been isolated from crabs. This member of the peptide family inhibits secretion of the terpenoid methyl farnesoate (MF) by the mandibular organ (see "Methyl farnesoate" below) and is called "mandibular organ-inhibiting hormone" (MOIH) (Wainwright et al. 1996; Liu et al. 1997). Several dozen members of the CHH/MIH/VIH/MOIH peptide family have been identified in several decapod species (see Chang 1997 for a review). A related peptide hormone that regulates water transport in the locust also has been isolated and sequenced (Audsley et al. 1992; Meredith et al. 1996).

Methyl farnesoate: A crustacean juvenile hormone?

Similarities between the endocrinology of molting by crustaceans and insects (mediation by 20E with regulation of ecdysone secretion from the molting gland by a neuropeptide) led to speculation that an analog to JH (Figure 2-6) may be present and functional in crustaceans.

JH or related compounds reportedly have biological activity in crustaceans, and extracts of crustacean tissues have been reported to have JH activity in insects. Several decades ago, JH activity was observed in eyestalks of the lobster using an insect bioassay (Schneiderman and Gilbert 1958). More recently, the normal development of larvae of the barnacles *Balanus galeatus* and *Elminius modestus* (Tighe-Ford 1977), the cladoceran *Daphnia magna* (Templeton and Laufer 1983), the crab *Rhithropanopeus harrisii* (Costlow 1976; Christiansen et al. 1977a; 1977b), and the lobster *H. americanus* (Hertz and Chang 1986) has been shown to be affected by additions of JH or analogs to the water in which these crustaceans were cultured.

Current data, however, indicate that JH probably is not present in crustaceans. It appears that the precursor to insect JH III, MF (Figure 2-6), may be acting as the crustacean JH. MF (and not JH) was isolated from several tissues of the crab *Libinia emarginata* (Laufer et al. 1987). Also, the secretory source of MF was determined to be the mandibular organ. Supporting data were obtained from the lobster *H. americanus*, the crayfish *Cambarus bartonii* and *Procambarus clarkii*, the crabs *Cancer borealis, Carcinus maenas*, and *Scylla serrata*, and the prawn *Macrobrachium rosenbergii* (see Homola and Chang 1997 for review).

Analogous to insect JH binding proteins, hemolymph MF binding proteins have been characterized in few crustacean species examined (see Homola and Chang 1997 for review). Intracellular binding proteins (i.e., receptors) for MF in crustaceans have not yet been characterized. Recent data suggest the presence of an inhibitor of MF secretion by the crustacean mandibular organ. Addition of crab (*L. emarginata*) sinus gland extracts to cultured mandibular organs resulted in a significant decline in secreted MF in vitro (Laufer et al. 1987). Injection of sinus-gland extracts in lobsters (*H. americanus*) resulted in a decline in circulating MF in vivo (Borst et al. 1991). Further work on this topic resulted in the characterization and sequencing of MOIH (Wainwright et al. 1996; Liu et al. 1997), which is functionally equivalent to the insect allatostatin.

Reproduction

Methyl farnesoate

Several lines of evidence indicate that MF may have a permissive or stimulatory effect upon reproduction in both female and male crustaceans. A role for MF-secreting mandibular organs in reproduction was first indicated through their implantation into juvenile female spider crabs, *L. emarginata*. Implanted females experienced an increased frequency of vitellogenic ovaries in the few animals that survived this difficult protocol. In mature females, mandibular organs secrete varying amounts of MF in vitro, depending upon the stage of the reproductive cycle (Laufer et al. 1987). A recent report presents strong evidence that MF does act as a female gonadotropin (Laufer et al. 1998).

Associations between MF concentrations and reproductive status have also been observed in male crustaceans. Male *L. emarginata* with worn epicuticles (indicative of a particular reproductive status) display mating behavior, have large reproductive systems, and have hemolymph with high MF titers (39 to 67 ng/ml). Males with intact epicuticles do not display mating behavior, have small reproductive systems, and have hemolymph with significantly lower MF titers (3 to 30 ng/ml) (Homola et al. 1991; Sagi et al. 1993, 1994).

Sex-specific differences in MF titer occur in several species of crustaceans. Male crab *L. emarginata*, lobster *H. americanus*, and crayfish *P. clarkii* have higher levels of MF in the hemolymph than females (Laufer et al. 1987; Borst et al. 1987; Landau et al. 1989). In *H. americanus*, the relative size of the mandibular organ increases in males after sexual maturity, but not in females, suggesting that the mandibular organ may function differently in males and females (Waddy et al. 1995). MF titers in females may be lower than in males because the mandibular organs are smaller and less active or because MF is sequestered by the gonads. MF is present in the lobster ovary at 2.7 ng/g (Borst et al. 1987), although the ovary is unable to synthesize MF. It is not certain whether these differences in MF titers mediate sex-specific physiology and morphology or whether these differences are the result of sexual differentiation.

The specific role of MF in crustacean reproduction is ambiguous. In the lobster *H. americanus*, mandibular organ ablation had no apparent effect upon the latter phases of female reproduction, including secondary vitellogenesis and oviposition (Byard 1975). MF injection had no effect upon vitellogenin titer in female *H. americanus* (Tsukimura et al. 1993) or in male freshwater prawns, *M. rosenbergii* (Wilder et al. 1994). However, it is possible that MF has a role in the early stages of ovarian development, which coincide with elevated MF titers (Waddy et al. 1995). Male *M. rosenbergii*, differentiate into alternative morphotypes, and MF titer in these animals varies from 0 to 37 ng/ml. The means, standard deviations, and ranges of titers are similar in males and females (Sagi et al. 1991). Differences in MF titer could not, however, be correlated with male morphotypes. Additionally, vitellogenesis does not appear to influence MF titers in female *M. rosenbergii* (Wilder and Aida 1995).

MF does appear to affect the duration of the molt cycle. When MF was added to the water in which lobster larvae were cultured, a significant increase in the time to metamorphosis was observed (Borst et al. 1987). Similar observations were made with larvae of *M. rosenbergii* (Abdu et al. 1998). These observations are consistent with the hypothesis that MF is acting as a crustacean JH in that it promotes the retention of juvenile characteristics. However, the morphologies of the larval lobsters were not significantly different.

Ecdysteroids

Ecdysteroids have been implicated in crustacean vitellogenesis as described above with some insects. Ovarian free-ecdysteroid concentrations fell with increasing stage of vitellogenesis in the prawns *Penaeus monodon* and *M. rosenbergii* (Young et al. 1993a, 1993b). Hemolymph concentrations of free ecdysteroid also were found to decrease with increasing stage of vitellogenesis in *P. monodon* (Young et al. 1993a), and whole-body ecdysteroid concentrations decreased in brine shrimp *Artemia salina* between primary and secondary vitellogenesis (Walgraeve et al. 1988). Conversely, ovarian and hemolymph ecdysteroid levels increased with increasing stage of vitellogenesis in the crab *Carcinus maenas* (Lachaise and Hoffman 1977).

Ecdysteroids have been shown to stimulate vitellogenesis in the ovaries of some crustaceans (Gohar and Souty 1984) and inhibit it in others (Fyhn et al. 1977). Thus, a role for ecdysteroids in crustacean vitellogenesis is clearly indicated, though the precise function of the steroid remains to be determined and may vary among taxa.

Ecdysteroids have been measured in gonads, developing oocytes, and eggs during the early stages of embryonic development (Quackenbush 1986; Spindler et al. 1987). These ecdysteroids are presumed to be of maternal origin and may contribute to the regulation of germinal vesicle breakdown and early embryonic maturation. Generally, ecdysteroid titers have been found to increase with increasing development of the embryo (McCarthy and Skinner 1979; Lanot and Cledon 1989; Young et al. 1991). This increase has been interpreted as evidence that the developing embryo is capable of ecdysteroid synthesis (Wilder and Aida 1995).

Molting and reproduction are highly coordinated in some crustaceans (e.g., branchiopods and brachyura). Accordingly, ecdysteroids can have a significant impact on reproductive success by virtue of their role in the molting process.

Development

Androgenic hormone

Sexual differentiation of malacostracan crustaceans (decapods, isopods, amphipods, etc.) is under the regulatory control of androgenic hormone. Androgenic hormone is the product of the androgenic gland, which is typically associated with the terminal region of the male gamete ducts. Two proteins, of similar molecular weights, were purified from the androgenic gland of the isopod *Armadillidum vulgare* and found to have androgenic hormone activity, as measured by their ability to cause elongation of the endopodites of the first pair of pleopods of female isopods (Martin et al. 1990).

Ablation of the androgenic gland caused feminization of male prawns *M. rosenbergii* (Nagamine et al. 1980a). Feminizing effects included the initiation of oogenesis and development of oviducts and female gonopores. Re-implantation of the androgenic gland to feminized males caused defeminization. Ablation of the androgenic gland from male penaeid shrimp *Penaeus indicus* caused loss of male secondary sex characteristics and blocked spermatogonial differentiation (Mohamed and Diwan 1991). In male isopods *A. vulgare*, ablation of the androgenic gland prevented elongation of the endopods (copulatory organs) (Suzuki et al. 1990).

Implantation of the androgenic gland into females causes masculinization. Consequences of androgenic gland implantation into females has included masculinization of pleopod morphology, development of male gonopore complexes, and masculinized chelipeds in *M. rosenbergii* (Nagamine et al. 1980b). Female isopods *A. vulgare* implanted with androgenic gland caused functional sex reversal of the gonads (Suzuki and Yamasaki 1997). Androgenic gland ablation and implantation experiments such as these have conclusively demonstrated that the development of primary and secondary male sex characteristics in malacostracan crustaceans is under the positive regulatory control of androgenic hormone. In the absence of androgenic hormone, these organisms develop into functional females.

Vitellogenin synthesis has also been shown to be under the negative regulatory control of androgenic hormone in the isopod *A. vulgare*. Ablation of the androgenic gland from male isopods initiated synthesis of vitellogenin by the fat body, whereas implantation of androgenic gland to female isopods inhibited vitellogenin synthesis by the fat body (Suzuki et al. 1990). This negative regulatory control of vitellogenin synthesis stands in contrast to the positive regulatory control of vitellogenesis by 17β-estradiol in oviparous vertebrates (Palmer and Palmer 1995).

Androgenic hormone appears to be common to all malacostracan crustaceans. This would include the mysid shrimp *Mysidopsis bahia*, an invertebrate species commonly used in toxicity testing. A functional role for androgenic hormones in lower crustaceans (i.e., Cladocera, Copepoda) has not been established. Additional research is needed to establish whether an endocrine pathway involving androgenic hormone also regulates processes of sexual differentiation and vitellogenin secretion by these species, which are used commonly for toxicity evaluations.

Vertebrate-type sex steroids

Vertebrate-type sex steroids such as 17β-estradiol (deLoof and DeClerk 1986), testosterone (Burns et al. 1984), and progesterone (Kanazawa and Teshima 1971; Yano 1985) have been measured in malacostracan crustaceans. The presence of these compounds in crustaceans, however, does not imply that the compounds function as hormones in these organisms. Fragmented evidence, however, does suggest that some of these compounds may function as hormones in crustaceans. For example, testosterone administration has been shown to stimulate the conversion of ovaries to testis in female ocypod crabs (Sarojini 1963), induce hypertrophy and hyperplasia of the androgenic gland of the penaeid shrimp (Nagabhushanam and Kulkarni 1981), and mobilize glycogen from the midgut to the testis of the penaeid shrimp (Nagabhushanam and Kulkarni 1981). Progesterone administration to the shrimp *Metapenaeus ensis* induced ovarian maturation and spawning (Yano 1985). The progesterone derivative 17β-hydroxyprogesterone stimulated vitellogenin secretion in the prawn *Penaeus japonicus* (Yano 1987).

Studies with the cladoceran *Daphnia magna* have shown their capability to extensively metabolize testosterone to a variety of oxido-reduced, hydroxylated, and conjugated derivatives (Baldwin and LeBlanc 1994a, 1994b). Testosterone is primarily eliminated from daphnids as a glucose conjugate, while several oxido-reduced derivatives of testosterone, which are variously androgenic in vertebrates, are preferentially retained by the organisms (Baldwin and LeBlanc 1994b). A functional role for testosterone or its metabolites in daphnids has not been established, though exposure of female daphnids to testosterone has been shown to interfere with normal embryo development (LeBlanc, personal observations). These limited data are suggestive but insufficient to conclude that vertebrate-type sex steroids function as hormones in crustaceans.

Pigmentation

There are 2 types of pigmentary effectors in the Crustacea: the chromatophores and the retinal pigment cells. The chromatophores are located primarily in the integument, and their activities result in the body pigmentation of the entire animal. The retinal pigments are located in the compound eyes. They regulate the amount of light impinging on the rhabdome, which is the light-sensitive portion of each ommatidium (the functional unit) of the compound eye. The physiology and morphology of these 2 types of pigmentary effectors are quite different, although both are subject to endocrine regulation. There are several excellent reviews on these topics, and they should be consulted for more detailed discussions (Kleinholz and Keller 1979; Fingerman 1985; Kleinholz 1985; Rao et al. 1985).

Chromatophores

The chromatophores are pigment-containing cells that exist not only on the surface of the crustacean but also in some internal tissues. Their function is to adjust the body coloration to its surroundings, depending upon the situation (e.g., protection, mating behavior, antagonistic displays). The chromatophores are single cells with numerous extensions. Pigment granules of various colors migrate into and out of these extensions. The most common pigment granules are black, red, white, and yellow.

When the pigment granules are spread out along the extensions of the chromatophore, they are described as being in a dispersed state. In this state, the granules are most visible and give the chromatophore its characteristic color. When the granules are withdrawn from the extensions of the chromatophore, they are in a concentrated state and are not as visible.

As noted by Fingerman (1989), the different body colors that arise from the differential activities of the chromatophores serve several important functions. First, they provide camouflage to match the background color and pattern (Brown 1950). This is useful for both prey capture and predator protection. Second, there is evidence that differential body coloration can affect body temperature. Thus light coloration on hot days and dark coloration on cold ones may assist in thermoregulation (Wilkens and Fingerman 1965). Third, pigment dispersion may protect the living cell layers from damage by ultraviolet irradiation (Coohill et al. 1970). Fourth, differential coloration may be utilized for intraspecific communication, e.g., courtship or antagonism (Crane 1944).

Endocrinology

In addition to precocious molting, it was seen many years ago that eyestalk ablation of crustaceans resulted in a dramatic alteration of body color (Koller 1925, 1927). In some species (e.g., *Crangon*, *Palaemon*, and *Palaemonetes*), the body color darkens following eyestalk ablation. In others (e.g., *Uca*), the body color whitens. An indication that regulation of body color may be under hormonal control was the demonstration that injection of eyestalk extracts could reverse the color changes induced by eyestalk ablation (Koller 1928; Perkins 1928).

Early observations indicated that the sinus gland (discussed above in reference to MIH) was a storage site for neurosecretory material that regulated color change (chromatophorotropins). However, it appears from a number of different observations that the chromatophorotropins are not only located in the X-organ/sinus gland complex, but in other neural tissue, both within the eyestalk and in other locations within the body as well (see Rao 1985 for a more thorough discussion).

Because of the number of different chromatophores, the diversity of suspected sources of the chromatophorotropins, and the likely presence of both concentrating and dispersing hormones in the same tissues, it is difficult to make any overall generalizations about the large number of publications dealing with the various physiological observations and partial purifications of the regulating factors (see Rao 1985 for review). However, there has been a great deal of recent work on 2 general classes of pigmentary effector hormones: the red pigment-concentrating hormones (RPCHs) and the pigment-dispersing/light-adapting hormones (PD/LAH). The chemistry of PD/LAH will be discussed in more detail in the following section on the retinal pigment hormones.

The first crustacean chromatophorotropin that was sequenced was RPCH from the eyestalks of the shrimp *Pandalus borealis* (Fernlund 1974a, 1974b; Fernlund and Josefsson 1968, 1972). Its amino acid sequence is shown in Figure 2-11. Some publications refer to this molecule as "erythrophore (red chromatophore) concentrating hormone," or ECH. In addition to the erythrophores, however, there are observations that other types of chromatophores are concentrated with RPCH (Josefsson 1975, 1983a, 1983b).

RPCH is a member of a family of structurally related peptides. Another member of this family, AKH, has been isolated from several different insect species. Interestingly, although there is biological crossreactivity between the insects and crustaceans (i.e., RPCH has some AKH activity in insects, and AKH has some RPCH activity in crustaceans), the functions of the hormones are quite distinct in the 2 taxonomic groups (i.e., AKH does not appear to alter lipid mobilization in crustaceans).

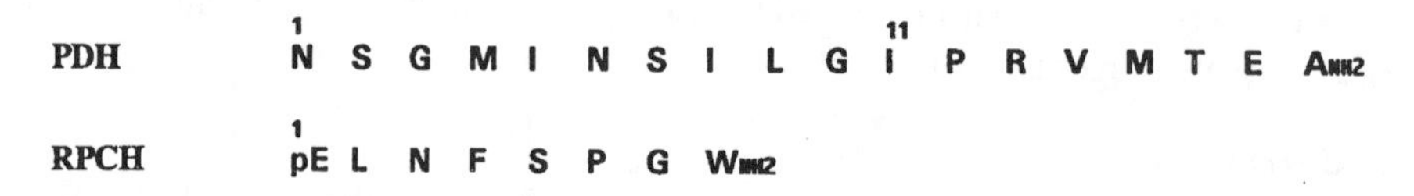

Figure 2-11 Amino acid sequences of one of the pigment-dispersing hormones (PDH; top) and red pigment-concentrating hormone (RPCH; bottom) (Fernlund 1976; Fernlund and Josefsson 1972)

Retinal pigments

There are 3 retinal pigments in the compound eyes of shrimp and other decapod crustaceans. The distal retinal pigment is found within cells that extend from the distal ends of the retinular (photoreceptor) cells to the cornea. The proximal retinal pigment is contained within the retinular cells. The reflecting retinal pigment is contained within cells that are variously located between the neighboring ommatidia, depending upon the species.

These 3 pigments alter their cellular locations in concert such that their combined activities result in either more or less light impinging upon the rhabdome, depending upon the ambient light intensity. In dim light or darkness, the light-sensitive rhabdome is exposed by the movement of the distal and proximal pigments. In addition, the reflecting pigment is located distally to the basement membrane in a position that permits light to be reflected onto the rhabdome. In bright light, the distal and proximal pigments move to screen the rhabdome from excess light. In addition, the reflecting pigment is positioned adjacent to the basement membrane. In this position, it does not reflect light onto the rhabdome.

PD/LAH was first isolated and its structure determined from the eyestalks of the shrimp *Pandalus borealis* (Figure 2-11; Fernlund 1976). It was initially referred to as the distal retinal pigment light-adapting hormone (DRPH) because of its action on the distal retinal pigment (Fernlund 1971). Later studies (Kleinholz 1975) demonstrated that this same molecule was able to disperse the pigment granules in several types of chromatophores. Since the chromatophore assay is easier to conduct than the characterization of the location of the retinal pigments, this hormone is presently often referred to as the pigment-dispersing hormone (PDH). A second PDH (β-PDH) has been isolated and characterized from the fiddler crab *Uca pugilator* (Rao 1985), and additional members of this neuropeptide family have been reported (Rao and Riehm 1989; Rao et al. 1985).

Limb regeneration

Crustaceans possess a remarkable ability to regenerate limbs and other appendages. The actual factors responsible for the growth of a new limb are unknown. However, it has been observed that there is a precise interplay between the molt cycle and regenerative events. Although the observation of multiple limb loss affecting the duration of the molt interval had been made by others (see review by Skinner 1985), this phenomenon was not thoroughly defined until the work of Skinner and Graham (1970, 1972). They observed in crabs (*Gecarcinus lateralis*) that multiple limb autotomy was in some ways a greater stimulus to molting than eyestalk removal and hypothesized that, when a sufficient number of limbs have been removed, a molt-promoting factor acts to initiate the molting process. Skinner (1985) termed this molt-promoting substance *limb autotomy factor, anecdysial* (LAF_{an}). Both the chemical nature and the source of this factor are unknown at present.

Holland and Skinner (1976) made additional observations on the interaction of regeneration and molting. They observed that the loss of limbs or primary regenerates (the limb buds of regenerating limbs) was able to inhibit the progress of premolt, as measured by the growth of the remaining primary regenerates. When the secondary regenerates (the limb buds formed following autotomy of the primary regenerates) reached a size that was comparable to the remaining primary regenerates, premolt events resumed. The factor responsible for this effect was termed "limb autotomy factor, proecdysial" (LAF_{pro}). The modes of action of both LAF_{an} and LAF_{pro} are not known at this time. LAF_{an} could either inhibit the secretion of MIH or antagonize MIH action at the level of the molting gland. LAF_{pro} could either promote the secretion of MIH or inhibit the secretion of ecdysone by the Y-organ. As with LAF_{an}, the chemical nature and the source of LAF_{pro} are not known.

Retinoic acid can disrupt limb regeneration in the fiddler crab *Uca pugilator* (Hopkins and Durica 1995). Presumably this action is mediated by the crab RXR that is present in these limbs (Chung et al. 1998), but neither the mechanism nor the possible normal role of retinoids in this regeneration are known.

Diapause

Some crustaceans are capable of existing in a state of quiescence and low metabolism called "diapause." This trait in crustaceans is primarily associated with those species (i.e., water fleas, fairy shrimp, brine shrimp) that inhabit environments that are periodically inhospitable due to freezing, desiccation, or other adversity. Crustaceans typically diapause as eggs that survive the inhospitable environment and hatch when favorable conditions return. Several environmental stimuli have been identified that initiate diapause in crustaceans including photoperiod, food quantity, and quality (Schwartz and Hebert 1987; Stross and Hill 1968a, 1968b). Diapause in crustaceans is presumably under neuro-endocrine control, as in insects, though the hormones responsible have not been identified.

Endocrine disruption

Molting

Except for the differences in the neuroendocrine regulation of the molting gland described above for insects and crustaceans, it is likely that any chemical that disrupts the normal endocrine control of molting in insects will also be active in crustaceans (see section on insects). These disrupters could either stimulate or inhibit the neurosecretory control of the Y-organ. At the cellular level, ecdysteroid disrupters could either positively or negatively compete with ecdysteroids for the ecdysteroid receptor (see section on insect ecdysteroid receptors). Little research has examined this question in crustaceans, especially the crabs, shrimp, and lobsters.

Most toxicological studies on crustacean physiology have not examined cellular effects or effects on hormone titers. These studies have been concerned primarily with morphological and temporal effects of pollutant exposure (see reviews by

Fingerman et al. 1996, 1998). Molting is inhibited by heavy metals (Weis et al. 1992), polychlorinated biphenyls (PCBs) (Fingerman and Fingerman 1977), benzene (Cantelmo et al. 1981), and vertebrate estrogens (Baldwin et al. 1995; Zou and Fingerman 1997a, 1997b). Because so much of a crustacean's physiology is directed toward molting, just about any xenobiotic that has a deleterious effect upon a crustacean will likely also disrupt molting. Therefore, one must exercise caution before strictly defining an exogenous chemical as an endocrine disrupter unless first certain that the chemical in question is not simply a general toxicant (see Chapter 3, section "Invertebrate full life-cycle as the 'gold standard,'" p 111).

Metabolism

The primary mediator of crustacean metabolism is CHH. This hormone mediates the breakdown of glycogen to glucose in the hepatopancreas (midgut gland) and muscles. Perturbations in normal CHH levels resulting from chemical exposure can be measured by an enzyme-linked immunosorbent assay (ELISA) (Chang et al. 1998).

Reproduction

Several processes that contribute to reproduction in crustaceans are under hormonal control. The hormonal regulation of some of these processes has been well described (e.g., the role of androgenic hormone in sex differentiation), while others require further evaluation (e.g., the role of MF in ovarian development).

Neither sex differentiation in males nor vitellogenin production in female malacostracans has been demonstrated to be directly altered by synthetic chemicals. However, intersex individuals have been identified in field populations of decapods (Farfante 1978; Sangalang and Jones 1997), isopods (Smith 1977; Korczynski 1985), and amphipods (Hasting 1981). Whether such occurrences represent aberrations in androgenic hormone action due to contaminant exposure remains to be determined. ED has been implicated in aberrations in sexual differentiation in cladocerans. Exposure of daphnids to p-*tert.*-pentylphenol during the 16- to 24-hour period following release from the brood chamber masculinized female daphnids as evidenced by morphological changes in the shape of the carapace and the development of setae on the carapace edge (Gerritsen et al. 1997). These morphological changes normally occur in male daphnids during sexual maturation. Interestingly, the elongation of the first antennae, a prominent aspect of sexual maturation of male daphnids, occurs at a relatively constant rate between the time of release from the brood chamber to the attainment of reproductive maturation (Olmstead and LeBlanc 1998). Masculinization of daphnids resulting from exposure to p-*tert.*-pentylphenol did not cause elongation of the first antennae. This suggests that morphological changes in the carapace and elongation of the first antennae are under distinct regulatory controls, and p-*tert.*-pentylphenol interfered with one of these processes.

Toxicant effects on the development or maintenance of female sexual characteristics have also been reported. Exposure of the amphipod *Melita nitida* to sediments contaminated with waste crankcase oil caused abnormalities in oostegite setae (Borst et al. 1987). The oostegite setae function to maintain developing embryos in the brood chamber and possibly provide movement and aeration to the embryos. Structural abnormalities in the oostegite setae could interfere with maintenance of viable embryos by the affected females. A functionally analogous structure in cladocerans is the abdominal process. This structure is absent in neonatal daphnids and adult males. The abdominal process develops in females during reproductive maturation (Olmstead and LeBlanc 1998). Exposure of female daphnids to the testosterone and estrone precursor, androstenedione, significantly inhibited the rate of development of the abdominal process in daphnids (*D. magna*) (Olmstead and LeBlanc 1998). Total length of these exposed daphnids was also reduced during these experiments, rendering it difficult to ascertain whether the steroid compound was eliciting a direct effect on the development of the sex characteristic.

Most daphnids utilize cyclic parthenogenesis as a reproductive strategy (Figure 2-12). Under favorable conditions, populations consist primarily of parthenogenetically reproducing females that are capable of high reproductive productivity. With the onset of less favorable conditions, such as those associated with the approach of winter, reduced food availability, or overcrowding, the female daphnids will produce males and enter a sexual phase of reproduction. Upon maturation, these males will mate with receptive females and produce diapause (resting) eggs. These eggs are encased in a protective ephippium that facilitates relocation of the eggs to more favorable environments.

The neuroendocrine signals that respond to the environmental stimuli resulting in the switch to sexual reproduction are not known. Reports of stimulation of sexual reproduction and diapause in response to environmental chemicals are rare, indicating that general toxic stress does not stimulate entry into diapause. However, exposure to some chemicals has been reported to stimulate aspects of diapause. For example, exposure of female daphnids to the herbicide atrazine stimulated the production of male offspring in a concentration-dependent manner (Dodson et al. 1999). Parthenogenetically reproducing daphnids exposed to the juvenile insect growth regulator pyriproxyfen were stimulated to produce diapause eggs (Trayler and Davis 1996). It must be emphasized that these studies were not designed to evaluate the mechanism of the apparent endocrine-disrupting action. Nonendocrine effects, such as reduced uptake or absorption of food, may also stimulate entry into sexual reproduction.

Simulated pulp mill effluent derived from balsam fir processes elicited acute toxicity (48-hour LC50) to *Ceriodaphnia* at an exposure concentration of between 10 and 20%. However, this same effluent reduced fecundity at exposure concentrations of 0.005 to 0.01% effluent. This reproductive toxicity was found to be due to dehydrojuvabione, a phytosesquijuvenoid structurally similar to JH and MF. The severe

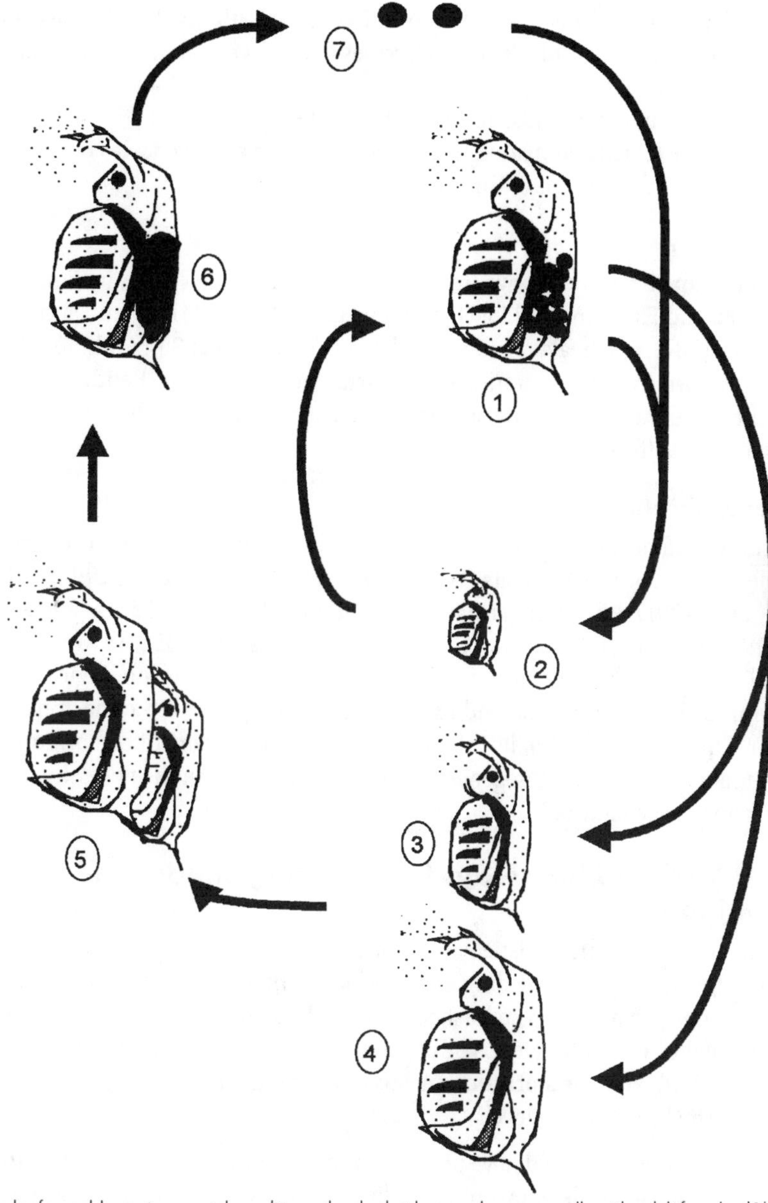

Under favorable environmental conditions, the daphnids reproduce asexually with adult females (1), producing female offspring (2) by parthenogenesis. These offspring then mature into parthenogenetically reproducing females (1). In response to stimuli denoting adverse environmental conditions, the adult females produce males (3). At about the time that the males attain reproductive maturity, the females enter a sexual reproductive phase (4), and males and females copulate (5). Fertilized eggs (diapause eggs) are encased in a protective ephippium (6), which allows survival during dessication or freezing. Upon the return of favorable environmental conditions, the eggs (7) hatch, releasing females (2) that will reproduce by parthenogenesis at maturity (1).

Figure 2-12 Reproductive cycle of the cladoceran *Daphnia magna*

reproductive toxicity of these compounds raised the possibility that JH-like compounds regulate critical reproductive processes in these crustaceans (O'Connor et al. 1992). The JH analogs methoprene and pyriproxyfen also were found to interfere with reproduction in daphnids, although their potency as reproductive toxicants, relative to acute toxicity, appeared to be lower than that observed with dehydrojuvabione (Trayler and Davis 1996; Hosmer et al. 1998).

Pigmentation

Xenobiotics that have been shown to alter coloration in fiddler crabs include PCBs (Fingerman and Fingerman 1978; Hanumante et al. 1981), PAHs (Staub and Fingerman 1984), and heavy metals (Reddy and Fingerman 1995). Cadmium has also been shown to affect migration of the distal retinal pigment (Reddy et al. 1997). All of these chemicals likely interfered with neurotransmitter-mediated release of the pigmentary effector hormone.

Limb regeneration

Deviations from normal regenerates have been observed in laboratory experiments involving the use of heavy metals (Weis 1977, 1978; Weis et al. 1986, 1992), TBT (Reddy et al. 1991), PCBs (Fingerman and Fingerman 1980), pesticides (Rao et al. 1978, 1979; Clare et al. 1992), and benzene (Cantelmo et al. 1981). As with all of the above discussion, care must be exercised in attributing some morphological or biochemical abnormality to an endocrine disrupter as opposed to some nonspecific stressor. This caveat is especially relevant because the involvement of hormones is largely unknown in the limb regeneration process (apart from the necessity of molting upon regeneration).

Biomarkers and bioassays for detecting endocrine disruption

Crustaceans have a characteristic pattern of ecdysteroid concentration during the course of the molt cycle, providing one measurable marker of hormone function. If baseline patterns have been established in control populations, it is possible that measurement of circulating levels of ecdysteroids by radioimmunoassay (Chang and O'Connor 1979) could be a practical means of assaying for ED through perturbations in normal ecdysteroid levels.

The ELISA CHH assay would have utility during controlled laboratory exposures to environmental chemicals but would have limited utility with field populations, since a number of environmental factors (such as temperature, salinity, and hypoxia) (Chang et al. 1998) as well as chemical pollutants can elevate CHH levels. Exposure to some heavy metals and organic chemicals has been shown to affect hemolymph levels of CHH (reviewed in Fingerman et al. 1998). As with molting, it is not certain whether these chemicals directly target CHH synthesis and/or secretion or whether effects on CHH levels are the consequence of general physiological stress.

Alteration of a population's normal coloration could potentially be a useful field indicator of an endocrine disrupter. However, since many species have some degree of natural color variation (e.g., color morphs of the green crab *Carcinus maenas* or reproductive coloration of male fiddler crabs), researchers would have to be cognizant of normal color patterns. Alteration of pigmentation could also serve as a useful laboratory-based assay of disrupters, since dramatic effects in whole animal coloration can be observed in only a few minutes following injection of test compounds.

Alterations in the normal appearance of regenerating limbs may serve as indicators of ED.

The use of daphnids in standardized bioassays and recent work on reproductive and developmental effects of EDCs make this group a promising candidate for developing ED-specific bioassays. The biology of daphnids is fairly well known, and the reproductive success, offspring morphology, and sexual/asexual reproduction switch are all potential endpoints for ED bioassays.

A Note on Minor Arthropod Groups

All arthropods have a cuticular exoskeleton, which they must shed periodically in order to grow. This periodic molting—the production of the new cuticle followed by the ecdysis of the old cuticle—is orchestrated by the ecdysteroids in all arthropods (Jegla 1990; Lachaise 1990; Spindler 1991; Figure 2-4). Most is known about the endocrinology of the insects and the crustaceans as described above, whereas relatively little is known about the endocrinology of the myriapods (Joly and Descamps 1988; Descamps 1991), the horseshoe crab *Limulus* (Walker 1988), and the arachnids (Stewart 1988). The first stage (trilobite) larva of *Limulus* has been used in bioassays for ecdysteroid activity (Jegla and Costlow 1979). Recent studies in ticks (arachnids) suggest that ecdysteroids are also used in reproduction and that a JH-like compound may be involved in reproductive maturation (Kaufman 1997; Lomas and Rees 1998).

Annelida

The phylum Annelida, the segmented worms, includes 3 classes of soft-bodied worms: the Polychaeta, Oligochaeta, and Hirudinea. The polychaetes are exclusively marine worms with parapodia, fleshy appendages that display a great diversity of body form and specialization, especially cephalic. The oligochaetes are smooth-bodied, mostly freshwater or terrestrial worms, including earthworms and *Tubifex* worms. The hirudineans or leeches are the most specialized of the annelids and inhabit freshwater, marine, and terrestrial environments. The segmented worms have nervous, excretory, and reproductive systems that repeat in each segment.

Unlike the other invertebrate phyla, most of the annelids have closed circulatory systems that are especially well developed in the polychaetes. All annelids possess a coelom, a true body cavity surrounding the digestive tract. The coelom is the location of the excretory structures and the site of gamete storage and maturation prior to release into aquatic environments. Hence, chemical signaling takes place in 2 fluid compartments, the blood and the coelomic fluid.

Endocrinology

The internal secretions of the annelids come from neuronal structures, rather than true endocrine glands of epithelial origin, a condition typical of the non-arthropod invertebrates. The primary hormones are neuropeptides released from neurosecretory cells organized in ganglionic masses that originate in the nervous system. The annelid simple nervous system consists of a head ganglion, the cerebral ganglion, and a single nerve chord running the length of the worm. The head contains neurosecretory cells, some of which end at a modified structure of the dorsal blood vessel. In addition, each segment has a ganglion emanating from the central nerve chord. There is a ganglionic mass in each of the segments of the worm, and often a neural connective with a separate ganglion in each segment. These ganglia end in secretory structures that are the most likely or identified sources of the neurosecretions (Highnam and Hill 1977).

More recently, neurosecretory cells or structures have been identified in the nervous system of the head and segmental ganglia, gut, and reproductive tissues of one or more classes of annelids. Neurosecretory cells in the cerebral ganglion of the head have been at least partially identified and characterized in the leech, using immunocytological methods (Verger-Bocquet et al. 1992). Interestingly, several different peptidergic cell types were identified using polyclonal antibodies raised against vertebrate hormones, indicating conservatism in the evolution of these molecules. Punin (1993) reported the presence of neuroendocrine structures in the digestive epithelium of annelids, and Kulkarni (1989) found evidence of steroid synthesis in leech testes.

Hormones

The identified hormones in annelids are all neuropeptides, though few have been definitively characterized (Porchet and Dainaut-Courtois 1988). The small cardioactive peptide, FMRFamide, first identified in mollusks (Greenberg and Price 1983; see next section for additional discussion) has been identified in the polychaetes *Nereis virens* (Krajniak and Price 1990) and *N. diversicolor* (Baratte et al. 1990). FMRFamide-like peptides have also been reported in both polychaetes (Baratte et al. 1990) and leeches (Evans et al. 1991), but none have yet been found in oligochaetes, although Krajniak (personal communication) found that the digestive tract is responsive to FMRFamide in a dose-dependent fashion. Pheromones are known to cue the release of gametes in mating behaviors in the polychaetes. Sperm maturation is regulated by a specific, though as yet unidentified, peptide.

Steroid and terpenoid hormones

Ecdysteroids reportedly have been found in some annelids, although a function for these putative hormones is not known (Barker et al. 1990). Juvenile hormone affects larval settlement in laboratory investigations (Biggers and Laufer 1996), yet this effect may simply be pharmacological, caused by some normal physiological mechanism. Nothing is known of vertebrate-type sex steroids in this phylum.

Processes under hormonal control

Growth, body regeneration, and reproductive processes are known or suspected to be under neurosecretory control in annelids. The maturation of sperm and oocytes is regulated by a peptide called "sperm maturation factor" (SMF), though this has primarily been identified in marine polychaetes. SMF may occur in oligochaetes (earthworms), but this is uncertain.

Reproduction

Several groups of marine polychaetes undergo various morphological transformations, particularly of the gamete-bearing segments, at spawning, generating a reproductive morph or "epitoke." Epitoke formation is under the control of neurosecretions, although the specific site of release is not known, nor is the mechanism of action.

Earthworms are hermaphroditic, and mutual insemination occurs at mating. Copulation requires secretion of a mucous ring secreted by the clitellum, and mucus production is likely under hormonal control.

Regeneration

Many, if not all, annelids are capable of regenerating at least some lost segments. Regeneration is under hormonal, neuroendocrine control by compounds known as morphogens, although this has not been extensively investigated in most worms. Two sites that regulate regeneration are the head, specifically the ganglion, and the terminal segment, the pygidium (see Highnam and Hill 1977; Gorbman and Davey 1991 for discussion).

Endocrine disruption

There is no information available on the action of endocrine-disrupting agents that modulate any of the processes known or suspected to be under hormonal control in annelids. The only evidence that ED may occur is the stimulation of larval settling and metamorphosis by insect JH and JH analogs (Biggers and Laufer 1992). These processes may be cued by chemical stimulation, similar to a pheromonal process (Pawlick 1992).

Biomarkers and bioassays for detecting endocrine disruption

Neither biomarkers nor bioassays have been developed for detecting ED in annelids. Both segment regeneration and larval settling are under hormonal control and have been used in laboratory experiments for other purposes. These 2 processes offer promise of providing useful bioassays for ED in annelids.

Mollusca

The mollusks are perhaps the most diverse of the invertebrate phyla and are second to the insects in number of identified species. Mollusks include the bivalves (clams, oysters, scallops, etc.), gastropods (snails, slugs, sea slugs), cephalopods (octopus, squid, etc.), limpets, slipper shells, and tusk shells. Because the morphologies and life histories of the various classes differ so greatly, the physiological and hormonal systems are understandably adapted to quite different conditions.

Endocrinology

The molluscan endocrine system centers mainly on neurosecretory loci of the central nervous system, namely the cerebral, pleural, pedal, and abdominal ganglia (Geraerts et al. 1988; Nassel 1996). The products of these neurosecretory sites are neuropeptides, about which substantial information is available. However, most work has been restricted to relatively few species of gastropods (*Aplysia* and *Lymnaea*), while there is some information on the bivalve *Mytilus* (Nassel 1996; Santama et al. 1996; Norekian and Satterlie 1997; de Lange and van Minnen 1998). Typical invertebrate steroids such as ecdysone have only rarely been reported, while there is good evidence for the presence and biosynthesis of vertebrate-type sex steroids such as progesterone and testosterone from precursors in the ovotestis (Wootton et al. 1995). This structure emphasizes the hermaphrodite nature of many mollusks and highlights the need for care and attention when interpreting gonad structure. Many species are protandric (testicular development precedes ovarian development), although in some taxa all species (e.g., cephalopods) or nearly all species (e.g., prosobranch gastropods) are strictly gonochoristic.

Steroid hormones

Ecdysteroids and juvenoids have been reported in a few molluscan species but remain unconfirmed in the majority. However, vertebrate-type sex steroids have been measured in several species of mollusks, and biosynthetic studies following the pathways taken by labeled precursors suggest the ability to synthesize a range of progestogens and androgens (Wootton et al. 1995). The ability of some molluscan species to convert vertebrate-type sex steroids to a variety of oxido-reduced and polar derivatives has been demonstrated (Ronis and Mason 1996; Oberdörster et al. 1998). Various studies have also implicated a function for steroidal androgens in sex differentiation of mollusks. For example, testosterone administration to castrated male slugs (*Euhadra prelionphala*) stimulated the production of male secondary sex characteristics (Takeda 1980). Similarly, administration of testosterone to female gastropods caused them to develop a penis (Spooner et al. 1991; Bettin et al. 1996).

Neuropeptides

By far the most well-documented neuroendocrine pathways in mollusks are those involving neuropeptides. Much of this work has been done with *Aplysia* and *Lymnaea* with the intent of identifying peptides that regulate reproductive activity and growth. Egg-laying hormone (ELH) is a product of the abdominal ganglion and controls gonad maturation, egg mass production, and the egg-laying behavior typical in both *Aplysia* and *Lymnaea.* Dorsal body hormones produced by the light-green cells (caudo-dorsal cells) of the cerebral ganglion control the development of female accessory sex organs, gonad maturation, and ovulation (Geraerts et al. 1988). Molluscan insulin-like peptides (MIPs) are also produced by the central nervous system (CNS) and are amongst the best-characterized peptides. At least 6 members of this structurally related family (Smit et al. 1996) bear a striking resemblance to the vertebrate insulins and insulin-like growth factors. They have established roles in growth, development, and metabolism (Pertseva et al. 1996).

The cardioacceleratory peptide FMRFamide was first characterized in mollusks (Greenberg and Price 1983). The FMRFamide family is now the best known, characterized and most widespread of all invertebrate neuroendocrine hormones. Its activities in mollusks are manifest in a range of physiological processes of which the regulation of heartbeat is the best characterized (Greenberg and Price 1983; Nassel 1996; Santama et al. 1996; Suzuki et al. 1997; Favrel et al. 1998; Tensen et al. 1998).

Endocrine disruption

The mollusks provide the best mechanistically defined example of chemically induced ED among the invertebrates. The biocide TBT has been shown to cause a pseudohermaphroditic condition in female neogastropods (i.e., *Nassarius, Ilyanassa, Ocenebra, Urosalpinx)* characterized by the development of a penis, vas deferens, and seminiferous tubules (Gibbs et al. 1991). This phenomenon has been termed "imposex" (Smith 1971). The widespread distribution of imposex in field population of gastropods is discussed in Chapter 4.

TBT exposure has been shown to elevate testosterone levels in the dogwhelk *Nucella lapillus* (Spooner et al. 1991), the netted whelk *Nassarius reticulatus* (Bettin et al. 1996), and the freshwater snail *Marisa cornuarietis* (Schulte-Oehlmann et al. 1995). Testosterone administration has been shown to stimulate the development of male sex characteristics in female and castrated male gastropods (Takeda 1980; Spooner et al. 1991; Bettin et al. 1996). These observations have led to the hypothesis that elevated testosterone levels following TBT exposure are mechanistically responsible for the development of imposex (Bettin et al. 1996). This hypothesis has been supported by reports that the vertebrate androgen antagonist cyproterone acetate prevents the development of imposex in TBT-exposed female gastropods (Bettin et al. 1996). These latter experiments also imply that the gastropods and vertebrates express a highly conserved androgen receptor.

Either increased testosterone synthesis or secretion, or reduced testosterone biotransformation or elimination, could cause the increased accumulation of testosterone in TBT-exposed gastropods. The effects of TBT on testosterone synthesis and secretion are not known; however, several lines of evidence suggest that TBT increases the biotransformation or elimination of testosterone. Spooner et al. (1991) first proposed that the increase in testosterone levels following exposure of *Nucella* to TBT was due to the inhibition of the enzyme aromatase CYP19, which is responsible for the metabolic conversion of testosterone to 17β-estradiol. This hypothesis was supported by observations that the aromatase inhibitors flavone and 1-methyl-1,4-androstadiene-3,17-dione both caused imposex in the whelk, and exposure to 17β-estradiol or estrone prevented TBT-induced imposex (Bettin et al. 1996). However, we are not aware of any data that directly demonstrate the inhibition of aromatase by TBT. Furthermore, an increase in serum testosterone levels by the inhibition of CYP19 should be accompanied by a stoichiometric decrease in 17β-estradiol levels. No such decrease has been observed (Spooner et al. 1991; Schulte-Oehlmann et al. 1995). Finally, aromatase inhibitors such as flavone tend to inhibit, and sometimes stimulate, multiple enzymes that contribute to the inactivation or elimination of testosterone (Ramos and Sultatos 1998). Thus, the inhibition experiments do not directly implicate the involvement of aromatase.

An alternative hypothesis to explain the increase in testosterone levels following exposure of gastropods to TBT was first postulated by Ronis and Mason (1996). These investigators observed that TBT inhibited the sulfation of testosterone by the periwinkle *Littorina littorea,* resulting in a decrease in the metabolic elimination of the steroid. As a result, [^{14}C] testosterone administered to TBT-exposed gastropods retained significantly more of the testosterone than did unexposed snails. These investigators hypothesized that TBT caused an elevation in testosterone levels by inhibiting the conversion of testosterone to the sulfate conjugates, thus impeding its elimination. Weaknesses in these experiments include the high concentrations of TBT to which the snails were exposed and the lack of analyses of endogenous testosterone levels.

The occurrence of imposex is most customarily identified and the degree of severity quantified by relative penis size index (RPSI) or the vas deferens sequence index (VDSI). The RPSI is a measure of the penis mass expressed as a percentage of the mass of penises of normal males collected with the females (Gibbs et al. 1987). The VDSI consists of staging the degree of development of the vas deferens using an established scale (Huet et al. 1995).

Biomarkers and bioassays for detecting ED

No bioassays or biomarkers have been developed for detecting ED in mollusks. However, given the fact that so many species of gastropods are affected by TBT, and given the endocrine nature of the response, this system is a likely candidate for assay development.

The clam heart has been used as a sensitive bioassay to detect cardio-active agents, especially those that exert their action on mygenic hearts. The FMRFamides were identified using that preparation.

Nematoda

Nematodes are important microscopic fauna of all habitats, where they play important roles in food chains or as parasites. They are of interest because most of their developmental processes involve specific sets of cells, the fate and distribution of which have been mapped for all developmental stages. The complete sequence of cell divisions (and deaths) that give rise to the 959 somatic cells of the adult hermaphrodite *Caenorhabditis elegans* have been described, and it is the first animal genome to be completely sequenced (*C. elegans* Sequencing Consortium 1998). As a result of these features, *C. elegans* has become a widely used model in developmental biology and molecular genetics. Because there is a considerable knowledge base about the fundamental biology of nematodes, they are potentially useful for detecting ED. However, knowledge of their endocrinology is currently limited.

Endocrinology

A number of developmental processes that are likely to be under endocrine control have been described for the nematodes. These include

1) four molts preceding the adult stage,
2) formation of a large gonad,
3) switching gonad production from sperm to eggs,
4) alternative formation of males, and
5) alternative formation of the specialized dispersal stage called "dauer larvae."

None of these processes has been definitively shown to be controlled by hormones, but results of several experiments provide evidence for hormonal processes. A few experiments suggest that there are JH-like controls on growth (Fodor et al. 1982; Fodor and Timar 1989), and all JH-like compounds thus far isolated are derivatives of the methyl ester of 10,11 epoxy-farnesoic acid (Figure 2-6). Neurosecretory cells have been identified by light and electron microscopy, and noradrenaline, dopamine, serotonin, and octopamine have been identified in nematode extracts. Furthermore, the biochemical pathways for catecholamine synthesis have been identified. JH and its analogs can cause the release of noradrenaline from neurosecretory cells in the cephalic papillary ganglion. This causes the release of ecdysial hormone, which causes the secretion of enzymes necessary for molting. Both ecdysone and 20E have been identified in some nematodes, but their biosynthesis has yet to be confirmed (Barker and Rees 1990; Barker et al. 1990). Finally, FMRFamide-like peptides have been isolated (Brownlee et al. 1996).

Endocrine disruption

Exposure of nematodes to exogenous JH and ecdysone had effects, but at concentrations considerably above those reported in insects. *C. elegans* has been reported to respond to the JH antagonist precocene III and the JH analog methoprene (Fodor et al. 1982; Fodor and Timar 1989). Many genes in the *C. elegans* genome encode members of the nuclear hormone receptor family, but none is an ortholog of the insect ecdysteroid receptor (Ruvken and Hobart 1998). One of these genes shows high similarity to a nuclear receptor DHR3 induced by ecdysteroids in insects. This gene is expressed in the epidermis and is necessary for the normal shedding of the old cuticle during molting (Kostrouchova et al. 1998).

Biomarkers and bioassays for detecting endocrine disruption

Nematodes hold promise for the development of assays for the detection of ED, since these organisms are amenable to toxicity testing and several processes in these organisms are under hormonal control.

Rotifera

Rotifers are microscopic invertebrates that possess a ciliated crown (corona) on the anterior end. The animals are usually found in abundance in fresh water and can withstand dehydrating conditions for years.

Some rotifer species have been exploited for use in evaluating the chronic toxicity of environmental chemicals (Chapter 3). While several aspects of the rotifer life cycle are likely to be under endocrine control (the life cycle of Rotifera is similar to that presented for Cladocera in Figure 2-12), the endocrinology of these organisms remains largely unknown. Likely targets of endocrine regulation include the initiation of egg production in females after a brief immature period, the rate of egg production, quantity and type of yolk supplied to eggs, the switch from asexual to sexual reproduction, sperm production in males, and changes in the yolk gland to produce the specialized resting egg.

Endocrinology

Neurosecretory cells have been identified in rotifers by light and electron microscopy (Clement and Wurdak 1991). Rotifers have cholinergic neurotransmission, and acetylcholine, acetylcholinesterase, and choline acetyltransferase are present in several species (Nogrady and Alai 1983). Catecholinergic structures have been mapped in male and female *Brachionus plicatilis* (Keshmirian and Nogrady 1988). Exogenous exposure to 10^{-5} M acetylcholine inhibited oviposition in *Philodina acuticornis* (Nogrady and Keshmirian 1986). Norepinephrine at 10^{-4} M caused increased ciliary motion, swimming speed, and foot flipping in *Brachionus plicatilis* (Keshmirian and Nogrady 1987), and norepinephrine, dopamine and octopamine at 1 to 5 mM induced sessile behavior (Keshmirian and Nogrady 1989).

Exogenous exposure of *B. plicatilis* to several vertebrate and invertebrate hormones caused changes in asexual and sexual reproduction (Gallardo et al. 1997). Aqueous exposure to the peptide hormones mammalian growth hormone (0.025 I.U./ml) and human chorionic gonadotropin (0.25 I.U./ml) increased asexual population growth rate (r) relative to controls. Other hormones, including the steroids 17β-estradiol and 20E, the terpenoid JH, and the thyroid hormone triiodothyroxine, had no effect on asexual population growth rate. Exposure of *B. plicatilis* to 0.0025 I.U./ml growth hormone increased the rate of mictic (sexual) female production by 218% relative to controls. JH, 17β-estradiol, and 20E at concentrations of 50 μg/L increased the rate of mictic female production by 205%, 173%, and 59%, respectively. Triiodothyronine and human chorionic gonadotropin had no effect on mictic female production. JH at 50 μg/L increased body length by about 10%. Triiodothyronine, 20E, and human chorionic gonadotropin all decreased rotifer length to 4% and 8%. Growth hormone and 17β-estradiol had no effect on rotifer body size.

Hatching of rotifer resting eggs is initiated by exposure to light. Resting eggs also can be induced to hatch in the dark by exposure to 3×10^{-12} M E1, E2, and F2 prostaglandins (Hagiwara et al. 1995). The mechanism of this effect is not known, but one hypothesis is that oxidation of endogenous prostaglandins inside the resting eggs may be the trigger for resting egg hatching.

Endocrine disruption

Exposure of the freshwater rotifer *Brachionus calyciflorus* to some suspected endocrine disrupters differentially inhibited sexual reproduction without affecting asexual reproduction (Snell and Carmona 1995). Pentachlorophenol (200 μg/L), cadmium (25 μg/L), naphthol (400 μg/L), and chlorpyrifos (300 μg/L) differentially inhibited sexual reproduction by reducing sexual female production. Chlorpyrifos also decreased sexual female fertilization rate and the number of males produced per sexual female. It is not known whether these effects are mediated through ED or simply represent reproductive toxicity through a nonendocrine mechanism.

Biomarkers and bioassays for detecting endocrine disruption

Too little is known of rotifer endocrinology to recommend their use for the detection of ED at this time.

Echinodermata

The Echinodermata is a marine invertebrate phylum comprising the sea stars, lilies, cucumbers, urchins, and brittle stars, and closely related to the Vertebrata (both are deuterostomes). The echinoderms may thus have more in common with the endocrine physiology of the vertebrates than the other major invertebrate phyla. Echinoderms may also share similar targets of ED with the vertebrates and be susceptible to many of the chemicals known to cause ED in vertebrates.

Endocrinology

Knowledge of the function of steroid hormones in echinoderms is fragmentary. No ecdysteroids or juvenoids have been found in members of this phylum, a feature that may be related both to their nonmolting life history and their deuterostome lineage. Processes known or suggested to be regulated by hormones, neurohormones, or locally released growth factors include gametogenesis (asteroids, ophiuroids, crinoids), spawning (asteroids), growth, and regeneration (asteroids, ophiuroids, crinoids).

Reproduction

The reproductive physiology of asteroids (sea stars; formerly named "starfish") has been studied in most detail. The majority of sea star species are broadcast breeders, producing large amounts of gametes, a process that involves a good deal of biosynthetic resource. In asteroids, nutrients for yolk formation (vitellogenesis) or for spermatogenesis can be stored in the pyloric caeca (digestive diverticula) during the resting stage (Oudejans and van der Sluis 1979; Lawrence and Lane 1982). This allows the reproductive cycle to become more independent of fluctuations in food availability and can also be regarded as an adaptation to life in environments that experience seasonal variations in productivity (e.g., in waters of temperate zones). During gametogenesis, the stored nutrients are released from the pyloric caeca (den Besten et al. 1994) and translocated to the gonads. In female sea stars, vitellogenin-like compounds are taken up by developing oocytes (Broertjes et al. 1984; Voogt et al. 1985). In males, the same glycolipoproteins seem to be present in the pyloric caeca. These glycolipoproteins are released during spermatogenesis, but they are not accumulated in the spermatozoa or in the testes and at present their precise function is not clear. The most likely function is that they serve as an energy source during spermatogenesis.

Steroid hormones

As a consequence of the transport of nutrients, the pyloric caeca show changes in size that are inversely related to gonadal growth (Giese 1966). There are no clear indications concerning which hormones are primarily responsible for the mobilization of the materials required for gametogenesis, although it has been suggested that cAMP has a role in the mobilization of stored triacylglycerols (Voogt and van

Rhenen 1984). Several studies have demonstrated the capacity of sea stars to synthesize at least 2 steroids that in vertebrates are known to be biologically active: progesterone and testosterone (Voogt et al. 1991), and key enzymes of steroidogenic pathways, including 3β-hydroxysteroid dehydrogenase and cytochromes P450, have been demonstrated in the pyloric caeca and in gonads (Schoenmakers 1980; den Besten 1998). In addition, the presence of cells with ultrastructural features characteristic of steroid biosynthesis has been described. Synthesis of estrone and estradiol could not be demonstrated, possibly due to the relatively low activities of aromatase (CYP 19). However, it is important to note that a receptor with specificity for estradiol has been described (de Waal et al. 1982) and that different investigators have demonstrated that 17β-estradiol stimulates vitellogenesis and ovary growth in *A. rubens*, *Asterina pectinefera* and in *Sclerastias mollis* (Schoenmakers and Dieleman 1981; Schoenmakers et al. 1981; Takahashi and Kanatani 1981; Barker and Xu 1993).

A role for steroid hormones in reproduction of echinoderms is further supported by the observation that the onset of ovarian growth in *A. rubens* is preceded by a marked increase in the activity of 3β-hydroxysteroid dehydrogenase, the enzyme that is generally regarded to be the key enzyme in steroid biosynthesis (Schoenmakers 1980). Furthermore, the levels of progesterone, estrone, and testosterone vary in a sex- and organ-specific manner. In female *A. rubens*, progesterone levels in the pyloric caeca drop after the start of oogenesis, remain low until the last phase, then increase again in the period just before spawning. Estrone levels in the pyloric caeca show a slight increase during oogenesis. In the ovaries, progesterone levels are highest during the resting stage and at the onset of oogenesis, after which levels drop. Estrone levels show a peak at the onset of oogenesis (Schoenmakers and Dieleman 1981). In male *A. rubens*, the level of progesterone in the pyloric caeca shows a trend comparable to that in females, except for the increase just before spawning. Estrone levels are relatively high at the onset of spermatogenesis, but they fluctuate less than progesterone. In testes, progesterone shows a clear peak at the onset of spermatogenesis, whereas estrone levels show little variation. The noted changes in steroid levels occur in female sea stars somewhat earlier in autumn (September and October) than in males (November and December) (Voogt and Dieleman 1984). Testosterone levels have not been measured throughout the entire cycle of sea stars, but existing data suggest that levels are highest in the pyloric caeca and attain a maximum level in November (den Besten, Elenbaas et al. 1991). In other sea star species, it has been shown that changes in the levels of estrogens, progesterone, and testosterone coincided with gametogenesis events (Xu and Barker 1990a, 1990b, 1990c; Xu 1991; Hines et al. 1992).

These observations strongly suggest that steroids are involved in the regulation of sea star reproductive processes. Information on the reproductive physiology of echinoderm groups other than the asteroids is scarce. In these echinoderm groups, digestive diverticula are lacking. Nutrient storage may take place in the gut or in the somatic and gametic cells of the gonads (see Lawrence and Lane 1982), but compared to the asteroids, these groups have a much lower capacity for the storage of nutrients under favorable conditions. There is no information on the control of the reproductive cycle in the ophiuroids, echinoids, holothuroids, or crinoids. The presence of a microsomal cytochrome P450 monooxygenase system has been demonstrated in sea urchins and sea cucumbers (den Besten 1998). However, nothing is known about a possible function in steroidal biosynthesis. A second function of steroid synthesis in asteroids is the production of saponins. Asterosaponins consist of a pregnene or cholestene steroid moiety with a sulfate and a carbohydrate group. They are found to be present in the oocytes, where they probably function as inhibitors of maturation or mitotic/meiotic cleavages, or provide defense against predators or pathogens (Voogt and Hulskamp 1979).

Neuropeptides

In sea stars, gonad-stimulating substance (GSS, also called "radial nerve factor" and possibly a neuropeptide) is known as a factor present in extracts of radial nerve cords, which can stimulate spawning (Caine and Burke 1984; Shirai et al. 1986; Smiley 1988). This has not yet been characterized, although several neuropeptides have been isolated and sequenced from echinoderm nervous systems (Elphick, Price et al. 1991; Elphick, Reeve et al. 1991). None of these have been shown to influence reproductive activity but some of these are known to be involved in other aspects of echinoderm physiology such as the regulation of feeding (Elphick et al. 1995; Potton and Thorndyke 1995).

Oocytes mature just before spawning and may be fertilized only during a critical time window of approximately 1 or 2 hours. GSS stimulates the local release of 1-methyl adenine (1-MA), which acts as a hormone by stimulating the release of maturation-promoting factor (MPF). MPF induces oocyte germinal vesicle breakdown, a process that is an essential prerequisite for fertilization.

Cadmium exposure during oogenesis delays oocyte maturation and reduces the proportion of oocytes responding to 1-MA (den Besten et al. 1989, 1991). This delay can have serious negative consequences for fertilization success and subsequent embryonic development, possibly due to an increased incidence of poly-spermy (den Besten et al. 1989).

Regeneration

Regeneration is a natural phenomenon found in brittlestars, featherstars, and asteroids. It can represent an important component of asexual reproduction, where individuals undergo fission and split to form new individuals (Emson and Wilkie 1980; Mladenov and Burke 1994). Alternatively, it comprises a rapid tissue-proliferative response, which allows the replacement of body parts (frequently arms or parts of arms) lost following predation. Pollutants, including PCBs, have been shown to have dramatic effects on regeneration in crinoids, where enhanced growth and substantial cell dedifferentiation and increased cell-cycle activity are prominent features (Candia Carnevali et al. 1999).

Endocrine disruption

There is limited information on endocrine-disrupting toxicity of chemicals in the echinoderms. Long-term exposure of sea stars to low levels of cadmium or a PCB mixture (Clophen A50 via the foodchain algae–mussels–sea star) resulted in lower steroid synthesis rates and significant changes in the steroid levels detectable in pyloric caeca and gonads (den Besten, van Donselaar et al. 1991). These changes were related to reductions in ovarian growth and an impaired quality of offspring observed after exposure to cadmium or PCBs under similar conditions (den Besten et al. 1989, 1990; den Besten, Elenbaas et al. 1991). In short-term exposure studies, it was later confirmed that steroid metabolism in pyloric caeca microsomal fractions can be impaired by exposure to benzo[a]pyrene and selected PCBs. As a possible mechanism for this effect, it was hypothesized that induction of P450 detoxification enzymes leads to a reduction in the steroid synthesizing capacity of the cytochrome P450 monooxygenase system (den Besten, van Donselaar et al. 1991; den Besten et al. 1993). Potential endpoints for assessing endocrine-mediated effects of chemicals in echinoderms may include gonadal growth (den Besten et al. 1990, 1991a), oocyte maturation (den Besten et al. 1989; den Besten, Elenbaas 1991), embryonic development (den Besten et al. 1989), and arm regeneration both in vitro and in vivo (Bonasoro et al. 1998, 1999; Candia Carnevali et al. 1998; Moss, Beesley et al. 1998; Moss, Hunter, Thorndyke 1998).

Biomarkers and bioassays for detecting endocrine disruption

The processes of regeneration, fertilization, gonadal growth, oocyte maturation, and early development offer possibilities for developing assays in the echinoderms. Endpoints such as regeneration and gonadal growth in sea stars could be highly relevant for the detection of ED, especially considering that this group relies on endogenous steroid hormones for regulating at least some of the same processes, as do vertebrates. In addition, fertilization and early embryology of urchins has been so well studied that these two could prove fruitful for identifying ED endpoints.

Chordata

The tunicates, or sea squirts, are members of the phylum Chordata and therefore represent a close phylogenetic (and perhaps functional) link with vertebrates, including humans. Like the Echinodermata, tunicates are found only in the marine environment.

Endocrinology

The adults are typical and rather unspecialized filter-feeding invertebrates with a simple neural complex comprising neurosecretory cells (Thorndyke and Dockray 1986). They are hermaphroditic, with clearly separated testis and ovary (protandrous), and there is some evidence for steroid metabolism and biosynthesis in the gonads (Colombo et al. 1977). The neuropeptide products of the neural ganglion include a gonadotropin releasing hormone-like molecule (Powell et al. 1996). Although the details of its interaction with gonad development are not clear, it has been found also in visceral parts of the nervous system and may well be involved in growth and regeneration of the neural complex (Bollner et al. 1997; Mackie 1995). The most well-characterized "endocrine tissue" in tunicates is the group of prethyroidal iodine-binding cells present in the endostyle. These cells produce and (presumably) release iodinated tyrosines, which are structurally and functionally identical with vertebrate thyroid hormones. However, their function in tunicates is obscure and in adults is most likely involved in the tanning process, which takes place during growth and tunic formation (Thorndyke 1973). An embryonic endostyle is present in the larval (tadpole) stage (see below), and there is some evidence to suggest that thyroid hormones are important in the regulation of metamorphosis and tail resorption (Ruppert and Barnes 1991).

Tadpole larva and development

The tunicates are chordates by virtue of their tailed, swimming, larval stage, "the tunicate tadpole," which is characterized by a dorsal, hollow CNS with anterior swelling ("brain") and notochord (Katz 1983; Nicol and Meinertzhagen 1991). Furthermore, the presence and expression sites of typical vertebrate patterning genes (Wada et al. 1998) clearly supports the idea of this larval form as a vertebrate ancestor. However, apart from the putative role of thyroid hormones in metamorphosis noted above, there are no known endocrine or neuroendocrine phenomena in this nonfeeding dispersive stage.

Other regulated processes

Many tunicates exist as colonies derived from asexual budding. In some planktonic forms, this also involves an alternation of generations (Ruppert and Barnes 1991). Adult budding is common and involves the asexual production of new adults from existing adult epidermal and mesodermal tissues. Retinoids, and thus retinoid receptor-mediated growth and differentiation, appear to be central to this activity (Hisata et al. 1998).

Biomarkers and bioassays for detecting endocrine disruption

No research indicates how tunicates might be used in assays, though there is every reason to believe that early development, especially of the tadpole stage, may provide a sensitive bioassay.

Cnidaria

The cnidarians represent one of the last taxa phylogenetically situated before the divergence of the protostomes and deuterostomes (Figure 2-2). These organisms thus can be viewed as precursors to both lineages. The endocrinology of these organisms may prove to represent those aspects of the system that are conserved among both protostomes and deuterostomes. Currently, too little is known of the endocrinology of this group to draw such conclusions.

Endocrinology

Cnidarians have a primitive nervous system where all cells seem to have combined properties of sensory, motor, and interneurons (Lesh-Laurie 1988), and these cells contain neuropeptides (Grimmelikhuijzen et al. 1996). These neuropeptides may act as neurotransmitters and also as locally released factors to regulate processes such as growth, regeneration, and possibly other processes (Lesh-Laurie 1988; Berking 1998). One such group of peptides is the glycine-leucine tryptophan amides (GLWamides), which are found in certain anterior neurosensory cells of the planula larva of *Hydractinia echinata* at the time of metamorphosis (Gajewski et al. 1996; Schmich et al. 1998). The GLWamides have been shown in bioassays to induce metamorphosis (Gajewski et al. 1996; Schmich et al. 1998). After metamorphosis, the peptide is no longer present in these cells but reappears in the hypostomal endoderm of the adult polyp. The function of the peptide in the adult is unknown. In several different cnidarian classes, metamorphosis could also be stimulated by protein kinase C activators and inhibited by specific protein kinase C inhibitors (Henning et al. 1998), indicating the involvement of a second messenger in the process (Berking 1998). Pattern formation during regeneration in *Hydra* is also dependent on the protein kinase C pathway, but whether this is activated by neuropeptides such as the "head-activating" factor or other types of cellular signals is currently uncertain (Berking 1998; Hassel 1998; Hassel et al. 1998).

Metagenesis occurs in many cnidarians and is characterized by an alternation of an asexual polyp form with the sexually reproducing medusa. The transformation into the medusa is called "strobilation" and is induced in the jellyfish *Aurelia* by iodide or thyroxine (Spangenberg 1967). Subsequent studies have shown that monoiodotyrosine, diiodotyrosine, or some other iodide derivative is necessary for this process, although some have questioned the effectiveness of thyroxine (reviewed in Lesh-Laurie 1988; Eales 1997). An RXR that can bind 9-*cis*-retinoic acid and also can heterodimerize with the thyroid hormone receptor, as it does in vertebrates (Mangelsdorf and Evans 1995), has recently been found in the jellyfish *Tripedalia cystophora* (Kostruch et al. 1998). This RXR is present during the polyp stage but is lost immediately after metamorphosis to the medusa. The RXR may be involved in the regulation of this process.

Summary

Knowledge of invertebrate endocrinology

Much of our knowledge of invertebrate endocrinology has been driven by the need to control invertebrate populations, particularly insect pests. Many insecticides interfere with some aspect of insect endocrinology, resulting ultimately in disruption of development and/or reproduction. Such insecticides generally have high specificity, since they are designed to target relatively unique aspects of the insect endocrine system. Hence, among invertebrates as a whole, most of our knowledge regarding invertebrate endocrinology pertains to insects and their susceptibility to environmental chemicals, particularly insecticides. There is also a significant amount of information available on the endocrinology of some crustaceans. Information on the endocrinology of other invertebrate groups, however, is often limited and fragmentary. It is important to note that even among groups where there is a significant amount of information, the information is usually limited to a few very well-studied species with little known about interspecies variability in endocrinology.

Unique and common aspects of invertebrate endocrinology

Knowledge of the comparative endocrinology of various invertebrate groups is an important consideration that should be taken into account when attempting to select reasonably representative invertebrate species and appropriate endpoints for laboratory testing (Chapter 3) and field monitoring studies (Chapter 4). Endocrine processes that are relatively conserved among invertebrate groups provide the greatest versatility for the extrapolation of study results to other invertebrate species (Chapter 3, section "Evaluation of endpoints and species for use in effects assessments" and Figure 3-2). The use of endpoints relevant to such endocrine processes may prove to be most informative in hazard and risk assessment approaches for identifying ED in invertebrates. Alternatively, unique aspects of the endocrinology of certain invertebrate groups may also provide targets of toxic action that render some species highly sensitive to ED.

The endocrine systems of invertebrates tend to regulate many of the same processes that are found in the vertebrates, i.e., processes associated with development, growth, and reproduction. Furthermore, the endocrine systems of both vertebrates and invertebrates, in general, tend to be an extension of the nervous system. The endocrine systems of invertebrates respond to a variety of environmental and physiological cues that are detected and processed by the nervous system. The nervous system responds to these stimuli by the secretion of neuropeptides that may directly elicit responses to the stimuli or may stimulate the secretion of additional hormones that mediate the response. Some of the endocrine-regulated processes associated with development, growth, and reproduction are unique to specific groups of invertebrates. For example, molting, diapause, and limb regeneration are endocrine-regulated processes that are associated with some invertebrate groups and are rare or absent among the vertebrates. Other endocrine-regulated processes, such as carbohydrate metabolism, gonadal development, and vitellogenin synthesis, tend to be common among the invertebrates (as well as relevant vertebrate taxa).

Some degree of conservation exists with respect to hormone usage by invertebrates. All invertebrates that periodically shed an exoskeleton use ecdysteroids to regulate the molting process. In addition, those species that use ecdysteroids as hormones also produce terpenoids. It is well established among the insects that the terpenoid JH functions in concert with ecdysteroids to regulate development and reproduction. The terpenoid MF may function similarly in crustaceans. A functional role for terpenoids and ecdysteroids also has been suggested in nematodes. All invertebrates extensively use neuropeptides to transduce neuroendocrine signals. Functional similarities have been noted among peptide hormones from different phyla (i.e., cardioactive/cardioacceleratory peptide). These functionally conserved peptides may represent common neuropeptide hormone families among the invertebrates.

Steroid hormone usage appears to have undergone significant divergence between the deuterostome and protostome invertebrates. Among some deuterostome invertebrates, strong evidence exists for the use of vertebrate-type sex steroids with no precedence for the use of ecdysteroids. Conversely, ecdysteroids are extensively used among many protostome invertebrates with little precedence for the use of vertebrate-type sex steroids. Exceptions to these generalities exist. For example, good evidence supports a role for vertebrate-type sex steroids in the regulation of sex differentiation in the protostome mollusks. Ecdysteroids and vertebrate-type sex steroids appear to have analogous roles in different species.

Clearly, there exist sufficiently common processes (i.e., ecdysteroid–receptor interactions) and functions (regulation of gametogenesis, vitellogenin synthesis, and energy metabolism) associated with invertebrate endocrinology that warrant consideration when assessing environmental ED. Many of the processes and functions are unique to the invertebrates, and information on these processes could not be derived from evaluations of ED in vertebrates. Finally, the endocrinologies of individual invertebrate phyla are sufficiently unique that no single species could adequately serve to represent invertebrate endocrinology as a whole in the assessment of environmental ED.

Endocrine processes, assays, and endpoints

The overview of invertebrate endocrinology presented in this chapter identifies many processes that could be exploited as endpoints for the detection and evaluation of chemically induced ED in invertebrates. Some examples are provided in Table 2-2. Some of these assays have been used historically to evaluate the endocrine properties of potential insecticides. These assays could be similarly used to evaluate

Table 2-2 Examples of endocrine-relevant endpoints and assays that may be susceptible to and serve as indicators of chemically induced endocrine disruption in invertebrates

Group	Hormone action or mechanism	Endpoints and biomarkers	Bioassays	Chemical examples
Insects	Anti-juvenile hormone (JH)	• Molting: premature metamorphosis (small adults) • Inhibition of vitrellogenin synthesis in some insects	• Expose early instar larvae (e.g., *Oncopeltus*, grasshoppers, *Manduca,* or other Lepidoptera); precocious metamorphosis • Expose normal *Manduca sexta*: if becomes black = JH antagonist • In vitro *Manduca sexta* epidermal (wound) assay	• Precocene (destroys CA gland that produces JH) • HMG CoA reductase inhibitor
Insects	JH agonist	• Molting: extra larval instars, or deformed intermediates formed, or absence of adults • Fecundity • Egg mortality	• Galleria wax wound assay • Black *Manduca sexta* larval test • *Tenebrio*: morphological assay • Locust adult: destroy CA; agonist to vitellogenin production • Cockroach: accessory gland development assay • Mosquito larvae: prevent emergence of adults, egg hatch	• Methoprene • Pyriproxifene • Fenoxycarb • JH III
Insects Crustaceans	Anti-ecdysteroids	• Inhibition of molting and metamorphosis • Inhibition of egg production in Diptera	• Inverse of *Calliphora* pupariation assay	• Azadirachtin
Insects Crustaceans	Ecdysteroid agonist	• Inhibition of feeding • Molting: premature incomplete molt with double head capsule in Lepidoptera, followed by lethality (deformed and molt susceptible to predation) • Vitellogenin synthesis in Diptera and some crustaceans	• *Calliphora* pupariation assay • Ecdysteroid-responsive cell lines • Ecdysone receptor binding assay	• RH5992 (Tebufenozide; specific for Lepidoptera) • 20-OH-Ecdysone • Ponasterone-A
Insects Crustaceans	Pheromones	• Mating behavior (sex attractants) • Disruption of behavior in social insects	• Various types	

the potential of other environmental chemicals to interfere with processes regulated by ecdysteroids or JH.

It is important to recognize that, among the invertebrates, the same process can be regulated by different endocrine controls. For example, vitellogenin production has been shown to be subject to endocrine control among the invertebrates thus far evaluated. However, vitellogenin production appears to be stimulated by 20E in some crustaceans and insects and inhibited by 20E in other crustaceans. In some crustaceans, androgenic hormone stimulates vitellogenesis, as does JH in some insects (i.e., Orthopterans) and vertebrate-type steroids in asteroids. Thus, the use of vitellogenin induction in one taxon to evaluate ED has limited value in predicting such effects in other taxa.

Even when a common mechanism is involved in regulating a physiological process among taxa, sufficient differences among organisms with respect to the mechanism may significantly alter levels of susceptibility to ED. An example of this is 20E and its receptors. The effects of 20E on insect molting are mediated by the ecdysone receptor. However, differences in the ligand-binding domain of the receptor among insect orders can result in significant differences in binding affinity of 20E analogs. This is clearly illustrated with the lepidopteran-selective insecticide, tebufenozide, which has the highest affinity for the lepidopteran ecdysone receptor relative to the ecdysone receptor from nonlepidopteran insects or other arthropods (Sundaram et al. 1998).

Test species

The insects and crustaceans could clearly be best exploited for use in evaluating chemically induced ED by virtue of the wealth of information that is available on the endocrinology of these organisms. In addition to more commonly used insect species such as *Chironomus* (Chapter 3), *Drosophila* should be considered among the potential insect test species due to its amenability to laboratory culture and experimentation and its environmental ubiquity, the existence of various strains, and the advanced level of understanding of the physiology and genetics of the organism. Most of our knowledge of the endocrinology of crustaceans stems from studies of economically important decapods (i.e., crabs, lobsters, shrimp, crayfish). These species can be readily used as indicators of environmental ED. However, they may have limited use as a laboratory test species due to life-cycle complexities. Mysid shrimp may serve as a viable surrogate for many crustaceans. Methods for evaluating chemical toxicity using this species exist (ASTM #E 1191-90; ASTM 1993) and could be modified to include endocrine-relevant endpoints (Chapter 3). Current toxicity-test methods with other commonly used test invertebrates (i.e., amphipods, ASTM #E 1367-90; daphnids, ASTM #E 1193-87; ASTM 1993) could also be modified to include endocrine-relevant indicators of toxicity (Chapter 3). However, limited knowledge of the endocrinology of these lower crustaceans would currently constrain their use for the detection of endocrine toxicity.

Key research needs

- Invertebrate hormone receptors must be identified, cloned, and characterized. This information will permit the assessment of hormone receptors that are common among invertebrate groups, thus allowing the use of common assay or surrogate species when chemical-receptor interactions are assessed. This information will also help to identify those receptors that are unique to some groups of invertebrates, thus limiting the use of surrogates.
- In vitro assays, such as receptor-binding assays, must be developed that are applicable to invertebrate endocrine processes. Such assays are necessary for use in screening chemicals for potential endocrine-disrupting activity and for evaluating specific mechanisms of ED.
- Endocrine processes that regulate physiological endpoints must be better defined so that these endpoints can be used as valid indicators of ED in laboratory tests or field evaluations.
- A database must be developed on the susceptibility of uniquely invertebrate processes (i.e., those regulated by JH, ecdysteroids, or invertebrate neuropeptides) to environmental chemicals.
- It must be established whether information on ED generated in vertebrates, as related to sex steroids (i.e., estrogen agonists, androgen antagonists), can be applied to deuterostome invertebrates (i.e., tunicates, echinoderms).

Overall conclusions

- Our ignorance of the endocrinology of invertebrates far outweighs our knowledge of the subject.
- Much of our knowledge of invertebrate endocrinology and invertebrate ED stems from efforts to generate target-specific insecticides.
- Physiological processes in invertebrates that are dependent on the endocrine system are as diverse as the endocrine systems among the various phyla. Therefore the potential for, and the consequences of, ED in invertebrates varies considerably.
- Physiological processes known to be under endocrine regulation provide insight into endpoints and biomarkers that could be developed for use in evaluating ED.
- The concept of a single "representative invertebrate" for the assessment of ED in all invertebrates is not scientifically valid.

References

Abdu U, Takac P, Yehezkel G, Chayoth R, Sagi A. 1998. Administration of methyl farnesoate through the *Artemia* vector, and its effect on *Macrobrachium rosenbergii* larvae. *Israeli J Aquacult* 50:73–81.

Abu-Hakima R, Davey KG. 1977. The action of juvenile hormone on the follicle cells of *Rhodnius prolixus. Can J Zool* 53:1187–1188.

Adams TS. 1997. Arthropoda: Insecta. In: Adiyodi KG, Adiyodi RG, editors. Reproductive biology of invertebrates. Chichester UK: J Wiley. p 277–338.

Adams TS. 1999. Hematophagy and hormone release. *Ann Ent Soc Amer* 92:1–13.

Ahl JS. 1986. Effects of juvenile hormone analog on molting and development in larval brine shrimp, Artemia. *Amer Zool* 26:29.

Ankley GT, Tietge JE, De Foe DL, Jensen KM, Holcombe GW, Durhan EJ, Diamond SA. 1998. Effects of ultraviolet light and methoprene on survival and development of *Rana pipiens. Environ Toxicol Chem* 17:2530–2542.

Arnold SF, Robinson MK, Notides A, Guillette Jr. LJ, McLachlan JA. 1996. A yeast estrogen screen for examining the relative exposure of cells to natural and xenoestrogens. *Environ Health Perspect* 104:544–548.

Ashburner M, Chihara C, Meltzer P, Richards G. 1974. On the temporal control of puffing activity in polytene chromosomes. *Cold Spring Harb Symp Quant Biol* 38:655–662.

Ashok M, Turner C, Wilson TG. 1998. Insect juvenile hormone resistance gene homology with the bHLH-PAS family of transcriptional regulators. *Proc Nat Acad Sci USA* 95:2761–2766.

[ASTM] American Society for Testing and Materials. 1993. ASTM standards on aquatic toxicology and hazard evaluation. Philadelphia PA: ASTM. 538 p.

Audsley N, McIntosh C, Phillips JE. 1992. Isolation of a neuropeptide from locust corpus cardiacum which influences ileal transport. *J Exp Biol* 173:261–274.

Audsley N, Weaver RJ, Edwards JP. 1999. Enzyme linked immunosorbent assay for *Manduca sexta* allostatin (Mas-AS), isolation and measurement of Mas-AS immunoreactive peptide in *Lacanobia oleracea. Insect Biochem Mol Biol* 28:775–784.

Baker ME. 1997. Steroid receptor phylogeny and vertebrate origins. *Mol Cell Endocrinol* 135:101–107.

Baldwin WS, LeBlanc GA. 1994a. Identification of multiple steroid hydroxylases in *Daphnia magna* and their modulation by xenobiotics. *Environ Toxicol Chem* 13:1013–1021.

Baldwin WS, LeBlanc GA. 1994b. In vivo biotransformation of testosterone by phase I and II detoxication enzymes and their modulation by 20-hydroxyecdysone in *Daphnia magna. Aquatic Toxicol* 29:103–117.

Baldwin WS, Milam DL, LeBlanc GA. 1995. Physiological and biochemical perturbation in *Daphnia magna* following exposure to the model environmental estrogen diethylstilbestrol. *Environ Toxicol Chem* 14:945–952.

Baratte B, Gras-Masse H, Ricart G, Bulet P, Dhainaut-Courtois N. 1990. Isolation and characterization of authentic Phe-Met-Arg-Phe-NH2 and the novel Phe-Thr-Arg-Phe-NH2 from *Nereis diversicolor. Eur J Biochem* 198:627–633.

Barker GC, Chitwood DJ, Rees HH. 1990. Ecdysteroids in helminths and annelids. *Invert Reprod Develop* 24:53–58.

Barker CG, Rees HH. 1990. Ecdysteroids in nematodes. *Parasitology Today* 6:384–387.

Barker MF, Xu RA. 1993. Effects of estrogens on gametogenesis and steroid levels in the ovaries and pyloric caeca of *Sclerasterias mollis* (Echinodermata: Asteroidea). *Invert Reprod Develop* 24:53–58.

Barton AE, Wing KD, Le DP, Slawecki RA, Feyereisen R. 1989. Arylpyridyl-thiosemicarbazones: A new class of anti-juvenile hormones active against Lepidoptera. *Experientia* 45:580–583.

Berking S. 1998. Hydrozoa metamorphosis and pattern formation. *Curr Top Dev Biol* 38:81–131.

Bettin C, Ohlmann J, Stroben E. 1996. Tributyltin-induced imposex in marine neogastropods is mediated by an increasing androgen level. *Helgol Meeresunters* 50:299–317.

Biggers WJ, Laufer H. 1992. Chemical induction of settlement and metamorphosis of *Capitella capitata* sp-1 (Polychaeta) larvae by juvenile hormone active compounds. *Invert Reprod Devel* 22:39–46.

Biggers WJ, Laufer H. 1996. Detection of juvenile hormone-active compounds by larvae of the marine annelid *Capitella sp. Arch Insect Biochem Physiol* 32:475–484.

Blomquist GJ, Tillman-Wall JH, Guo L, Quiliar DR, Gu P, Schal C. 1993. Hydrocarbon and hydrocarbon-derived sex phermones in insects: Biochemistry and endocrine regulation. In: Stanley-Samuelson DW, Nelson DR, editors. Insect lipids: chemistry, biochemistry and biology. Lincoln NE: Univ Nebr Pr. p 317–335.

Bollner T, Beesley PW, Thorndyke MC. 1997. Investigation of the contribution from peripheral GnRH-like immunoreactive "neuroblasts" to the regenerating central nervous system in the protochordate *Ciona intestinalis Proc R Soc Lond B* 264:1117–1123.

Bonasoro F, Candia Carnevali MD, Sala F, Patruno M, Thorndyke MC. 1999. Regenerative potential of arm explants. In: Candia Carnevali MD, editor. Echinoderms. Milano: Balkema, Rotterdam, The Netherlands.

Bonasoro F, Candia Carnevali MD, Thorndyke MC. 1998. Epimorphic versus morphallactic mechanisms in arm regeneration of crinoids and asteroids: Pattern of cell proliferation/ differentiation and cell lineage. In: Mooi R, Telford M, editors. Echinoderms. San Francisco. p 13–18.

Borst DW, Laufer H, Landau M, Chang ES, Hertz WA, Baker FC, Schooley DA. 1987. Methyl farnesoate and its role in crustacean reproduction and development. *Insect Biochem* 17:1123–1127.

Borst DW, Tsukimura B, Chang ES. 1991. The regulation of methyl farnesoate levels in crustaceans. La Londe les Maures, France.

Bowers WS, Ohta T, Cleere JS, Marsella PA. 1976. Discovery of insect anti-juvenile hormones in plants. *Science* 193:542–547.

Broertjes JJS, de Waard P, Voogt PA. 1984. On the presence of vitellogenic substances in the starfish, *Asterias rubens* (L.). *J Mar Biol Ass UK* 64:261–269.

Brown FA Jr. 1950. Studies on the physiology of *Uca* red chromatophores. *Biol Bull* 98:218–226.

Brownlee DJA, Fairweather I, Holden-Dye L, Walker RJ. 1996. Nematode neuropeptides: localization, isolation, and functions. *Parasitology Today* 12:343–351.

Burns BG, Sangalang GB, Freeman HC, McMenemy M. 1984. Isolation of testosterone from the serum and testes of the American lobster (*Homarus americanus*). *Gen Comp Endocrin* 54:429–435.

Byard EH. 1975. The female specific protein and reproduction in the lobster, *Homarus americanus*. [Ph.D. dissertation]. London, Ontario: Univ Western Ontario.

Caine GD, Burke RD. 1984. Immunohistiochemical localization of gonad stimulating substance in the seastar *Pycnopodia helianthoides*. In: Keegan BF, O'Connor BDS, editors. Galway, Ireland. Balkema, Rotterdam.

Candia Carnevali MD, Bonasoro F, Patruno M, Thorndyke MC. 1998. Cellular and molecular mechanisms of arm regeneration in crinoid echinoderms: The potential of arm explants. *Dev Genes Evol* 208:421–430.

Candia Carnevali MD, Galassi F, Bonasoro F, Dina G, Terlizzi MA, Patruno M, Thorndyke MC. 1999. PCB-induced environmental stress and –the regenerative response in crinoids. In: Candia Carnevali MD, editor. Milano. Balkema, Rotterdam.

Cantelmo AC, Lazell RJ, Mantel LH. 1981. The effects of benzene on molting and limb regeneration in juvenile *Callinectes sapidus*. *Mar Biol Lett* 2:333–343.

C. elegans Sequencing Consortium. 1998. Genome sequence of the nematode *C. elegans*: A platform for investigating biology. *Science* 282:2012–2018.

Champlin DT, Truman JW. 1998. Ecdysteroid control of cell proliferation during optic lobe neurogenesis in the moth, *Manduca sexta*. *Development* 125:269–277.

Chang ES. 1997. Chemistry of crustacean hormones that regulate growth and reproduction. In: Fingerman M, Nagabhushanam R, Thompson M-F, editors. Recent advances in marine biotechnology. New Delhi, India: Oxford & IBH Publ. p 163–178.

Chang ES, Keller R, Chang SA. 1998. Quantification of crustacean hyperglycemic hormone by ELISA in haemolymph of the lobster, *Homarus americanus*, following various stresses. *Gen Comp Endocrinol* 111:359–366.

Chang ES, O'Connor JD. 1977. Secretion of a-ecdysone by crab Y-organs in vitro. *Proc Nat Acad Sci USA* 74:615–618.

Chang ES, O'Connor JD. 1979. Arthropod molting hormones. In: Jaffe BM, Beherman HR, editors. Methods of hormone radioimmunoassay. New York: Academic Pr. p 797–814.

Chang ES, Prestwich GD, Bruce MJ. 1990. Amino acid sequence of a peptide with both molt-inhibiting and hyperglycemic activities in the lobster, *Homarus americanus*. *Biochem Biophys Res Commun* 171:818–826.

Cherbas P, Cherbas L. 1996. Molecular aspects of ecdysteroid hormone action. In: Gilbert LI, Tata JR, Atkinson BG, editors. Metamorphosis: Postembryonic reprogramming of gene expression in amphibian and insect cells. San Diego CA: Academic Pr. p 175–221.

Cherbas P, Cherbas L, Lee SS, Nakanishi K. 1988. 26-[125 I]Iodoponasterone A is a potent ecdysone and a sensitive radioligand for ecdysone receptors. *Proc Nat Acad Sci USA* 85:2096–2100.

Chippendale GM. 1983. Larval and pupal diapause. In: Downer RGH, Laufer H, editors. Endocrinology of insects. New York: Alan R. Liss. p 343–356.

Christiansen ME, Costlow JD, Monroe RJ. 1977a. Effects of the JH mimic ZR-512 (Altozar) on larval development of the mud-crab *Rhithropanopeus harrisii* at various cyclic temperatures. *Mar Biol* 39:281–288.

Christiansen ME, Costlow JD, Monroe RJ. 1977b. Effect of the JH mimic ZR-515 (Altosid) on larval development of the mud-crab *Rhithropanopeus Harrissii* in various salinities and cyclic temperatures. *Mar Biol* 39:269–279.

Chung ACK, Durica DS, Clifton SW, Roe BA, Hopkins PM. 1998. Cloning of crustacean ecdysteroid receptor and retinoid-X receptor gene homologs and elevation of retinoid-X receptor mRNA by retinoic acid. *Mol Cell Endocr* 139:209–227.

Clare AS, Costlow JD, Bedair HM, Lumb G. 1992. Assessment of crab limb regeneration as an assay for developmental toxicity. *Can J Fish Aquat Sci* 49:1268–1273.

Clement P, Wurda KE. 1991. Rotifera: Microscopic anatomy of invertebrates. Wiley-Liss. p 219–297.

Clever U. 1964. Actinomycin and puromycin: Effects on sequential gene activation by ecdysone. *Science* 146:794–795.

Clever U, Karlson P. 1960. Indukton von Puff-Veranderungen in den Speicheldrusen-chromosomen von Chironomus tentans durch Ecdyson. *Exp Cell Res* 20:623–626.

Coast GM. 1998. The regulation of primary urine production in insects. In: Coast GM, Webster SG, editors. Recent advances in arthropod endocrinology. Cambridge: Cambridge Univ Pr. Soc. Exp. Biol. Seminar Series 65. p 289–209.

Coast GM, Webster SG, editors. 1988. Recent Advances in Arthropod Endocrinology. Cambridge: Cambridge Univ Pr. Soc. Exp. Biol. Seminar Series 65.

Colombo L, Colombo-Belvedere P, Breda L. 1977. Progesterone metabolism by the gonads of the ascidian *Styela plicata* Lesueur. *Comp Biochem Physiol* 56B:49–54.

Coohill TP, Bartell CK, Fingerman M. 1970. Relative effectiveness of ultraviolet and visible light in eliciting pigment dispersion directly in melanophores of the fiddler crab, *Uca pugilator*. *Physiol Zool* 43:232–239.

Costlow JD. 1976. The effect of JH mimics on development of the mud crab, *Rhithropanopeus harrisii*. In: Calabrese A, editor. Proceedings of Symposium on Pollution and Physiology of Marine Organisms. New York: Academic Pr. p 439–459.

Crane J. 1944. On the color changes of fiddler crabs (genus *Uca*) in the field. *Zoologica* 29:161–168.

de Lange RPJ, van Minnen J. 1998. Localization of the neuropeptide AGPWamide in gastropod mollusks by in situ hybridization and immunocytochemistry. *Gen Comp Endocrinol* 109:166–174.

deLoof A, DeClerk D. 1986. Vertebrate-type steroids in arthropods: Identification, concentrations, and possible functions. In: Porchet M, Andriesm JC, Dhainaut A, editors. Advances in invertebrate reproduction 4. Amsterdam: Elsevier. p 117–123.

den Besten PJ. 1998. Cytochrome P450 monooxygenase system in echinoderms. *Comp Biochem Physiol* 121C:139–146.

den Besten PJ, Elenbaas JML, Maas EJR, Dielman SJ, Herwig HJ, Voogt PA. 1991. Effects of cadmium and polychlorinated biphenyls (Clophen A50) on steroid metabolism and cytochrome P-450 monooxygenase system in the sea star *Asterias rubens* L. *Aquatic Toxicol* 20:95–110.

den Besten PJ, Herwig HJ, Smaal AC, Zandee DI, Voogt PA. 1990. Interference of polychlorinated biphenyls (Clophen A50) with gametogenesis in the sea star *Asterias rubens* L. *Aquatic Toxicol* 18:231–246.

den Besten PJ, Herwig HJ, Zandee DI, Voogt PA. 1989. Effects of cadmium and PCBs on reproduction of the sea star *Asterias rubens*: Aberrations in the early development. *Ecotoxicol Environ Safety* 18:173–180.

den Besten PJ, Lemaire P, Livingstone DR, Woodin B, Stegeman JJ, Herwig HJ, Seinen W. 1993. Time-course and dose-response of the apparent induction of the cytochrome P450 monooxygenase system of pyloric caeca microsomes of the female sea star *Asterias rubens* L. by benzo[a]pyrene and polychlorinated biphenyls. *Aquatic Toxicol* 26:23–40.

den Besten PJ, van Donselaar EG, Herwig HJ, Zandee DI, Voogt PA. 1991. Effects of cadmium on gametogenesis in the sea star *Asterias rubens* L. *Aquatic Toxicol* 20:83–94.

den Besten PJ, van Donselaar EG, Voogt PA, Herwig HJ. 1994. Ultrastructural changes in the storage cells of the pyloric caeca of the sea star *Asterias rubens* L. (Echinodermata: Asteroidea) during the reproductive cycle. *The Netherlands J Zool* 44:65–76.

Denlinger DL. 1985. Hormonal control of diapause. In: Kerkut GA, Gilbert LI, editors. Comprehensive insect physiology, biochemistry, and pharmacology. Oxford UK: Pergamon Pr. p 353–412.

De Ruijter A, Vandersteen J. 1987. Felldversuche zum Effekt einer insecgar (fenoxycarb)-Spritzung wahrend der Apfelblute auf die Honigbiene. *Apidologie* 18:355–356.

Descamps M. 1991. Roles of morphogenetic hormones in spermatogenesis in Myriapoda. In: Gupta AP, editor. Morphogenetic hormones of arthropods. New Brunswick NJ: Rutgers Univ Pr. p 567–592.

de Waal M, Poortman J, Voogt PA. 1982. Steroid receptors in invertebrates. A specific 17b-estradiol binding protein in a sea star. *Mar Biol Lett* 3:317–323.

De Wilde J. 1983. Endocrine aspects of diapause in the adult stage. In: Downer RGH, Laufer H, editors. Endocrinology of insects. New York: Alan R. Liss. p 357–367.

Dhadialla TS, Carlson GR, Le DP. 1998. New insecticides with ecdysteroidal and juvenile hormone activity. *Ann Rev Entomol* 43:545–569.

Dodson SI, Merrit CM, Shanrahan JP, Shults CM. 1999. Low exposure concentrations of atrazine increase male production in *Daphnia pulicaria*. *Environ Toxicol Chem:* In press.

Donly BC, Ding Q, Tobe SS, Bendena WG. 1993. Molecular cloning of the gene for the allatostatin family of neuropeptides from cockroach *Diploptera punctata Proc Nat Acad Sci USA* 90:8807–8811.

Duve H, Thorpe A, Johnsen AH, Maestro JL, Scott AG, East PD. 1998. The dipteran Leu-callatostatins: Structural and functional diversity in an insect neuroendocrine peptide family. In: Coast GM, Webster SG, editors. Recent advances in arthropod endocrinology. Cambridge UK: Cambridge Univ Pr. p 229–2447.

Eales JG. 1997. Iodine metabolism and thyroid-related functions in organisms lacking thyroid follicles: Are thyroid hormones also vitamins? *Proc Soc Exp Biol Med* 214:302–316.

Edwards JP. 1975. The effects of a juvenile hormone analog on laboratory colonies of Pharoah's ant, *Monomorium pharaonis* (L.) (Hymenoptera, Formicidae). *Bull Ent Res* 65:75–80.

Edwards JP. 1987. Activity of optical and geometrical isomers of the juvenile hormone analog hydroprene on endogenous juvenile hormone levels in larval *Manduca sexta. Pestic Sci* 21:203–210.

Edwards JP, Cerf DC, Staal GB. 1985. Inhibition of ootheca production in *Periplaneta americana* with the anti juvenile hormone fluoromevalonate. *J Insect Physiol* 31:723–728.

Edwards JP, Price NR. 1983. Effects of three compounds with anti juvenile hormone activity and a juvenile hormone analog on endogenous juvenile hormone levels in the tobacco hornworm, *Manduca sexta. J Insect Physiol* 29:83–89.

Elphick MR, Newman SJ, Thorndyke MC. 1995. Distribution and action of SALMFamide neuropeptides in the starfish *Asterias rubens. J Exp Biol* 198:2519–2525.

Elphick MR, Price DA, Lee TD, Thorndyke MC. 1991. The SALMFamides: a new family of neuropeptides isolated from an echinoderm. *Proc Roy Soc Lond B* 243:121–127.

Elphick MR, Reeve JR, Burke RD, Thorndyke MC. 1991. Isolation of the neuropeptide SALMFamide-1 from starfish using a new antiserum. *Peptides* 12:455–459.

Emson RH, Wilkie IC. 1980. Fission and autonomy in echinoderms. *Oceanogr Mar Biol A Rev* 18:155–250.

Evans BD, Pohl J, Kartsonis NA, Calabrese RL. 1991. Identification of Rfamide neuropeptides in the medicinal leech. *Peptides* 12:897–908.

Ewer J, De Vente J, Truman JW. 1994. Neuropeptides induction of cGMP increases in the insect CNS: Resolution at the level of single identifiable neurons. *J Neurosci* 14:7704–7712.

Ewer J, Gammie SC, Truman JW. 1997. Control of insect ecdysis by a positive-feedback endocrine system: Roles of eclosion hormone and ecdysis triggering hormone. *J Exp Biol* 200:869–881.

Fain MJ, Riddiford LM. 1975. Juvenile hormone titers in the haemolymph during late larval development of the tobacco hornworm, *Manduca sexta* (L.). *Biol Bull* 149:506–521.

Farfante IP. 1978. Intersex anomalies in shrimp of the genus *Penaeopsis* (Crustacea: Penaeidae). *US Nat Mar Fish Serv Fish Bull* 76:687–691.

Favrel P, Lelong C, Mathieu M. 1998. Structure of the cDNA encoding the precursor for the neuropeptide FMRFamide in the bivalve mollusk *Mytilis edulis. Neuroreport* 9:2961–2965.

Fernlund P. 1971. Chromactivating hormones of *Pandalus borealis*. Isolation and purification of a light-adapting hormone. *Biochem Biophys Acta* 237:519–529.

Fernlund P. 1974a. Structure of the red-pigment-concentrating hormone of the shrimp, *Pandalus borealis. Biochem Biophys Acta* 371:312–322.

Fernlund P. 1974b. Synthesis of the red-pigment-concentrating hormone of the shrimp, *Pandalus borealis. Acta* 371:17–25.

Fernlund P. 1976. Structure of a light-adapting hormone from the shrimp, *Pandalus borealis. Biochem Biophys Acta* 439:17–25.

Fernlund P, Josefsson L. 1968. Chromactivating hormones of *Pandalus borealis.* Isolation and purification of the 'red pigment-concentrating hormone'. *Biochem Biophys Acta* 158:262–273.

Fernlund P, Josefsson L. 1972. Crustacean color-change hormone: Amino acid sequence and chemical synthesis. *Science* 177:173–175.

Fingerman M. 1985. The physiology and pharmacology of crustacean chromatophores. *Amer Zool* 25:233–252.

Fingerman M. 1989. Pigment hormones in Crustacea. In: Laufer H, Downer RGH, editors. Endocrinology of selected invertebrate types. New York: Alan R. Liss. p 357–374.

Fingerman M, Devi M, Reddy PS, Katyayani R. 1996. Impact of heavy metal exposure on the nervous system and endocrine-mediated processes in crustaceans. *Zool Stud* 35:1–8.

Fingerman M, Jackson NC, Nagabhushanam R. 1998. Hormonally-regulated functions in crustaceans as biomarkers of environmental pollution. *Comp Biochem Physiol* 120C:343–350.

Fingerman SW, Fingerman M. 1977. Effects of a polychlorinated biphenyl and polychlorinated dibenzofuran on molting of the fiddler crab, *Uca pugilator. Bull Environ Contam Toxicol* 18:138–142.

Fingerman SW, Fingerman M. 1978. Influence of the polychlorinated biphenyl preparation aroclor 1242 on color changes of the fiddler crab, *Uca pugilator. Mar Biol* 50:37–45.

Fingerman SW, Fingerman M. 1980. Inhibition by the polychlorinated biphenyl Aroclor 1242 of limb regeneration in the fiddler crab, *Uca pugilator*, in different salinities from which different numbers of limbs have been removed. *Bull Environ Contam Toxicol* 25:744–750.

Fodor A, Deak P, Kiss I. 1982. Competition between juvenile hormone antagonist precocene III and juvenile hormone analog methoprene in the nematode *Caenorhabditis elegans. Gen Comp Endocrin* 74:32–44.

Fodor A, Timar T. 1989. Effects of precocene analogs on the nematode *Caenorhabditis remanei.* (var. Bangaloreiensis) 2. Competitition with a juvenile hormone analog (methoprene). *Gen Comp Endocrin* 74:32–44.

Forman BM, Goode E, Chen J, Oro AE, Bradley DJ, Pearlmann T, Noonan DJ, Burke LT, McMorris T, Lamph WW, Evans RM, Weinberger C. 1995. Identification of a nuclear receptor that is activated by farnesol metabolites. *Cell* 81:687–693.

Fraenkel G. 1935. A hormone causing pupaion in the blowfly *Calliphora erythrocephala. Proc Roy Soc B* 118:1–12.

Fyhn UEH, Fyhn HJ, Costlow JD. 1977. Cirripede vitellogenesis: Effect of ecdysterone in vitro. *Gen Comp Endocrinol* 32:266–271.

Gade G, Hoffmann KH, Spring JH. 1997. Hormonal regulation in insects: Facts, gaps, and future directions. *Physiol Rev* 77:963–1032.

Gajewski M, Leitz T, Schlossherr J, Plickert G. 1996. LWamides from Cnidaria constitute a novel family of neuropeptides with morphogenetic activity. *Wilh Roux Arch Dev Biol* 205:232–242.

Gallardo WG, Hagiwara A, Tomita Y, Soyano K, Snell TW. 1997. Effect of some vertebrate and invertebrate hormones on the population growth, mictic female production, and body size of the marine rotifer *Brachionus plicatilis* Muller. *Hydrobiologia* 358:113–120.

Gammie SC, Truman JW. 1999. Eclosion hormone provides a link between ecdysis-triggering hormone and crustacean cardioactive peptide in the neuroendocrine cascade that controls ecdysis behavior. *J Exp Biol* 202:343–352.

Garside CS, Nachman RJ, Tobe SS. 1997. Catabolism of insect neuropeptides: Allatostatins as models for peptidomimetic design. In: Kawashima S, Kikuyama S, editors. Advances in comparative endocrinology. Bologna Italy: Monduzzi Editore. p 169–173.

Geraerts WPM, Vreugdenhil E, Ebberink RHM. 1988. Bioactive peptides in mollusks. In: Thorndyke MC, Goldsworthy GJ, editors. Neurohormones in invertebrates. Cambridge UK: Cambridge Univ Pr. p 261–281.

Gerritsen A, van der Hoeven N, Seinen W. 1997. Masculinization of female *Daphnia magna* due to treatment with p-*tert*-pentylphenol (TPP). Doctoral thesis. The Netherlands: Univ of Utrecht.

Gibbs PE, Bryan GW, Pascoe PL. 1991. Tributyltin-induced imposex in the dogwhelk, *Nucella lapillus*: Geographical uniformity of the response and effects. *Mar Environ Res* 32:79–87.

Gibbs PE, Bryan GW, Pascoe PL, Burt GR. 1987. The use of the dog-whelk, *Nucella lapillus*, as an indicator of tributyltin (TBT) contamination. *J Mar Biol Assoc UK* 67:507–523.

Giese AC. 1966. On the biochemical constitution of some echinoderms. In: Boolootian RA, editor. Physiology of Echinodermata. New York: Interscience. p 757–796.

Gilbert LI, Rybczynski R, Tobe SS. 1996. Endocrine cascade in insect metamorphosis. In: Gilbert LI, Tata JR, Atkinson BG, editors. postembryonic reprogramming of gene expression in amphibian and insect cells. San Diego CA: Academic Pr. p 59–107.

Gilbert LI, Schneiderman HA. 1960. The development of a bioassay for the juvenile hormone of insects. *Trans Am Microsc Soc* 79:38–67.

Gohar M, Souty C. 1984. Action temporelle d'ecdysteroides sur la synthese proteique ovarienne in vitro chez le Crustace isopode terrestrial *Porecllio dilatatus*. *Reprod Nutr Dev* 24:137–145.

Goldsworthy GJ. 1983. The endocrine control of flight metabolism in locusts. *Adv Insect Physiol* 17:149–204.

Gorbman A, Davey K. 1991. Endocrines. In: Prosser CL, editor. Neureal and integrative animal physiology. New York: Wiley-Liss. p 693–754.

Greenberg MJ, Price DA. 1983. Invertebrate neuropeptides: Native and naturalized. *Ann Rev Physiol* 45:271–288.

Grimmelikhuijzen CJP, Leviev I, Carstensen K. 1996. Peptides in the nervous systems of cnidarians: Structure, function, and biosynthesis. *Int Rev Cytol* 167:37–89.

Guo X, Harmon MA, Jin X, Laudet V, Mangelsdorf DJ, Palmer MJ. 1998. Isolation of 2 functional retinoid X receptor subtypes from the Ixodid tick, *Amblyoma americanum* (L.). *Mol Cell Endocrinol* 139:45–60.

Guo X, Harmon MA, Laudet V, Mangelsdorf DJ, Palmer MJ. 1998. Isolation of a functional ecdysteroid receptor homologue from the Ixodid tick, *Amblyomma americanum* (L.). *Insect Biochem Mol Biol* 27:945–962.

Hagiwara A, Hoshi N, Kawahara F, Tominaga K, Hirayama K. 1995. Resting eggs of the marine rotifer *Brachionus plicatilis* Muller: Development, and effect of irradiation on hatching. *Hydrobiologia* 313/314:223–229.

Hampshire R, Horn DHS. 1966. Structure of crustecdysone, a crustacean molting hormone. *Chem Commun* 1966:37–38.

Hanumante MM, Fingerman SW, Fingerman M. 1981. Antagonism of the inhibitory effect of the polychlorinated biphenyl preparation, Aroclor 1242, on color changes of the fiddler crab, *Uca pugilator*, by norepinephrine and drugs affecting noradrenergic neurotransmission. *Bull Environ Contam Toxicol* 26:479–484.

Happ GM. 1992. Maturation of the male reproductive system and its endocrine regulation. *Ann Rev Entomol* 37:303–320.

Hardie J, Lees AD. 1985. Endocrine control of polymorphism and polyphenism. In: Kerkut GA, Gilbert LI, editors. Comprehensive insect physiology, biochemistry, and pharmacology. Oxford UK: Pergamon. p 441–490.

Harmon MA, Boehm MF, Heyman RA, Mangelsdorf DJ. 1995. Activation of mammalian retinoid X receptors by the insect growth regulator methoprene. *Proc Nat Acad Sci USA* 92:6157–6160.

Hassel M. 1998. Upregulation of a *Hydra vulgaris* cPKC gene is tightly coupled to the differentiation of head structures. *Dev Genes Evol* 207:489–501.

Hassel M, Bridge DM, Stover NA, Kleinholz H, Steele RE. 1998. The level of expression of a protein kinase C gene may be an important component of the patterning process in *Hydra*. *Dev Genes Evol* 207:502–514.

Hasting MH. 1981. Intersex specimens of the amphipod *Ampelisca brevicornis* (Costa). *Crustaceana* 41:199–205.

Henning G, Hofmann DK, Benayahu Y. 1998. Metamorphic processes in the soft corals *Heteroxenia fuscescens* and *Xena umbellata*: The effect of protein kinase C activators and inhibitors. *Invert Reprod Dev* 34:35–45.

Henrich VC, Szekely AA, Kim SJ, Brown NE, Antoniewski C, Hayden MA, Lepesent JA, Gilbert LI. 1994. Expression and function of the ultraspiracle (usp) gene during development of *Drosophila melanogaster. Dev Biol* 165:38–52.

Henrick CA. 1991. Juvenoids and anti-juvenile hormone agents: Past and present. In: Hrdy I, editor. Insect chemical ecology. Prague: Academia Praha. p 429–452.

Hertz WA, Chang ES. 1986. Juvenile hormone effects on metamorphosis of lobster larvae. *J Invert Reprod Develop* 10:71–77.

Highnam KC, Hill L. 1977. The Comparative endocrinology of the invertebrates. series. New York: American Elsevier. 270 p.

Hines GA, Watts SA, Sower SA, Walker CW. 1992. Sex steroid levels in the testes, ovaries, and pyloric caeca during gametogenesis in the sea star *Asterias vulgaris. Gen Comp Endocrin* 87:451–460.

Hiruma K, Yagi S, Endo A. 1983. ML-236B (Compactin) as an inhibitor of juvenile hormone biosynthesis. *Appl Entomol Zool* 18:111–115.

Hisata K, Fujiwara S, Tsuchida Y, Ohashi M, Kawamura K. 1998. Expression and function of a retinoic acid receptor in building ascidians. *Dev Genes Evol* 208:537–546.

Holland C, Skinner D. 1976. Interactions between molting and regeneration in the land crab. *Biol Bull* 150:222–240.

Homola E, Chang ES. 1997. Methyl farnesoate: Crustacean juvenile hormone in search of functions. *Comp Biochem Physiol* 117B:347–356.

Homola E, Sagi A, Laufer H. 1991. Relationship of claw form and exoskeleton condition to reproductive system size and methyl farnesoate in the male spider crab, *Libinia emarginata*. *Invert Reprod Develop* 20:219–225.

Hopkins PM, Durica DS. 1995. Effects of all-trans retinoic acid on regenerating limbs of the fiddler crab, *Uca pugilator*. *J Exp Zool* 272:455–463.

Horn DHS, Middleton EJ, Wunderlich JA, Hampshire F. 1966. Identity of the molting hormones of insects and crustaceans. *Chem Commun* 1966:339–340.

Hosmer AJ, Warren LW, Ward TJ. 1998. Chronic toxicity of pulse-dosed fenoxycarb to *Daphnia magna* exposed to environmentally realistic concentrations. *Environ Toxicol Chem* 17:1860–1866.

Huet M, Fioroni P, Oehlmann J, Stroben E. 1995. Comparison of imposex response in three Prosobranch species. *Hydrobiologia* 309:29–35.

Ishaaya, I. 1992. Insect resistance to benzylphenylureas and other insect growth regulators. In: Mullin CA and Scott JG, editors. Molecular mechanisms of insecticide resistance: Diversity amongst insects. Washington DC: American Chemical Society. ACS Symposium Series 505. p 231–246.

Ishizaki H, Suzuki A. 1992. Brain secretory peptides of the silkmoth *Bombyx mori*: Prothoracicotropic hormone and bombyxin. *Prog Brain Res* 92:1–14.

Jansons I, Cusson M, McNeil JN, Tobe SS, Bendena WG. 1996. Molecular characterization of a cDNA from *Pseudaletia unipuncta* encoding the *Manduca sexta* allatostatin peptide (Mas-AST). *Insect Biochem Molec Biol* 26:767–773.

Jegla TC. 1990. Evidence for ecdysteroids as molting hormones in Chelicerata, Crustacea, and Myriapoda. In: Gupta AP, editor. Morphogenetic hormones of arthropods. New Brunswick NJ: Rutgers Univ Pr. p 229–273.

Jegla TC, Costlow JD. 1979. The Limulus bioassay for ecdysteroids. *Biol Bull* 156:103–114.

Jindra M, Huang JY, Malone F, Asahina M, Riddiford LM. 1997. Identification and mRNA profiles of 2 Ultraspiracle isoforms in the epidermis and wings of *Manduca sexta. Insect Mol Biol* 6:41–53.

Jindra M, Malone F, Hiruma K, Riddiford LM. 1996. Developmental profiles and ecdysteroid regulation of the mRNAs for 2 ecdysone receptor isoforms in the epidermis and wings of the tobacco hornworm, *Manduca sexta Dev Biol* 180:258–272.

Joly R, Descamps M. 1988. Endocrinology of myriapods. In: Laufer H, Downer RGH, editors. Endocrinology of selected invertebrates. New York: Alan R. Liss. p 429–449.

Jones G. 1995. Molecular mechanisms of action of juvenile hormone. *Ann Rev Entomol* 40:140–169.

Jones G, Sharp PA. 1997. Ultraspiracle: An invertebrate nuclear receptor for juvenile hormones. *Proc Natl Acad Sci USA* 94:13499–13503.

Josefsson L. 1975. Structure and function of crustacean chromatophorotropins. *Gen Comp Endocrinol* 25:199–202.

Josefsson L. 1983a. Chemical properties and physiological actions of crustacean chromatophorotropins. *Amer Zool* 23:507–515.

Josefsson L. 1983b. Invertebrate neuropeptide hormones. *Internat J Pept Prot Res* 21:459–470.

Kamimura M, Tomita S, Kiuchi M, Fujiwara H. 1997. Tissue-specific and stage-specific expression of 2 silkworm ecdysone receptor isoforms - Ecdysteroid-dependent transcription in cultured anterior silk glands. *Eur J Biochem* 248:786–793.

Kanazawa A, Teshima S. 1971. In vivo conversion of cholesterol to steroid hormones in the spiny lobster, *Panulirus japonicus*. *Bull Japan Soc Sci Fish* 37:891–898.

Kapitskaya M, Wang S, Cress DE, Dhadialla TS, Raikhel AS. 1996. The mosquito ultraspiracle homologue, a partner of ecdysteroid receptor heterodimer: Cloning and characterization of isoforms expressed during vitellogenesis. *Mol Cell Endocrinol* 121:119–132.

Kataoka H, Toschi A, Li JP, Carney RL, Schooley DA, Kramer SJ. 1989. Identification of an allatotropin from adult *Manduca sexta*. *Science* 243:1481–1483.

Katz M. 1983. Comparative anatomy of the tunicate tadpole, *Ciona intestinalis*. *Biol Bull* 164:1–27.

Kaufman WR. 1997. Arthropoda: Chelicerata. In: Adiyodi KG, Adiyodi RG, editors. Reproductive biology of invertebrates. Volume 8, Progress in reproductive endocrinology. Chichester: J Wiley p 211–245.

Kelce WR, Stone CR, Laws SC, Gray LE, Kemppainen JA, Wilson EM. 1995. Persistent DDT metabolite *p.p'*-DDE is a potent androgen receptor antagonist. *Nature* 375:581–585.

Keller R, Schmid E. 1979. In vitro secretion of ecdysteroids by Y-organs and lack of secretion by mandibular organs of the crayfish following molt induction. *J Comp Physiol* 130:347–353.

Keshmirian J, Nogrady T. 1987. Rotifer neuropharmacology. III. Adernergic drug effects on *Brachionus plicatilis*. *Comp Biochem Physiol* 86C:329–332.

Keshmirian J, Nogrady T. 1988. Histofluorescent labeling of catecholaminergic structures in rotifers (Aschelminthes) II. Males of *Brachionus plicatilis* and structures from sectioned females. *Histochemistry* 89:189–192.

Keshmirian J, Nogrady T. 1989. Rotifer neuropharmacology. IV. Involvement of aminergic neurotransmitters in the abnormal sessile behavior of *Brachionus plicatilis* (Rotifera, Aschelminthes). *Hydrobiologia* 174:213–216.

Kleinholz LH. 1975. Purified hormones from the crustacean eyestalk and their physiological specificity. *Nature* 258:256–257.

Kleinholz LH. 1985. Biochemistry of crustacean hormones. In: Bliss DE, Mantel LH, editors. Volume 9, The biology of crustacea. New York: Academic Pr.

Kleinholz LH, Keller R. 1979. Endocrine regulation in Crustacea. In: Barrington EJW, editor. Hormones and evolution. New York: Academic Pr. p 159–213.

Koller G. 1925. Farbwechsel bei *Crangon vulgris*. *Verh Deutsch Zool Ges* 30:128–132.

Koller G. 1927. Uber Chromatophorensystem, Farbensinn und Farbechsel bei *Crangon vulgaris*. *Z Vergl Physiol* 5:191–246.

Koller G. 1928. Versuche uber die inkretorischen Voragange beim Garneenlenfarbwechsel. *Z Vergl Physiol* 8:601–612.

Korczynski RE. 1985. Intersexuality in the arctic isopod *Mesidotea (=Saduria) sibirica*. *Arctic* 38:68–69.

Kostrouchova M, Krause M, Kostrouch Z, Rall JE. 1998. CHR3: a *Caenorhabditis elegans* orphan nuclear hormone receptor required for proper epidermal development and molting. *Development* 125:1617–1626.

Kostruch Z, Kostruchova M, Love W, Jannini E, Piatigorsky J, Rall JE. 1998. Retinoic acid X receptor in the diploblast, *Tripedalia cystophora*. *Proc Nat Acad Sci USA* 95:13442–13447.

Kothapalli R, Palli SR, Ladd TR, Sohi SS, Cress DE, Dhadialla TS, Tzertzinis G, Retnakaran A. 1995. Cloning and developmental expression of the ecdysone receptor gene from the spruce budworm, *Choristoneura fumiferana*. *Dev Genet* 17:319–330.

Krajniak KG, Price DA. 1990. Authentic FMRFamide is present in the polychaete *Nereis virens*. *Peptides* 11:75–77.

Kramer SJ, Toschi A, Miller CA, Kataoka H, Quistad GB, Li JP, Carney RL, Schooley DA. 1991. Identification of an allatostatin from the tobacco hornworm, *Manduca sexta*. *Proc Natn Acad Sci USA* 88:9458–9462.

Kulkarni GK. 1989. Histochemical evidence for the occurrence of steroid synthesis in the testis of a freshwater leech *Hirudo birmanica* Blanchard. *J Anim Morphol Physiol* 36:201–208.

Kuwano E, Takeya R, Eto M. 1983. Terpenoid imidazoles: New anti-juvenile hormones. *Agric Biol Chem* 47:921–923.

Kuwano E, Takeya R, Eto M. 1984. Synthesis and anti-juvenile hormone activity of 1-citronellyl-5-substituted imidazoles. *Agric Biol Chem* 48:3115–3119.

La Clair JJ, Bantle JA, Dumont J. 1998. Photoproducts and metabolites of a common insect growth regulator produce developmental deformalities in Xenopus. *Environ Sci Technol* 32:1453–1461.

Lachaise F. 1990. Synthesis, metabolism and effects on molting of ecdysteroids in Crustacea, Chelicerata, and Myriapoda. In: Gupta AP, editor. Morphogenetic hormones of arthropods. New Brunswick NJ: Rutgers Univ Pr.

Lachaise F, Carpentier G, Somme G, Colardeau J, Beydon P. 1989. Ecdysteroid synthesis by crab Y-organs. *J Exp Zool* 252:283–292.

Lachaise F, Hoffman JA. 1977. Ecdysone et development ovarien chez un Decapod, *Carcinus maenas*. *CR Acad Sci (Paris)* 285:701–704.

Lagueux M, Hetru C, Goltzene F, Kappler C, Hoffmann JA. 1979. Ecdysone titer and metabolism in relation to cuticulogenesis in embryos of *Locusta migratoria*. *J Insect Physiol* 25:709–723.

Landau M, Laufer H, Homola E. 1989. Control of methyl farnesoate synthesis in the mandibular organ of the crayfish *Procambarus clarkii*: Evidence for peptide neurohormones with dual functions. *Invert Reprod Develop* 16:165–168.

Lanot R, Cledon P. 1989. Ecdysteroids and meiotic reinitiation in *Palaemon serratus* (Crustacea Decapoda Natantia) and in *Locust migratoria* (Insecta Orthoptera). A comparative study. *Invert Reprod Devel* 16:169–175.

Laufer H, Biggers WJ, Ahl JSB. 1998. Stimulation of ovarian maturation in the crayfish *Procambarus clarkii* by methyl farnesoate. *Gen Comp Endocrinol* 111:113–118.

Laufer H, Borst D, Baker FC, Carrasco C, Sinkus M, Reuter CC, Tsai LW, Schooley DA. 1987. Identification of a juvenile hormone-like compound in a crustacean. *Science* 235:202–205.

Lawrence JM, Lane JM. 1982. The utilization of nutrients by post-metamorphic echinoderms. In: Jangoux M, Lawrence JM, editors. Echinoderm nutrition. Rotterdam: Balkema.

Lee KY, Horodyski FM, Chamberlin ME. 1998. Inhibition of midgut ion transport by allatotropin (Mas-AT) and Manduca FLRFamides in the tobacco hornworm *Manduca sexta*. *J Exp Biol* 201:3067–3074.

Lesh-Laurie GE. 1988. Coelenterate endocrinology. In: Laufer H, Downer RGH, editors. Endocrinology of selected invertebrate types. New York: Alan R. Liss. p 3–29.

Liu L, Laufer H, Wang YJ, Hayes T. 1997. A neurohormone regulating both methyl farnesoate synthesis and glucose metabolism in a crustacean. *Biochem Biophys Res Commun* 237:694–701.

Lomas LO, Rees HH. 1998. Endocrine regulation of development and reproduction in acarines. In: M. CG, Webster SG, editors. Recent advances in arthropod endocrinology. Cambridge UK: Cambridge Univ Pr. p 91–124.

Mackie GO. 1995. On the "visceral nervous system" of *Ciona. J Mar Biol Assoc UK* 75:141–151.

Mangelsdorf DJ, Evans RM. 1995. The RXR heterodimers and orphan receptors. *Cell* 83(841–850).

Martin G, Juchault P, Sorokine O, A. VD. 1990. Purification and characterization of androgenic hormone from the terrestrial isopod *Armadillidum vulgare* Latr. (Crustacea, Oniscidea). *Gen Comp Endocrin* 80:349–354.

Mattson MP, Spaziani E. 1985. 5-Hydroxytryptamine mediates release of molt-inhibiting hormone activity from isolated crab eyestalk ganglia. *Biol Bull* 169:246–255.

McCarthy JF, Skinner DM. 1979. Changes in ecdysteroids during embryogenesis of the blue crab *Callinectes sapidus. Dev Biol* 69:627–633.

McLachlan JA. 1993. Functional toxicology: A new approach to detect biologically active xenobiotics. *Environ Health Perspect* 101:386–387.

Meredith J, Ring M, Macins A, Marschall J, Cheng NN, Theilmann D, Brock HW, Phillips JE. 1996. Locust ion transport peptide (ITP): Primary structure, cDNA and expression in a baculovirus system. *J Exp Biol* 199:1053–1061.

Mittler TE. 1991. Juvenile hormone and aphid polymorphism. In: Gupta AP, editor. Morphogenetic hormones of arthropods. New Brunswick NJ: Rutgers Univ Pr. p 453–474.

Mladenov PV, Burke RD. 1994. Echinodermata: asexual reproduction. In: Adiyodi KG, Adiyodi RG, editors. Reproductive biology of invertebrates, VI B, Asexual propogation and reproductive strategies. India: Oxford and IBH. p 339–383.

Mohamed KS, Diwan AD. 1991. Effect of androgenic gland ablation on sexual characters of the male Indian while prawn *Panaeus indicus* H. Milne Edwards. *Ind J Exp Biol* 29:478–480.

Monger DJ, Lim WA, Kezdy FG, Law JH. 1982. Compactin inhibits insect HMG-CoA reductase and juvenile hormone biosynthesis. *Biochem Biophys Res Comm* 105:1374–1380.

Mordue W, Luntz AJ, Blackwell A. 1993. Azadirachtin: An update. *J Insect Physiol* 39:903–924.

Moss C, Beesley PW, Thorndyke MC, Bollner T. 1998. Preliminary observations on ascidian and echinoderm neurons and neural explants in vitro. *Tissue and Cell.*

Moss C, Hunter AJ, Thorndyke MC. 1998. Patterns of bromodeoxyuridine incorporation and neuropeptide immunoreactivity in the regenerating arm of the starfish *Asterias rubens. Phil Trans Roy Soc* 353:421–436.

Mouillet JF, Delbecque JP, Quennedey B, Delachambre J. 1997. Cloning of 2 putative ecdysteroid receptor isoforms from *Tenebrio molitor* and their developmental expression in the epidermis during metamorphosis. *Eur J Biochem* 248:856–863.

Nachman RJ, Moyana G, Williams HJ, Garside CS, Tobe SS. 1997. Active conformation and peptidase resistance of conformationally restricted analogs of the insect allatostatin neuropeptide family. In: Kawashima S, Kikuyama S, editors. Advances in comparative endocrinology. Bologna Italy: Monduzzi Editore. p 1353–1359.

Nagabhushanam R, Kulkarni GK. 1981. Effect of exogenous testosterone on the androgenic gland and testis of a marine penaeid prawn, *Parapenaeopsis hardwickii* (Miers) (Crustacea, Decapoda, Penaeidae). *Aquaculture* 23:19–27.

Nagamine C, Knight AW, Maggenti A, Paxman G. 1980a. Effects of androgenic gland ablation on male primary and secondary sexual characteristics in the Malaysian prawn, *Macrobrachium rosenbergii* (de man) (Decapoda, Palaemonidae),with first evidence of induce feminization in a nonhermaphroditic decapod. *Gen Comp Endocrin* 41:423–441.

Nagamine C, Knight AW, Maggenti A, Paxman G. 1980b. Masculinization of female *Macrobrachium rosenbergii* (de Man) (Decapoda, Palaemonidae) by androgenic gland implantation. *Gen Comp Endocrin* 41:442–457.

Nassel DR. 1996. Peptidergic neurohormonal control systems in invertebrates. *Current Opinion in Neurobiology* 6:842–850.

Nassel DR, Lundquist CT, Muren JE, Winther AME. 1998. An insect peptide family in search of functions: The tachykinin-related peptides. In: Coast GM, Webster SG, editors. Recent advances in arthropod endocrinology. Cambridge UK: Cambridge Univ Pr. p 248–277.

Nicol D, Meinertzhagen IA. 1991. Cell count and maps in the larval central nervous system of the ascidian *Ciona intestinalis*. *J Comp Neurol* 309:415–429.

Nijhout HF. 1994. Insect hormones. Princeton NJ: Princeton Univ Pr.

Nijhout HF, Wheeler DE. 1982. Juvenile hormone and the physiological basis of insect polymorphisms. *Quart Rev Biol* 57:109–133.

Nisbet AJ, Mordue AJ, Mordue W. 1995. Detection of [22, 23-3H2]dihydroazadirachtin binding sites on *Schistocerca gregaria* (Forskal) testes membranes. *Insect Biochem Mol Biol* 25:551–557.

Nogrady T, Alai M. 1983. Cholinergic neurotransmission in rotifers. *Hydrobiologia* 104:149–153.

Nogrady T, Keshmirian J. 1986. Rotifer neuropharmacology. I. Cholinergic drug effects on oviposition of *Philodina acuticornis* (Rotifera, Aschelminthes). *Comp Biochem Physiol* 83C:335–338.

Noirot C, Bordereau C. 1991. Termite polymorphism and morphogenetic hormones. In: Gupta AP, editor. Morphogenetic hormones of arthropods. New Brunswick NJ: Rutgers Univ Pr. p 293–324.

Norekian TP, Satterlie RA. 1997. Distribution on myomodulin-like and buccalin-like immunoreactivities in the central nervous system and peripheral tissues of the mollusk *Clione limacina*. *J Comp Neurobiol* 381:41–52.

Oberdörster E, Rittschof D, McClellan-Green P. 1998. Testosterone metabolism in imposex and normal *Ilyanassa obsoleta*: A comparison of field and tributyltin Cl-induced imposex. *Mar Pollut Bull* 36:144–151.

O'Connor BI, Kovacs TG, Voss RH. 1992. The effect of wood species composition on the toxicity of simulated mechanical pulping effluents. *Environ Toxicol Chem* 11:1259–1270.

Olmstead AW, LeBlanc GA. 1998. Effects of endocrine active chemicals on the development of sex characteristics of *Daphnia magna*. 19th Annual Meeting of the Society of Environmental Toxicology and Chemistry; Charlotte NC.

Orchard I, Lange AG. 1998. The distribution, biological activity, and pharmacology of SchistoFLRFamide and related peptides in insects. In: Coast GM, Webster SG, editors. Recent advances in arthropod endocrinology. Cambridge UK: Cambridge Univ Pr. p 278–301.

Orchard I, Ramirez JM, Lange AG. 1993. A multifunctional role for octopamine in locust flight. *Ann Rev Entomol* 38:227–249.

Oudejans RCHM, van der Sluis I. 1979. Storage and depletion of lipid components in the pyloric caeca and ovaries of the sea star *Asterias rubens* during its annual reproductive cycle. *Mar Biol* 53:239–247.

Palmer BD, Palmer SK. 1995. Vitellogenin induction by xenobiotic estrogens in the red-eared turtle and African clawed frog. *Environ Health Perspect* 103(Suppl.4):19–25.

Pawlick JR. 1992. Chemical ecology of the settlement of benthic marine invertebrates. *Oceanogr Mar Biol Ann Rev* 30:273–335.

Pener MP. 1991. Locust phase polymorphism and its endocrine relations. *Adv Insect Physiol* 23:1–80.

Pener MP, Yerushalmi Y. 1998. The physiology of locust phase polymorphism: An update. *J Insect Physiol* 44:365–377.

Perkins EB. 1928. Color changes in crustaceans, especially in *Palaemonetes. J Exp Zool* 50:7–105.

Pertseva MN, Plensneva SA, Kuznetsova LA, Shpakov AO, Derkach KV. 1996. On the tyrosine kinase mechanism of the novel effect of insulin and insulin-like growth factor 1- stimulation of the adrenyl cyclase system in muscle tissue. *Biochem Pharmacol* 52:1867–1874.

Phillips JE. 1983. Endocrine control of salt and water balance: Excretion. In: Downer RGH, Laufer H, editors. endocrinology of insects. New York: Alan R. Liss. p 411–425.

Porchet M, Dhainaut-Courtois N. 1988. Neuropeptides and monoamins in annelids. In: Thorndike M, Goldsworthy GJ, editors. Neurohormones in invertebrates. Cambridge UK: Cambridge Univ Pr. p 219–234.

Postlethwait JH. 1974. Juvenile hormone and the adult development of Drosophila. *Biol Bull* 147:119–135.

Potton DJ, Thorndyke MC. 1995. S2-regulator of feeding in *Asterias rubens*? In: Emson RH, Smith A, Campbell AC, editors. Echinoderm research. Rotterdam The Netherlands: Balkema. p 57–61.

Powell J, Reskaskinner SM, Prakash MO, Fischer W, Park M, Rivier JE, Craig AG, Mackie GO, Sherwood NM. 1996. 2 new forms of gonadotrophin-releasing hormone in a protochordate and the evolutionary implications. *Proc Natl Acad Sci USA* 93:10461–10464.

Pratt GE, Jennings RC, Hamnett AF, Brooks GT. 1980. Lethal metabolism of precocene-I to a reactive epoxide by locust corpora allata. *Nature* 284:320–323.

Price DA, Greenberg MJ. 1977. Structure of a molluscan cardioexcitatory neuropeptide. *Science* 197:670–671.

Punin MY. 1993. The system of regulatory elements in the digestive tract epithelium of invertebrates (in Russian, from the Abstract). *Tsitoloya* 35:3–26.

Quackenbush LS. 1986. Crustacean endocrinology, a review. *Can J Fish Aquat Sci* 43:2271–2282.

Quistad GB, Cerf DC, Kramer SJ, Bergot BJ, Schooley DA. 1985. Design of novel insect anti-juvenile hormones: Allylic alcohol derivatives. *J Agric Food Chem* 33:47–50.

Quistad GB, Cerf DC, Schooley DA, Staal GB. 1981. Fluoromevalonate acts as an inhibitor of insect juvenile hormone biosynthesis. *Nature* 289:176–177.

Raikhel AS, Lea AO. 1985. Hormone-mediated formation of the endocytic complex in mosquito oocytes. *Gen Comp Endocrinol* 57:422–433.

Raina AK. 1993. Neuroendocrine control of sex pheromone biosynthesis in Lepidoptera. *Ann Rev Entomol* 38:329–349.

Rajfer B, Sikka SC, Rivera F, Handelsman DJ. 1986. Mechanism of inhibition of human testosterone steroidogenesis by oral ketoconazole. *Clin Endocrin Met* 63:1193–1198.

Ramos S, Sultatos L. 1998. Flavonoid-induced alterations in cytochrome P450-dependent biotransformation of the organophosphorus insecticide parathion in the mouse. *Toxicology* 131:155–167.

Ramsay JA. 1954. Active transport of water by the Malpighian tubules of the stick insect, *Dixippus morosus* (Orthoptera, Phasmidae). *J Exp Biol* 31:104–113.

Rao KR. 1985. Pigmentary effectors. In: Bliss DE, Mantel LH, editors. Volume 9, The biology of crustacea. New York: Academic Pr. p 395–462.

Rao KR, Conklin PH, Branno AC. 1978. Inhibition of limb regeneration in the grass shrimp, *Palaemonetes pugio*, by sodium pentachlorophenate. In: Rao KR, editor. Pentachlorophenol. New York: Plenum. p 193–203.

Rao KR, Fox FR, Conklin PJ, Cantelmo AC, Brannon AC. 1979. Physiological and biochemical investigations of the toxicity of pentachlorophenol to crustaceans. In: Vernberg WB, Thurberg FP, Calabrese A, Vernberg FJ, editors. Marine pollution: functional responses. New York: Academic Pr. p 307–340.

Rao KR, Riehm JP. 1989. The pigment-dispersing hormone family: Chemistry, structure-activity relations, and distribution. *Biol Bull* 177:225–229.

Rao KR, Riehm JP, Zahnow CA, Kleinholz LH, Tarr GE, Norton S, Semmes OJ, Jorenby WH. 1985. Characterization of a pigment-dispersing hormone in the eyestalks of the fiddler crab *Uca pugilator*. *Proc Natl Acad Sci USA* 82:5319–5322.

Reagan JD. 1994. Expression cloning of an insect diuretic hormone receptor: A member of the calcitonin/secretin receptor family. *J Biol Chem* 269:9–12.

Reddy PS, Fingerman M. 1995. Effect of cadmium chloride on physiological color changes of the fiddler crab, *Uca pugilator*. *Ecotoxicol Environ Safety* 31:69–75.

Reddy PS, Nguyen LK, Obih P, Fingerman M. 1997. Effect of cadmium chloride on the distal retinal pigment cells of the fiddler crab, *Uca pugilator*. *Bull Environm Contam Toxicol* 58:504–510.

Reddy PS, Sarojini R, Nagabhushanam R. 1991. Impact of tributyltin oxide (TBTO) on limb regeneration of the prawn, *Caridina rajadhari*, after exposure to different time intervals of amputation. *J Tiss Res* 1:35–39.

Restifo LL, Wilson TG. 1998. A juvenile hormone agonist reveals distinct developmental pathways mediated by ecdysone-inducible Broad Complex transcription factors. *Dev Genet* 22:141–159.

Reynolds SE. 1985. Hormonal control of cuticular mechanical properties. In: Kerkut GA, Gilbert LI, editors. Volume 8, Comprehensive insect physiology, biochemistry, and pharmacology. Oxford UK: Pergamon Pr. p 335–351.

Riddiford LM. 1985. Hormone action at the cellular level. In: Kerkut GA, Gilbert LI, editors. Volume 8, Comprehensive insect physiology, biochemistry, and pharmacology. Oxford UK: Pergamon. p 37–84.

Riddiford LM. 1993. Hormones and Drosophila development. In: Bate M, Martinez-Arias A, editors. The development of Drosophila. Cold Spring Harbor: Cold Spring Harbor Laboratory Pr. p 899–939.

Riddiford LM. 1994. Cellular and molecular actions of juvenile hormone. I. General considerations and premetamorphic actions. *Adv Insect Physiol* 24:213–274.

Riddiford LM, Ashburner M. 1990. Role of juvenile hormone in larval development and metamorphosis in *Drosophila melanogaster*. *Gen Comp Endocrin* 82:172–183.

Robinow S, Talbot WS, Hogness DS, Truman JW. 1993. Programmed cell death in the Drosophila CNS is ecdysone-regulated and coupled with a specific ecdysone receptor. *Development* 119:1251–1259.

Robinson GE. 1987. Hormonal regulation of age polyethism in the honeybee *Apis mellifera*. In: Menzel R, Mercer A, editors. Neurobiology and behavior of honeybees. Berlin: Springer-Verlag. p 266–279.

Roller H, Dahm KH, Sweeley CC, Trost BM. 1967. The structure of the juvenile hormone. *Angew Chem Int Ed* 6:179–180.

Ronis MJJ, Mason AZ. 1996. The metabolism of testosterone by the periwinkle (*Littorina littorea*) in vitro and in vivo: Effects of tributyltin. *Mar Environ Res* 42:161–166.

Rudolph PH, Stay B. 1997. Cockroach allatostatin-like immunoreactivity in the central nervous system of the freshwater snails *Bulinus globosus* (Planorbidae) and *Stagnicola elodes* (Lymnaeidae). *Gen Comp Endocrinology* 106:241–250.

Ruppert EE, Barnes RD. 1991. Invertebrate zoology 6th ed. Orlando: Saunders. p 1056.

Ruvken G, Hobart O. 1998. The taxonomy of developmental control in *Caenorhaditis elegans*. *Science* 282:2033–2041.

Saegusa HA, Mizoguchi A, Kitahora H, Nagasawa H, Suzuki A, Ishizaki H. 1992. Changes in the titer of bombyxin. Immunoreactive material in the haemolymph during postembryonic development of the silkmoth *Bombyx mori*. *Dev Growth Differ* 34:595–605.

Safranek L, Riddiford LM. 1975. The biology of the black larval mutant of the tobacco hornworm, *Manduca sexta*. *J Insect Physiol* 21:1931–1938.

Sagi A, Ahl JSB, Danaee H, Laufer H. 1994. Methyl farnesoate levels in the male spider crabs exhibiting active reproductive behavior. *Horm Behav* 28:261–272.

Sagi A, Homola E, Laufer H. 1991. Methyl farnesoate in the prawn *Macrobrachium rosenbergii*: synthesis by the mandibular organ in vitro, and titers in the haemolymph. *Comp Biochem Physiol* 99B:879–882.

Sagi A, Homola E, Laufer H. 1993. Distinct reproductive types of male spider crabs *Libinia emarginata* differ in circulating and synthesizing methyl farnesoate. *Biol Bull* 185:168–173.

Sangalang G, Jones G. 1997. Oocytes in testis and intersex in lobsters (*Homarus americanus*) from Nova Scotian sites: Natural or site-related phenomenon? *Can Tech Rep Fish Aquat Sci* 2163:46.

Santama N, Li KW, Geraerts WPM, Benjamin PR, Burke JF. 1996. Post-translational processing of the alternative neuropeptide precursor encoded by the FMRFamide gene in the pulmonate snail *Lymnaea stagnalis. Eur J Neurosci* 8:968–977.

Sarojini S. 1963. Comparison of the effects of androgenic hormone and testosterone propionate on the female ocypod crab. *Curr Science* 33:55–56.

Sbrenna G. 1991. Roles of morphogenetic hormones in embryonic cuticle deposition in arthropods. In: Gupta AP, editor. Morphogenetic hormones of arthropods. New Brunswick NJ: Rutgers Univ Pr. p 44–80.

Schmich J, Trep E, Leitz T. 1998. The role of GLWamides in metamorphosis of *Hydractinia echinata. Dev Genes Evol* 208:267–273.

Schneiderman H, Gilbert L. 1958. Substances with juvenile hormone activity in Crustacea and other invertebrates. *Biol Bull* 115:530–535.

Schoenmakers HJN. 1980. The variation of 3b-hydroxysteroid dehydrogenase activity of the ovaries and pyloric caeca of the starfish *Asterias rubens* during the annual reproductive cycle. *J Comp Physiol* 138:27–30.

Schoenmakers HJN, Dieleman SJ. 1981. Progesterone and estrone levels in the ovaries, pyloric caeca and peri visceral fluid during the annual reproductive cycle of starfish, *Asterias rubens. Gen Comp Endocr* 43:63–70.

Schoenmakers HJN, van Bohemen CG, Dieleman SJ. 1981. Effects of oestradiol-17b on the ovaries of the starfish *Asterias rubens. Devl Growth Differ* 23:125–135.

Schooley DA, Edwards JP. 1996. Anti juvenile hormones: From precocenes to peptides., Farnham UK: British Crop Protection Council.

Schubiger M, Wade AA, Carney GE, Truman JW, Bender M. 1998. Drosophila EcR-B ecdysone receptor isoforms are required for larval molting and for neuron remodeling during metamorphosis. *Development* 125:2053–2062.

Schulte-Oehlmann U, Bettin C, Fioroni P, Oehlmann J, Stroben E. 1995. *Marisa cornuarietis* (Gastropods, Prosobranchia): A potential tributyltin bioindicator for freshwater environments. *Ecotoxicology* 4:372–384.

Schwartz SS, Hebert PDN. 1987. Methods for activation of resting eggs of Daphnia. *Freshwater Biol* 17:373–379.

Seol W, Choi HS, Moore DD. 1995. Isolation of proteins that interact specifically with retinoid X receptor: 2 novel orphan receptors. *Mol Endocrin* 9:72–85.

Shirai H, Bulet P, Kondo N, Isobe M, Imai K, Goto T, Kubota I. 1986. Endocrine control of oocyte maturation and spawning in starfish. In: Porchet M, Andries J-C, Dhainaut A, editors. Advances in invertebrate reproduction 4. New York: Elsevier. p 249–256.

Skiebe P, Schneider H. 1994. Allatostatin peptides in the crab stomatogastric nervous system: Inhibition of the pyloric motor pattern and distribution of allatostatin-like immunoreactivity. *J Exp Biol* 194:195–208.

Skinner D, Graham D. 1970. Molting in land crabs: Stimulation by leg removal. *Science* 169:383–385.

Skinner D, Graham D. 1972. Loss of limbs as a stimulus to ecdysis in Brachyura (true crabs). *Biol Bull* 143:222–233.

Skinner DM. 1985. Molting and regeneration. In: Bliss DE, Mantel LH, editors. The biology of crustacea. New York: Academic Pr. p 43–146.

Smart D, Johnston CF, Curry WJ, Williamson R, Maule AG, Skuce PJ, Shaw C, Halton DW, Buchanan KD. 1994. Peptides related to the *Diploptera punctata* allatostatins in non arthropod invertebrates: an immunocytochemical survey. *J Comp Neurol* 347:426–432.

Smart D, Johnston CF, Maule AG, Halton DW, Hrckova G, Shaw C, Buchanan KD. 1995. Localisation of *Diploptera punctata* allatostatin-like immunoreactivvity in helminths: An immunocytochemical study. *Parasitology* 110:87–96.

Smiley S. 1988. Investigation into purification and identification of the oocyte maturation hormone of *Stichopus californica* (Holothuroidea: Echinodermata). In: Burke RD, Mladenov PV, Lambert P, Parsley RL, editors. Echinoderm biology. Rotterdam: Balkema. p 541–549.

Smit AB, Spijker S, van Minnen J, Burke JF, de Winter F, van Elk R, Geraerts WPM. 1996. Expression and characterization of molluscan insulin-related peptide VII from the mollusk *Lymnaea stagnalis. Neuroscience* 70:589–596.

Smith BS. 1971. Male characteristics on female mud snails caused by antifouling bottom paints. *J Appl Toxicol* 1:22–25.

Smith DG. 1977. An occurrence of a female intersex of *Asellus communis* Say (Isopoda, Asellidae). *Crustaceana* 32:89–90.

Snell TW, Carmona MJ. 1995. Comparative toxicant sensitivity of sexual and asexual reproduction in the rotifer *Brachionus calyciflorus. Environ Toxicol Chem* 14:415–420.

Soumoff C, O'Connor JD. 1982. Repression of Y-organ secretory activity by molt inhibiting hormone in the crab *Pachygrapsus crassipes. Gen Comp Endocrinol* 48:432–439.

Soyez D, Le Caer JP, Noel PY, Rossier J. 1991. Primary structure of 2 isoforms of vitellogenesis inhibiting hormone from the lobster *Homarus americanus. Neuropeptides* 20:25–32.

Spangenberg DB. 1967. Iodine induction of metamorphosis in *Aurelia. J Exp Zool* 165:441–450.

Spaziani E, Rees HH, Wang WL, Watson RD. 1989. Evidence that Y-organs of the crab *Cancer antennarius* secrete 3-dehydroecdysone. *Molec Cell Endocrinol* 66:17–25.

Spindler KD. 1991. Roles of morphogenetic hormones in the metamorphosis of arthropods other than insects. In: Gupta AP, editor. Morphogenetic hormones of arthropods. New Brunswick NJ: Rutgers Univ Pr. p 131–149.

Spindler KD, Van Wormhoudt A, Sellos D, Spindler-Barth M. 1987. Ecdysteroid levels during embryogenesis in the shrimp *Palaemon serratus* (Crustacea Decapoda): Quantitative and qualitative changes. *Gen Comp Endocrinol* 66:116–122.

Spooner N, Gibbs PE, Bryan GW, Goad LJ. 1991. The effects of tributyltin upon steroid titres in the female dogwhelk, *Nucella lapillus*, and the development of imposex. *Marine Environ Res* 32:37–49.

Staal GB. 1975. Insect growth regulators with juvenile hormone activity. *Ann Rev Entomol* 20:417–460.

Staal GB. 1986. Anti juvenile hormone agents. *Ann Rev Entomol* 31:391–429.

Staal GB, Henrick CA, Bergot BJ, Cerf DC, Edwards JP, Kramer SJ. 1981. Relationships and interactions between JH and anti JH analogs in Lepidoptera. In: Sehnal F, Zabza A, Menn JJ, Cymborowski B, editors. Regulation of insect development and behavior: Wroclaw Technical Univ Pr. p 323–340.

Staub GC, Fingerman M. 1984. Effect of naphthalene on color changes in the sand fiddler crab, *Uca pugilator. Comp Biochem Physiol* 77C:7–12.

Stay B, Tobe SS, Bendena WG. 1994. Allatostatins: Identification, primary structure, functions and distribution. *Adv Insect Physiol* 25:267–338.

Stewart DM. 1988. Endocrinology of arachnids. In: Laufer H, Downer RGH, editors. Endocrinology of selected invertebrates. New York: Alan R. Liss.

Stross RG, Hill JC. 1968a. Photoperiod control of winter diapause in freshwater crustacean Daphnia. *Biol Bull* 134:176–198.

Stross RG, Hill JC. 1968b. Diapause induction in Daphnia requires 2 stimuli. *Science* 150:1462–1464.

Sundaram M, Palli SR, Krell PJ, Sohi SS, Dhadialla TS, Retnakaran A. 1998. Basis for selective action of a synthetic molting hormone agonist, RH-5992 on lepidopteran insects. *Insect Biochem Mol Biol* 28:693–704.

Suzuki H, Kimura T, Sekiguchi T, Mizukami A. 1997. FMRFamide-like immunoreactive primary sensory neurons in the olfactory system of the terrestrial mollusk, *Limax marginatus. Cell Tiss Res* 289:339–345.

Suzuki S, Yamasaki K. 1997. Sexual bipotentiality of developing ovaries in the terrestrial isopod *Armadillidium vulgare* (Malacostraca, Crustacea). *Gen Comp Endocrin* 107:136–146.

Suzuki S, Yamasaki K, Katakura Y. 1990. Vitellogenin synthesis in andrectomized males of the terrestrial isopod, *Armadillidium vulgare* (Malacostracan Crustacea). *Gen Comp Endocrin* 77:283–291.

Takahashi N, Kanatani H. 1981. Effect of 17b-estradiol on growth of oocytes in cultured ovarian fragments of the starfish, *Asterina pectinefera. Dev Growth Differ* 23:565–569.

Takeda N. 1980. Hormonal control of head-wart development in the snail *Euchadra peliomphala. J Embryol Exp Morph* 60:57–69.

Talbot WS, Swyryd EA, Hogness DS. 1993. Drosophila tissues with different metamorphic responses to ecdysone express different ecdysone receptor isoforms. *Cell* 73:1323–1337.

Taylor PAI, Bhatt TR, Horodyski FM. 1996. Molecular characterization and expression analysis of *Manduca sexta* allatotropin. *Eur J Biochem* 239:588–596.

Templeton NS, Laufer H. 1983. The effects of a juvenile hormone analog (Altosid ZR-515) on the reproduction and development of *Daphnia magna* (Crustacea: Cladocera). *J Invert Reprod* 6:99–110.

Tensen CP, Cox KJA, Smit AB, Vanderschors RC, Meyerhof W, Richter D, Planta RJ, Hermann PM, van Minnen J, Geraerts WPM, Knol JC, Burke JF, Vreygdenhil E, van Heerikhuizen H. 1998. The *Lymnaea* cardioexcitatory peptide (LyCEP) receptor: A G-protein coupled receptor for a novel member of the Rfamide neuropeptide family. *J Neurosci* 18:9812–9821.

Tensen CP, de Kleijn DPV, Van Herp F. 1991. Cloning and sequence analysis of cDNA encoding 2 crustacean hyperglycemic hormones from the lobster *Homarus americanus. Eur J Biochem* 200:103–106.

Thorndyke MC. 1973. An in-vivo stimulatory effect of 3,3,5-tri-iodo-L-thyronine on polyphenol oxidase activity in an ascidian. *J Endocrinol* 58:679.

Thorndyke MC, Dockray GJ. 1986. Identification and localisation of material with gastrin-like immunoreactivity in the neural ganglion of a protochordate *Ciona intestinalis. Regul Peptides* 16:269–279.

Thummel CS. 1996. Flies on steroids—Drosophila metamorphosis and the mechanisms of seroid hormone action. *Trends Genet* 12:306–310.

Tighe-Ford DJ. 1977. Effects of juvenile hormone analogs on larval metamorphosis in the barnacle *Elminius modestus* Darwin (Crustacea: Cirripeda). *J Exp Mar Biol Ecol* 26:163–176.

Tobe SS, Stay B. 1979. Modulation of juvenile hormone synthesis by an analog in the cockroach. *Nature* 281:481–482.

Trayler KM, Davis JA. 1996. Sensitivity of *Daphnia carinata* sensu lato to the insect growth regulator, pyriproxyfen. *Ecotoxicol Environ Safety* 33:154–156.

Truman JW, Talbot WS, Fahrbach SE, Hogness DS. 1994. Ecdysone receptor expression in the CNS correlates with stage-specific responses to ecdysteroids during Drosophila and Manduca development. *Development* 120:219–234.

Tsukimura B, Waddy S, Burrow CW, Borst DW. 1993. Characterization and regulation of lobster vitellogenin. *Am Zool* 33:122a.

Van Marrewijk WJA, Van der Horst DJ. 1998. Signal transduction of adipokinetic hormone. In: Coast GM, Webster SG, editors. Recent advances in arthropod endocrinology. Cambridge UK: Cambridge Univ Pr. p 172–188.

Verger-Bocquet M, Wattez C, Salzet M, Malecha J. 1992. Immunocytochemical identification of peptidergic neurons in compartment 4 of the supraesophogeal ganglion of the leech *Theromyzon tessulatum* O.F.M. *Can J Zool* 70:856–865.

Voogt PA, Broertjes JJS, Oudejans RCHM. 1985. Vitellogenesis in sea star: Physiological and metabolic implications. *Comp Biochem Physiol* 80A:141–147.

Voogt PA, den Besten PJ, Jansen M. 1991. Steroid metabolism in relation to the reproductive cycle in *Asterias rubens* L. *Comp Biochem Physiol* 99B:77–82.

Voogt PA, Dieleman SJ. 1984. Progesterone and oestrone levels in the gonads and pyloric caeca of the male sea star *Asterias rubens*: A comparison with the corresponding levels in the female sea star. *Comp Biochem Physiol* 79A:635–639.

Voogt PA, Hulskamp R. 1979. Sex dependence and seasonal variation of saponins in the gonads of the starfish *Asterias rubens*: Their relation to reproduction. *Comp Biochem Physiol* 62A:1049–1055.

Voogt PA, van Rhenen JW. 1984. Is the level of cyclic AMP in the pyloric caeca of the starfish *Asterias rubens* related to the reproductive cycle? *Comp Biochem Physiol* 78B:719–722.

Wada H, Saiga H, Satoh N, Holland PWH. 1998. Tripartite organisation of the ancestral chordate brain and the antiquity of placodes: Insights from ascidian *Pax-2/5/8, Hox* and *Otx* genes. *Development* 125:1113–1122.

Waddy SL, Aiken DE, de Klein DPV. 1995. Control of growth and reproduction. In: Factor JR, editor. Biology of the Lobster *Homarus americanus*. San Diego CA: Academic Pr. p 217–266.

Wainwright G, Webster SG, Wilkinson MC, Chung JS, Rees HH. 1996. Structure and significance of mandibular organ-inhibiting hormone in the crab, *Cancer pagurus*. *J Biol Chem* 271:12749–12754.

Walgraeve H, Criel GR, Sorgeloos P, DeLeenheer AP. 1988. Determination of ecdysteroids during the molt cycle of adult *Artemia*. *J Insect Physiol* 34:597–602.

Walker RJ. 1988. Endocrinology of merostomates. In: Laufer H, Downer RGH, editors. Endocrinology of selected invertebrates. New York: Alan R. Liss. p 395–413.

Watkinson IA, Clark BS. 1973. The insect molting hormone as a possible target site for insecticidal action. *Proc Nat Acad Sci USA* 14:488–506.

Weaver RJ, Edwards JP. 1990. The role of corpora allata and associated nerves in the regulation of ovarian cycles in the oviparous cockroach *Periplaneta americana*. *J Insect Physiol* 36:51–59.

Weaver RJ, Edwards JP, Bendena W, Tobe SS. 1998. Structures, functions and occurrence of insect allatostatic peptides. In: Coast GM, Webster SG, editors. recent advances in arthropod endocrinology. Cambridge: Cambridge Univ Pr. p 3–32.

Weaver RJ, Freeman ZA, Pickering MG, Edwards JP. 1994. Identification of 2 allatostatins from CNS of the cockroach *Periplaneta americana* - Novel members of a family of neuropeptide inhibitors of insect juvenile hormone biosynthesis. *Comp Biochem Physiol* 107C:119–127.

Weis JS. 1977. Limb regeneration in fiddler crabs: species differences and effects of methylmercury. *Biol Bull* 152:263–274.

Weis JS. 1978. Interactions of methylmercury, cadmium, and salinity on regeneration in the fiddler crabs *Uca pugilator, U. pugnax* and *U. minax*. *Mar Biol* 49:119–124.

Weis JS, Cristini A, Rao KR. 1992. Effects of pollutants on molting and regeneration in Crustacea. *Amer Zool* 32:495–500.

Weis JS, Donellon N, Barnwell FH. 1986. Regeneration of tubercles on the limbs of *Uca pugilator* and effects of mercury and cadmium on their growth. *J Crust Biol* 6:648–651.

Wilder MN, Aida K. 1995. Crustacean ecdysteroids and juvenoids: Chemistry and physiological roles in 2 species of prawn, *Macrobrachium rosenbergii* and *Penaeus japonicus*. *Israeli J Aquacult* 47:129–136.

Wilder MN, Okumura T, Suzuki Y, Fusetani N, Aida K. 1994. Vitellogenin production induced by eyestalk ablation in juvenile giant freshwater prawn *Macrobrachium rosenbergii* and trial methyl farnesoate administration. *Zool Sci* 11:45–53.

Wilkens JL, Fingerman M. 1965. Heat tolerance and temperature relationships of the fiddler crab *Uca pugilator*, with reference to body coloration. *Biol Bull* 128:133–141.

Williams CM. 1956. The juvenile hormone of insects. *Nature* 178:212–213.

Williams CM. 1967. Third-generation pesticides. *Sci Amer* 217:13–17.

Wilson TG, Fabian J. 1986. A *Drosophila melanogaster* mutant resistant to chemical analog of juvenile hormone. *Dev Biol* 118:190–201.

Wolfner M. 1997. Tokens of love: Functions and regulation of Drosophila male accessory gland products. *Insect Biochem Mol Biol* 27:179–192.

Woodhead AP, Stay B, Seidel SL, Khan MA, Tobe SS. 1989. Primary structure of four allatostatins: Neuropeptide inhibitors of juvenile hormone biosynthesis. *Proc Nat Acad Sci USA* 86:5997–6001.

Wootton AN, Herrig C, Spry A, Wiseman A, Livingstone DR, Goldfarb PS. 1995. Evidence for the existence of cytochrome-P450 gene families (Cyp 1A, 3A, 4A, 11A) and modulation of gene-expression (cyp 1A) in the mussel *Mytilus* spp. *Mar Env Res* 39:21–26.

Wyatt GR, Davey KG. 1996. Cellular and molecular actions of juvenile hormone in adult insects. *Adv Insect Physiol* 26:1–155.

Xu RA. 1991. Annual changes in the steroid levels in the testis and the pyloric caeca of *Sclerasterias mollis* (Hutton) (Echinodermata: Asteroidea) during the reproductive cell. 20:147–152.

Xu RA, Barker MF. 1990a. Annual changes in the steroid levels in the ovaries and the pyloric caeca of *Sclerasterias mollis* during the reproductive cycle. *Comp Biochem Physiol* 95A:127–133.

Xu RA, Barker MF. 1990b. Effect of diets on the steroid levels and reproduction in the starfish, *Sclerasterias mollis. Comp Biochem Physiol* 96A:33–39.

Xu RA, Barker MF. 1990c. Photoperiodic regulation of oogenesis in the starfish *Sclerasterias mollis* (Hutton 1872) (Echinodermata: Asteroides). *J Exp Mar Biol Ecol* 141:159–168.

Yamashita O. 1996. Diapause hormone of the silkworm, *Bombyx mori*: Structure, gene expression and function. *J Insect Physiol* 42:669–679.

Yamashita O, Suzuki K. 1991. Roles of morphogenetic hormones in embryonic diapause. In: Gupta AP, editor. Morphogenetic hormones of arthropods. New Brunswick NJ: Rutgers Univ Pr. p 81–128.

Yano I. 1985. Induced ovarian maturation and spawning in greasyback shrimp *Metapenaeus ensis*, by progesterone. *Aquaculture* 47:223–229.

Yano I. 1987. Effect of 17a-hydroxy-progesterone on vitellogenin secretion in Kuruma prawn, *Penaeus japonicus*. *Aquaculture* 47:223–229.

Yao TP, Forman BM, Jiang Z, Cherbas L, Chen JD, Mckeown M, Cherbas P, Evans RM. 1993. Functional ecdysone receptor is the product of EcR and Ultraspiracle genes. *Nature* 366:476–479.

Yao TP, Segraves WA, Oro AE, McKeown M, Evans RM. 1992. Drosophila ultraspiracle modulates ecdysone receptor function via heterodimer formation. *Cell* 71:63–72.

Yin CM, Chippendale GM. 1979. Diapause of the southwestern corn borer, *Diatraea grandiosella*: Further evidence showing juvenile hormone to be the regulator. *J Insect Physiol* 25:513–523.

Young NJ, Webster SG, Jones DA, Rees HH. 1991. Profile of embryonic ecdysteroids in the decapod crustacean, *Macrobrachium rosenbergii*. *Invert Reprod Devel* 20:201–212.

Young NJ, Webster SG, Rees HH. 1993a. Ecdysteroid profiles and vitellogenesis in *Penaeus monodon* (Crustacea: Decapoda). *Invert Reprod Devel* 24:107–118.

Young NJ, Webster SG, Rees HH. 1993b. Ovarian and haemolymph ecdysteroid titers during vitellogenesis in *Macrobrachium rosenbergii*. *Gen Comp Endocrin* 90:183–191.

Zeleny C. 1905. Compensatory regulation. *J Exp Zool* 2:1–102.

Zitnam D, Kingan TG, Hermesan JL, Adams ME. 1996. Identification of ecdysis-triggering hormone from an epitracheal endocrine system. *Science* 271:88–91.

Zou E, Fingerman M. 1997a. Effects of estrogenic xenobiotics on molting of the water flea, *Daphnia magna*. *Ecotoxicol Environ Safety* 38:281–285.

Zou E, Fingerman M. 1997b. Synthetic estrogenic agents do not interfere with sex differentiation but do inhibit molting of the cladoceran *Daphnia magna*. *Bull Environ Contam Toxicol* 58:596–602.

CHAPTER 3

Laboratory Toxicity Tests for Evaluating Potential Effects of Endocrine-disrupting Compounds

Christopher G. Ingersoll (chair), Tom Hutchinson (reporter), Mark Crane, Stanley Dodson, Ted DeWitt, Andreas Gies, Marie-Chantal Huet, Charles L. McKenney Jr., Eva Oberdörster, David Pascoe, Donald J. Versteeg, Oliver Warwick

Introduction

Scope

The scope of the Laboratory Testing Work Group was to evaluate methods for testing aquatic and terrestrial invertebrates in the laboratory. Specifically, discussions focused on the following objectives:

1) assess the extent to which consensus-based standard methods and other published methods for conducting toxicity tests with invertebrates in the laboratory can be used to detect and assess endocrine-disrupting compounds (EDCs);
2) identify the strengths and weaknesses in these methods for use in the ecological risk assessment of EDCs;
3) suggest potential improvements to existing methods and endpoints, and where appropriate, recommend the development of new approaches; and
4) develop a prioritized list of short-term and long-term research needs.

During the workshop, 2 additional objectives were identified:

5) evaluate the use of reference compounds to help identify species, life stages, and endpoints potentially responsive to EDCs and to help in the development of standardized tests; and
6) identify groups of organisms that could be evaluated in transgenerational (multigenerational) exposures to detect ecologically relevant effects of EDCs on progeny of exposed organisms.

Endocrine Disruption in Invertebrates: Endocrinology, Testing, and Assessment. Peter L. deFur et al., editors.
 ISBN 1-880611-27-9

Before the start of the workshop, the Laboratory Testing Work Group developed the following questions to address:

1) What standard methods are available for conducting laboratory toxicity tests with invertebrates to evaluate reproductive effects?
2) Is there a need for further development of existing standard methods to detect EDCs?
3) Are these methods used in a regulatory context in Europe, Japan, North America, or other locations?
4) What are the groups of invertebrates for which there are few or no existing methods, but which may be at risk through exposure to EDCs?
5) What further development of methods, including validation, standardization, and harmonization, should be considered?
6) What are the key research needs?

During the workshop, 3 additional questions were identified:

7) What factors should be considered in selecting test organisms to evaluate EDCs?
8) What endpoints (endocrine-specific versus integrative) are appropriate for measuring effects of EDCs in laboratory tests?
9) Are there case studies that can be used to help identify factors that should be considered in selecting appropriate species, life stages, and endpoints for measuring effects associated with EDCs?

Background

Growing evidence from wildlife and laboratory observations indicate that environmental EDCs may be having adverse effects on wildlife populations. Such observations relate to a variety of ecosystems and include impacts upon birds, fish, and reptiles (Chapters 1 and 4). Even though invertebrates are vital to the structure and function of ecosystems, the focus of attention on effects of EDCs to date has been directed toward vertebrates, as summarized by a series of previous workshops (Tattersfield et al. 1997; Ankley et al. 1998).

Nonetheless, a substantial body of research suggests that EDCs in the environment may also cause adverse effects on a variety of nontarget invertebrates (Pinder and Pottinger 1998). Relevant examples include insects, crustaceans, and mollusks (Payen and Costlow 1977; Moore and Stevenson 1991, 1994; Matthiessen and Gibbs 1998; Chapters 2 and 4). Therefore, the need exists for organizations to develop screening and testing program for evaluating effects of EDCs on invertebrates, both in the field and in the laboratory, in order to protect ecologically and economically important species from potential effects of this high-profile class of environmental contaminants. This need includes not only aquatic invertebrates but also extends to terrestrial invertebrates that may also be exposed to hormonally active substances (Chapter 4; Pinder and Pottinger 1998).

Approach

Definition of an endocrine disrupter

At the beginning of the workshop, the Laboratory Testing Work Group made a practical decision to use an established definition of an endocrine disrupter as "a substance which causes adverse effects in an intact organism, or its progeny, subsequent to changes in endocrine function" (Holmes et al. 1997). Key considerations in the decision to use this definition included

1) the overall ecological context of the discussions on EDCs and invertebrates, with an emphasis on whole-organism responses and
2) a recognition of the importance of potential effects associated with transgenerational exposures of invertebrates to contaminants. While there was some discussion over the exact ecological interpretation of the term "adverse effects," there was agreement that this definition of endocrine disruption (ED) was sufficient for current practical purposes.

There is a need to use accurate terms to describe results of tests that do not provide information on specific mechanisms of ED. For example, if a substance produces adverse effects on fecundity in a toxicity test, then the best term to describe such a substance would be a "reproductive toxicant." Alternatively, the term "developmental toxicant" might best be used to describe a chemical that reduces growth or delayed sexual maturation. In both examples, if specific endocrine measurements are not reported, then the term "endocrine disrupter" would not be justified. Where reproductive or developmental toxicity has been observed, but supporting mechanistic data are lacking, then the term "potential endocrine disrupter" would be the best term to use. The misuse of the term "endocrine disruption" undervalues the importance of using this term for substances where the experimental evidence indicates this mode of action.

Integrated effects versus mechanistic data

A second critical aspect of the initial discussion by the Laboratory Testing Work Group was the identification of the need to define both ecologically relevant effects (Chapter 4) and a mechanism of endocrine system dysfunction (Chapter 2). Overall, adverse effects on whole-organism endpoints (e.g., development or reproduction) represent a greater level of potential ecological risk than does mechanistic evidence of endocrine activity *per se* (Pinder and Pottinger 1998). In brief, the Laboratory Testing Work Group agreed that, since ecological risk assessments are driven by evidence of effects rather than mechanistic data, the discussion on invertebrate testing for EDCs in the laboratory should give priority to such integrative effects. Ecologically relevant endpoints that may be responsive to EDCs might include behavior, development, feeding rate, growth, reproduction, or long-term survival (Pinder and Pottinger 1998). It should be noted that none of these endpoints are specific to effects associated with EDCs.

Structure–activity and in vitro methods

There is a need to screen thousands of compounds rapidly for the potential to affect invertebrates adversely through an endocrine mode of action (Tattersfield et al. 1997). While screening was not a topic to be specifically addressed during the workshop, it is appropriate briefly to discuss approaches that could be used to evaluate a large number of compounds. In vitro screening is also useful in confirming the toxic mode of action in cases where whole-organism testing or field studies provide empirical evidence that suggests ED. Some in vitro assays suggested as screening methods would be useful in confirming or rejecting ED as a cause of an observed effect. The use of screening tools and in vitro assays is briefly outlined in the appendix at the end of this chapter, "Molecular and biochemical techniques" (p 168; see also Tattersfield et al. 1997 and Ankley et al. 1998 for a discussion of these approaches for vertebrates).

Structure-based tools include both structure–activity relationships (SARs; Chen and Schüürmann 1997; Ankley et al. 1998) and common reactivity patterns (COREPAs). These 2 approaches relate biological effects to parameters derived from optimized 2- and 3-dimensional structures. The statistically derived relationships are based on structural and mechanistic understanding of the active site and the correlation of effects with relevant structural parameters. Ideally, these relationships are based on the biological system of interest (i.e., model the endocrine effect in the whole organism as opposed to modeling the response of an in vitro screen) and are based on a large number ($n > 30$) of compounds that adequately cover the chemistry of the compounds for which predictions are needed. The advantage of SARs and COREPAs is the capability to predict effects without synthesis of the compound of interest and subsequent laboratory testing. The disadvantages include the need for large amounts of data for construction and validation and the need for inclusion of all relevant structural properties of the parent compound as well as a knowledge of the site of action. This is a sizeable task due to the range of chemistries that may need to be covered, the number of different species for which predictions may be needed, and the number of distinct modes of ED that potentially operate within any one organism. Despite these drawbacks, standard approaches to data development should be considered so that structurally based predictive tools can be developed in the future. The appendix section "Molecular and biochemical techniques" (p 168) outlines several in vitro techniques with invertebrates that could be refined for use as screening tools for detecting potential effects associated with EDCs (Peel and Milner 1992; Lee and Noone 1995; Cottam and Milner 1997, 1998; Cherbas et al. 1988; Audsley et al. 1999).

Interspecies mechanisms

If a substance acts on the endocrine system of one species or phylum, this substance may not affect other species or other phyla by the same mechanism (Chapter 2). For example, nonylphenol is one of the best known xenoestrogens in mammals and fish (Jobling and Sumpter 1993). However, there is no evidence that the reproductive toxicity of nonylphenol in crustaceans is specifically due to an estrogenic mechanism (Baldwin et al. 1997, 1998). Despite uncertainty regarding mechanism of action, it is notable that in terms of the ecological risk assessment of nonylphenol to aquatic organisms, the most sensitive endpoint and species available to date is for reproductive impairment in the mysid shrimp (Tattersfield et al. 1997). While nonylphenol is only one case of a reproductive toxicant to invertebrates having a well-described endocrine activity in fish and mammals, this example illustrates the importance of developing ecologically relevant methods and identifying the toxic mode of action in the assessment of specific effects of EDCs on invertebrates.

Invertebrate full life-cycle as the "gold standard"

The sensitivity of developmental stages and critical periods of endocrine function in invertebrates exposed to potential EDCs are not yet known. Therefore, it is currently difficult to identify the specific life stages or endpoints that should be incorporated into methods for evaluating potential effects of EDCs. An alternative approach to consider in evaluating the utility of full life-cycle tests with invertebrates is outlined in "Factors to consider in selecting organisms to evaluate potential EDCs" (p 112) and in the appendix section "Groups of organisms for use in laboratory toxicity testing" (p 149). These full life-cycle tests could include embryolarval stages, periods of gonadal development, and reproduction in the assessment of EDCs (Figure 3-1). In contrast, it is difficult to conduct full life-cycle tests with most vertebrates because of their relatively long life cycles. Moreover, depending on the species of concern or the purpose of the investigation, exposures with some invertebrates can be extended over more than one generation to include some or all of the key life stages in the next generation (transgenerational exposures). Where relevant to a particular group of invertebrates, such an approach may be considered as a "gold standard" for the purposes of testing EDCs in the context of ecological risk assessment.

Given the charge set out in the "Scope" (see p 107), the Laboratory Testing Work Group first developed a list of factors that should be considered when groups of invertebrates are selected to evaluate sublethal responses of toxicants ("Factors to consider in selecting organisms to evaluate potential EDCs," p 112). Then, a second list of traditional and unique endpoints that could be used to measure potential effects of EDCs in toxicity tests with invertebrates was developed (appendix section "Groups of organisms for use in laboratory toxicity testing," p 149). Issues of experimental design and statistical analysis of data for these endpoints were outlined ("Statistical analysis of data," p 119). Research needs for the development of new methods, the evaluation of endpoint and species sensitivity, the use of in situ toxicity tests, and the development of consensus-based standard methods were then identified (from "Further development of methods and use of surrogate species," p 123, to "Further development of consensus-based standard methods," p 130). Finally, 3 case studies were described to illustrate the use of toxicity tests to detect potential effects of EDCs on invertebrates (effluent monitoring, chemical testing of methoprene and tributyltin [TBT]; "Case studies," p 132).

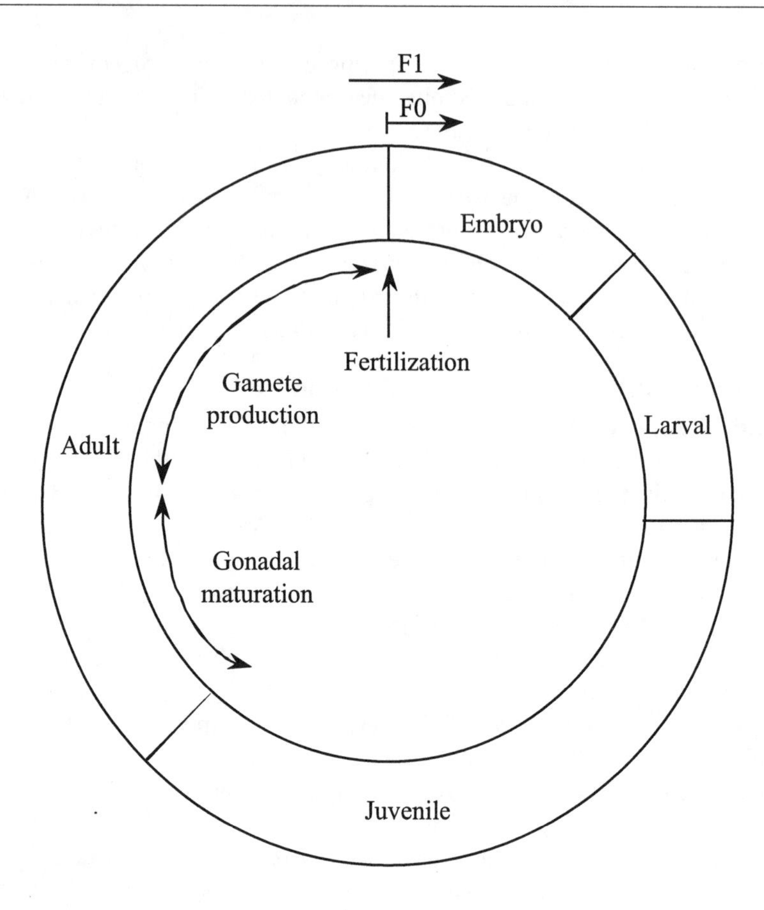

Figure 3-1 Evaluating effects of potential endocrine-disrupting compounds on invertebrates: A life-cycle context

Factors to Consider in Selecting Organisms to Evaluate Potential Endocrine-disrupting Compounds

Although there is an enormous variety of invertebrates from which to select test organisms, the choice will depend on practical constraints and the need to choose surrogate test organisms with endpoints relevant to ED. While the number of potentially exposed species is large, only a relatively small number of species are ever evaluated directly using toxicity tests (Glennon 1982). Ideally, a subset of these surrogate species is identified for testing, and the results of tests with these species are then used for effects assessment to protect a range of species.

In the assessment of potential effects associated with EDCs, there are clearly advantages in using, where possible, surrogate species that have been described in standard methods and test organisms for which there is evidence from field studies for adverse effects on the population. Since various components of the sexual reproductive cycle are regulated by endocrine systems, it is important that the endpoints measured include a variety of life stages, as illustrated in Figure 3-1. However, organisms that reproduce asexually should not be excluded from consideration, since their life cycles may well include relevant endocrine-regulated processes.

Table 3-1 was developed to address the following attributes that should be considered when test organisms are selected for conducting toxicity tests to evaluate potential effects of EDCs.

- Primary mode of reproduction: This selection criterion includes means by which the organism produces progeny (asexual, sexual, or both) and whether gender is fixed or predetermined (i.e., hermaphroditic). EDCs (as well as other environmental factors) can influence or modify sexuality and reproduction.
- Culture in the laboratory: The ability to culture an organism in the laboratory is desirable in order to provide test organisms at the required stages of development. Culturing also provides a measure of quality control (i.e., avoidance of previous exposure to EDCs, parasites, or disease). The ability to evaluate effects in the progeny of the exposed animals in transgenerational exposures is also desirable to detect effects that might be transmitted from one generation to the next.
- Generation time: Information on the generation time is useful as an indicator to evaluate the practicability of conducting full life-cycle (i.e., egg to egg) or transgenerational exposures.
- Size: Size of organisms is important for determining measurable tissue concentrations of hormones. Organisms with a short generation time will typically be small organisms. Hence, it may be difficult to measure hormones or other chemical biomarkers in organisms with a short generation time. There is a trade-off between short generation time and the size of the organism for testing.
- Knowledge of endocrinology: The extent of information available on endocrine regulation of molting and growth, sexual determination, and reproduction are all critical factors when organisms are selected for testing of potential EDCs (Chapter 2).
- Standard methods: Availability of existing consensus-based standard methods is a useful factor to consider when a test organism is selected for evaluating potential effects of EDCs. A consensus-based standard is a method that has been developed through a process opened to all interested parties and includes a mechanism for due process (i.e., a reviewer can seek clarification or appeal to the standards developer; Thomas 1997). The following international and national organizations have developed standards using some form of a consensus-based process: American Society for Testing and Materials (ASTM), Organization for Economic Cooperation and Develop-

Table 3-1 Factors to consider in selecting organisms for laboratory testing of potentially endocrine-disrupting compounds

Group	Primary mode of reproduction[1]	Culture in laboratory or transgenerational exposures[2]	Generation time[3]	Size[4]	Endocrinology known[5]			Standard methods[6]
					Molting or growth	Sexual determination	Reproduction	
Cnidarians								
Hydra spp.	B	Y	d – m	< 1 mg	L	L	L	ND
Marine hydrozoans	B	Y	d – y	mg	L	L	L	ND
Anemones	B	Y	y	> g	L	L	L	ND
Coral	B	Y	y	> g	L	L	L	ND
Turbellarians	B	Y	m	< 10 mg	L	L	L	ND
Rotifers								
Brachionus sp.	B	Y	d – w	< 1 mg	L	L	L	A,P
Nematodes								
Caenorhabdites elegans	S, H	Y	d – w	< 1 mg	L	L	L	ND
Mollusks								
Bivalves								
Crassostrea gigas	S	Y	y	mg – g	L	L	L	A
Crassostrea virginica	S	Y	y	mg – g	L	L	L	A
Macoma baltica	S	N	y	mg – g	L	L	L	A
Macoma nausta	S	N	y	mg – g	L	L	L	A
Mercenaria mercenaria	S	Y	y	mg – g	L	L	L	A
Mytilus edulis	S	Y	y	mg – g	L	L	L	A
Voldia limatula	S	N	y	mg – g	L	L	L	A
Gastropods	S	N	y	mg – kg	L	L	L	ND
Cephalopods	S	Y	y	g – kg	L	L	L	ND
Annelids								
Polychaetes	B	Y	d – m	mg – g	L	L	L	A
Capitella capaitata	S	Y	w	< 1 mg	L	L	L	A
Dinophilus gyrociliatus	B	Y	d	μg	L	L	L	A
Neanathes spp.	S	Y	w	< 1 mg	L	L	L	A

Table 3-1 *continued*

Group	Primary mode of reproduction[1]	Culture in laboratory or transgenerational exposures[2]	Generation time[3]	Size[4]	Endocrinology known[5]			Standard methods[6]
					Molting or growth	Sexual determination	Reproduction	
Annelids *(continued)*								
Nereis diversicolor	S	Y	w	mg	L	L	L	A
Ophryotrocha diadema	B	Y	d	µg	L	L	L	A
Oligochaetes	B	Y	w – m	mg	L	L	L	A
Aporrectodea claiginosa	A	N	w	mg	L	L	L	ND
Cognettia sphagnetorum	A	N	w	mg	L	L	L	ND
Eisenia spp.	S	Y	8 w	g	L	L	L	A
Lumbricus variegatus	A	Y	2 w	mg	L	L	L	A
Tubifex tubifex	S	Y	4 w	mg	L	L	L	A
Mites	S	N	m – y	mg	L	L	L	I
Millipedes and centipedes								
Centipedes	S	N	m – y	< 1 mg	L	L	L	NA
Millipedes	S	N	m – y	< 1 mg	L	L	L	ND
Crustaceans								
Amphipods								
Ampelisca abdita	S	N	4 – 8 w	< 1 mg	L	L	L	A,E
Amphiporeia virginiana	S	N	1 y	< 1 mg	L	L	L	C
Corophium spp.	S	Y	3 – 6 w	< 1 mg	L	L	L	C
Diporeia spp.	S	N	1 y	< 1 mg	L	L	L	A
Eohaustorius spp.	S	N	1 y	< 1 mg	L	L	L	A,C,E
Foxiphalus xixmeus	S	N	1 y	< 1 mg	L	L	L	C
Gammarus spp.	S	Y	6 – 8 m	< 6 mg	L	L	L	ND
Grandidierella japonica	S	Y	4 w	< 1 mg	L	L	L	A
Hyalella azteca	S	Y	5 w	< 1 mg	L	L	L	A,C,E
Leptochirus spp.	S	Y	4 w	< 1 mg	L	L	L	A,C,E

Table 3-1 *continued*

Group	Primary mode of reproduction[1]	Culture in laboratory or transgenerational exposures[2]	Generation time[3]	Size[4]	Endocrinology known[5]			Standard methods[6]
					Molting or growth	Sexual determination	Reproduction	
Amphipods *(continued)*								
Rhepoxinius abronius	S	N	1 y	< 1 mg	L	L	L	A,C,E
Cladocerans	B	Y	w	< 1 mg	L	L	L	A,C,E,O,I
Isopods	S	Y	3 m	mg	L	L	L	I
Fairy shrimp								
Artemia spp.	S	Y	d	mg	L	L	L	A
Thamnocephalus spp.	S	Y	d	mg	L	L	L	ND
Copepods	S	Y	d – w	< 1 mg	L	L	L	I
Mysid shrimp								
Holmesimysis costata	S	Y	3 m	< 1 mg	L	L	L	A
Mysidopsis bahia	S	Y	3 w	< 10 mg	L	L	L	A,E
Neomysis mercedis	S	Y	3 m	< 10 mg	L	L	L	A
Decapods								
Rhithropanopeus harrissii (mud crab)	S	Y	5 m	0.5 g	E	E	M	ND
Callinects sapidus (blue crab)	S	Y	6 m	10 g	E	M	E	ND
Palaemonetes pugio (grass shrimp)	S	Y	5 m	50 mg	E	M	M	ND
Homarus (lobster)	S	Y	y	g – kg	E	E	M	ND
Procambarus (crayfish)	S	Y	y	g	E	E	M	A
Balanus (barnacles)	S	Y	m – y	mg	L	M	M	ND
Insects								
Aquatic beetles	S	Y	1 y	< 1 mg	E	E	E	ND
Aquatic bugs	S	N	m – 1 y	< 1 mg	E	E	E	ND
Bees	S	Y	1 y	< 1 mg	E	E	E	A,O
Blackflies	S	N	m	< 1 mg	E	E	E	ND
Caddisflies	S	N	m – y	< 1 mg	E	E	E	ND
Chironomus spp.	S	Y	21 – 28 d	< 1 mg	E	E	E	A,C,E,O

Table 3-1 *continued*

Group	Primary mode of reproduction[1]	Culture in laboratory or transgenerational exposures[2]	Generation time[3]	Size[4]	Endocrinology known[5]			Standard methods[6]
					Molting or growth	Sexual determination	Reproduction	
Insects *(continued)*								
Dragonflies	S	N	1 – 3 y	mg – 1 g	E	E	E	ND
Drosophila spp.	S	Y	w	< 1 mg	E	E	E	ND
Hexagenia spp.	S	N	1 y	< 1 mg	E	E	E	A
Mosquitoes	S	Y	d	< 1 mg	E	E	E	ND
Springtails	S	Y	w	< 1 mg	E	E	E	I
Staphylinid beetles	S	Y	m	mg	E	E	E	I
Stoneflies	S	N	1 – 3 y	< 1 mg	E	E	E	ND
Echinoderms								
Echinoids	S	Y	y	g	L	L	L	A
Arbacia punctulata	S	U	m – y	g	L	L	M	A,C
Dendraster excentricus	S	U	m – y	g	L	L	L	A,C
Lytechnius pictus	S	U	m – y	g	L	L	L	A
Strongylocentrotus spp.	S	Y	y	g	L	L	L	A,C
Sea stars	S	Y	y	g	L	L	L	ND
Brittle stars	S	Y	m	mg	L	L	L	ND
Tunicates	B	Y	m – y	mg – g	L	L	L	ND

[1] Mode of reproduction: S = Sexual, A = Asexual, H = Hermaphrodite, B = Both sexual or asexual
[2] Ability to culture in laboratory or to conduct transgenerational exposures: Y = Yes, N = No, I = Impractical, U = Unknown
[3] Generation time: d = days, w = weeks, m = months, y = years
[4] Approximate adult body size in milligrams, grams, or kilograms
[5] Knowledge of endocrine system: L = Limited, M = Moderate, E = Extensive
[6] Standardized methods: A = ASTM, O = OECD, C = Environment Canada, E= USEPA, P = American Public Health Association, I = International Organization for Standardization, ND = Not developed

ment (OECD), U.S. Environmental Protection Agency (USEPA), Environment Canada, American Public Health Association (APHA), and International Organization for Standardization (ISO). Other organizations that have developed or are developing standard methods for toxicity testing include Deutsche Institut für Normung (DIN) in Germany, British Standards Institute (BSI) in the United Kingdom, and the Association Française de Normalization (AFNOR) in France. Information from studies published in the peer-reviewed literature is also useful for selection of potential endpoints or test organisms when procedures outlined in consensus-based standard methods are not considered suitable or adequate.

In addition to the factors outlined above, other factors that could be considered for selecting a test organism might include (Pinder and Pottinger 1998; ASTM 1999a)

1) relative sensitivity to and discrimination between a range of chemicals of concern,
2) existence of a database for interlaboratory comparisons of procedures,
3) ecological or economic importance,
4) compatibility with selected exposure methods and endpoints,
5) ability to expose organisms using in situ tests, and
6) availability of data comparing responses of laboratory-exposed and field-collected organisms

The appendix at the end of this chapter summarizes endpoints that can be measured in laboratory toxicity tests with a variety of different invertebrates. Endpoints that may be responsive to effects of EDCs but are not necessarily characteristic of them include survival, fecundity, sexual maturation, sex ratio, mating behavior, biomass, growth rate, molt time and success, embryonic or larval development, and feeding behavior (Table A-1). The primary source of information for this summary in the appendix was consensus-based standard methods (ASTM, OECD, USEPA, Environment Canada, APHA, and ISO). Methods described in the peer-reviewed literature were also evaluated when standard methods either did not exist or were limited in scope relative to potential endpoints of interest. This list of potential endpoints in Table A-1 was used to summarize procedures by major group (phylum, class, order) or by individual species. Additional endpoints that would require minor or major development are also identified in Table A-1 (i.e., endpoints not described in either standard methods or other publications and that would require additional research before routine use). Endpoints that are either impractical or impossible to measure with a particular organism or group of organisms are also identified. Finally, the Laboratory Testing Work Group identified the need for additional research on methods for under-represented groups of organisms.

There was sufficient information available to describe methods in the appendix and in appendix Table A-1 for laboratory toxicity tests with cnidarians, turbellarians, mollusks, annelids, mites, millipedes, centipedes, crustaceans (amphipods, cladocerans, fairy shrimp, isopods, copepods, mysids, crabs, shrimp, lobsters, crayfish, barnacles), insects (midges, other aquatic insects, and terrestrial insects), echinoderms, and tunicates (in "Further development of methods and use of surrogate species," p 123). The Laboratory Testing Work Group was unaware of information that could be used to describe adequately methods for conducting toxicity tests with sponges, jellyfish, anemones, bryozoans, leeches, spiders, sea stars, brittle stars, or sea cucumbers. The need for additional research on methods for these underrepresented groups of organisms is discussed in the section titled "Further development of methods and use of surrogate species" (p 123).

Statistical Analysis, Method Development, Validation, and Standardization

The importance of invertebrate communities in ecosystems, coupled with the observation of endocrine-mediated effects in gastropods argues for consideration of the effects of EDCs on invertebrates (Chapter 4). Few laboratory methods have been developed for invertebrates that are known to be responsive to EDCs. Hence, additional methods for a variety of species need to be developed and validated. The most useful of these methods should be standardized for routine assessment and screening of EDCs. The following section summarizes the application of statistical analysis of data from studies of potential EDCs. Research needs are also outlined for

1) developing methods to implement life-cycle toxicity tests more fully with invertebrates listed in Table 3-1 and in the appendix,
2) evaluating approaches to determine the sensitivity of endpoints and species for use in effects assessments of EDCs,
3) using toxicity tests to evaluate effects of EDCs in the field, and
4) further developing consensus-based standard methods.

Statistical analysis of data

Table 3-2 outlines the types of data that might be collected from laboratory toxicity tests using endpoints identified in the appendix (Table A-1). It is clear that a diversity of data types could be generated, requiring a diversity of quantitative analyzes to maximize the amount of information that can be gained from each test.

Analysis of results from toxicity tests

Three main approaches to the analysis of toxicity data have been identified (OECD 1998f):

1) Hypothesis testing to determine a no-observed-effects concentration (NOEC): The NOEC is the highest concentration in a toxicity test producing a response that does not differ from the control when compared in a statistical significance test. An NOEC is usually calculated from concentration-response data by using analysis of variance (ANOVA), followed by a multiple comparison

Table 3-2 Types of data collected for endpoints identified in Table 3-1 and appendix Table A-1

Endpoints	Data type
Survival	Quantal (dead or alive) Continuous (time to death)
Fecundity	Count (number of offspring)
Sexual maturation	Quantal (mature or immature) Continuous (time to maturation)
Sex ratio	Quantal (male or female) Continuous (percent population female) Nominal (male, female, or hermaphrodite)
Size	Continuous (weight, length, area)
Growth rate	Continuous (size change per unit time)
Molt timing and success	Quantal (molted or not) Continuous (time to first molt, intermolt time) Count (molt frequency)
Regeneration	Quantal (regeneration or not) Nominal or ordinal (class of regeneration) Continuous (time to reach specific stage of regeneration, length of regeneration after specific time)
Embryonic development	Quantal (development to specific stage or not) Continuous (time to reach developmental stage)
Larval development	Quantal (development to specific stage or not) Continuous (time to reach developmental stage)
Gross morphology	Quantal (abnormal or not) Ordinal (severity of abnormalities) Count (number of abnormalities) Continuous (size of abnormalities)
Pigmentation	Continuous (change in color)
Histopathology	Quantal (pathology present or not) Nominal (class of pathology) Count (number of pathologies) Continuous (size of each pathology)
Mating	Quantal (mated or not) Nominal (class of mating behavior) Continuous (time to find mate, mating duration)
Feeding	Quantal (feeding or not) Ordinal (severity of feeding inhibition) Count (number of items consumed) Continuous (weight of food consumed)
Osmoregulation	Continuous (concentration of each ion) Multivariate (concentrations of all ions)
Biochemical	Continuous (concentration of 1 biochemical) Multivariate (concentrations of several biochemicals)
Molecular	Quantal (gene expressed or not) Continuous (amount of each gene expressed) Multivariate (amounts of several genes expressed)

test such as Dunnett's or Tukey's test (Newman 1995). This can only be done if there is replication of each treatment. ANOVA is a parametric technique, so data that are highly skewed or have heterogeneous variances should be transformed before analysis. Count data and proportional data should also be transformed (Zar 1984). NOECs could be calculated for endpoints such as fecundity or growth (Table 3-2). The perceived advantage of the NOEC is that it is easy to understand (OECD 1998f). However, there are many disadvantages to its use that have been discussed elsewhere (Bruce and Versteeg 1992; Chapman et al. 1996).

2) Regression analysis to estimate a time-specific effect concentration such as a median lethal (LC50) or median effective (EC50) concentration: The estimation of an LCx or ECx value overcomes most of the problems associated with hypothesis testing (Chapman et al. 1996) and is the usual form of analysis for acute tests. Data from fixed times of observation (usually 24, 48, 72, or 96 hours) are transformed so that least-squares fits can be made to linear models. Linearity is usually achieved for quantal data by taking the log of the exposure concentration and converting the response to its probit (Bliss 1935) or logit (Berkson 1944). ECx values are then estimated for the magnitude of effect of interest. This is usually an EC50 or LC50 for acute data, because more precise estimates are possible at this median point. An LC or EC could be estimated for most of the quantal, count, or continuous data listed in Table 3-2. Generalized linear models have also been proposed as a method of analyzing data without the need for an initial linearizing transformation (Kerr and Meador 1996), and techniques are available for modeling continuous, rather than quantal, toxicity data (Bruce and Versteeg 1992; Moore and Caux 1997). Although there is general agreement that the use of a dose-response curve to estimate an ECx has many advantages over the derivation of an NOEC, the calculation of an ECx at specific time intervals still uses data in a suboptimal way. This is because most investigators will take some measurements during the course of a toxicity test, especially if survival or reproduction are the endpoints. However, these data from intermediate observation periods are not often used in the final estimation or risk assessment.
3) Time-to-event analysis: Sun et al. (1995) recommend the use of survival time modeling and accelerated life testing in toxicity testing. Survival time models have also been used as a method for integrating time, concentration, response, and ecologically important co-variables such as organism weight and sex (e.g., Newman et al. 1994). These approaches are commonly used in engineering, where "failure time analysis" is used to analyze the failure of a component; in medicine, where "survival analysis" is used to analyze the deaths of patients; and in ecology, where "time-to-event" analysis can be used to analyze the arrival of a migrant or parasite, the germination of a seed, the death of an organism, or the time to a particular behavioral display (Muenchow 1986). Parametric models can be based upon any theoretical

distribution. For example, Newman et al. (1994) chose from normal, log normal, Weibull, gamma and log-logistic distributions by comparing the fit of each of these to the data. Sun et al. (1995) chose a Weibull distribution. Nonparametric tools, such as life tables and the Kaplan-Meier method (Kaplan and Meier 1958), are also available for analyzing time-to-event data when the underlying distribution remains unknown. The life table approach groups occurrence times into intervals, while the Kaplan-Meier, or "product-limit," approach produces an estimate at each individual occurrence time (Newman 1995). Time-to-event analysis is likely to be the most powerful method for the endpoints listed in Table 3-2, which occur only once in time (e.g., death or development to a particular stage) and for which observations have been made at several time points during the experiment.

Analysis of additional types of data from toxicity tests for potential endocrine-disrupting compounds

Further types of data that may be obtained from toxicity tests for potential EDCs include nominal or ordinal data. Nominal data are those in which a response is classified as a member of a group, with all classes weighted equally. For example, pathologies may be divided into several discrete classes, each of equal or unknown seriousness. Ordinal data are those in which a response is ranked. For example, a pathology may be ranked minor, severe, or life threatening. These types of data are often analyzed using a contingency table (Zar 1984).

Multivariate data may also be obtained from biochemical or molecular analyses (in the appendix section "Molecular and biochemical techniques," p 168, and in Chapter 2). These take the form of a data matrix in which, for example, concentrations of several hormones or the levels of expression of several genes are recorded. Many multivariate techniques are available to analyze these data; such techniques may help show overall trends that are not obvious from analyzing single variables. Among the best known are principal components analysis (PCA) and multidimensional scaling (MDS), which can be used to produce an ordination of the data in 2 dimensions. Such ordinations may reveal patterns in chemical concentrations or gene expression that are associated with exposure to EDCs. Clarke and Warwick (1994) provide a highly readable summary of multivariate approaches specifically written for biologists with only limited mathematical knowledge.

Finally, if full life-cycle or transgenerational tests are conducted (Figure 3-1), population-level summaries are a very useful way of summarizing data (e.g., Meyer et al. 1986, 1987). If age-specific survival and fecundity are known, population parameters such as the intrinsic rate of population increase (r) can be estimated. These parameters allow experimenters and risk assessors to judge the hazard of a chemical to population processes, rather than just processes that operate at the individual level. Populations are generally considered to be one of the most important foci for environmental toxicologists (Caswell 1996), and most EDC impacts on invertebrates can be expected to affect development or reproduction. Population-level summaries are therefore a simple and obvious way of summarizing the hazard of EDCs.

Further development of methods and use of surrogate species

Table 3-3 summarizes the degree of method development required before various groups of organisms are used to evaluate effects of potential EDCs in full life-cycle or transgenerational exposures. Organisms requiring limited method development to conduct full life-cycle or transgenerational exposures were placed in this first category based on the availability of either consensus-based standards or other publications. Organisms requiring moderate to extensive method development were placed in this second category if there is a reasonable potential to develop additional endpoints or exposure methods in the near future. Finally, organisms that are not practical for development of methods for conducting full life-cycle or transgenerational exposures were placed in this third category if there were unique aspects of the life cycle that makes laboratory culturing or testing difficult or impossible. Table 3-3 also identifies organisms that are not practical for use in routine full life-cycle testing but that have been used to conduct partial life-cycle tests to assess potential effects of EDCs (i.e., crabs, shrimp, lobster, crayfish as described in the appendix section "Groups of organisms for use in laboratory toxicity testing," p 149).

A variety of organisms listed in Tables 3-1 and A-1 can be used with limited method development to conduct full life-cycle or transgenerational exposures (Table 3-3). For example, methods are well developed for conducting life-cycle exposures with cladocerans. However, it is unknown whether cladocerans would be useful surrogates for evaluating potential EDCs in all crustaceans (particularly sexually reproducing organisms). For most of the groups of organisms listed in Tables 3-3 and A-1, there is limited knowledge of the endocrinology controlling molting, growth, reproduction, or sexual determination (e.g., rotifers, nematodes, polychaetes, oligochaetes, amphipods, *Artemia* spp., cladocerans, and copepods). To the best of our knowledge, insects and crustaceans are the only groups of organisms in which there are surrogates available for life-cycle testing with a relatively well-understood endocrine system (Chapter 2). An important research need will be to determine whether surrogates of a particular group (e.g., *Chironomus* spp.) are representative of the range of responses in a larger group (i.e., insects). Specifically, "Evaluation of endpoints and species for use in effects assessments" (p 125) and Figure 3-2 outline an approach that could be used to validate endpoint and species sensitivity to a series of potential reference compounds suspected of having an ED mode of action. To the best of our knowledge, the echinoderms are the only major group of organisms that would require moderate to extensive method development but are also important organisms for monitoring EDCs in the field (Table 3-3; Chapter 4).

Table 3-3 Degree of method development required to conduct life-cycle or transgenerational exposures for surrogate species within a group of organisms

	Method development required	
Little	Moderate to extensive	Not practical in near future
Rotifers[1]	*Hydra* spp.	Sponges
Nematodes[1]	Planarians	Jellyfish, coral
Polychaetes[1]	Bivalves	Cephalopods
Oligochaetes[1]	Gastropods	Leeches
Amphipods[1]	Mites, spiders	Barnacles
Artemia spp.[1]	Isopods	Lobster[3], crayfish[3]
Cladocerans[1]	Ostracods	Mud crabs[3]
Copepods[1]	Millipedes	Aquatic beetles and bugs
Mysids[1]	Centipedes	Bees[3]
Grass shrimp[1]	Moths, butterflies	Blackflies
Drosophila melanogaster	Sea urchins[2]	Caddisflies
Chironomus spp.	Sea stars[2]	Dragonflies
Mosquitoes		Stoneflies
Springtails		Brittle stars, sea cucumbers
Staphylid beetles		Tunicates

[1] Require little method development, but knowledge of the endocrinology is limited (Chapter 2).

[2] Require moderate to extensive method development or are not practical to develop methods in the near future, but were identified as important organisms for monitoring EDCs in the field (Chapter 4).

[3] Have been used to conduct partial life-cycle exposures to evaluate effects of EDCs.

If the mode of action of an EDC is known, testing of effects on invertebrates may be more expeditiously directed at those targeted processes. In these cases, entire life-cycle testing may require more time, effort, and cost to identify effective concentrations than would a focus on those more susceptible life stages. This particularly holds true in longer-lived species for which entire life-cycle testing is impractical given the length of the life cycle, while those particularly susceptible life stages would experience relatively longer duration of exposure than possible in shorter-lived species. For these reasons, partial life-cycle exposures with particularly sensitive life stages of crabs, shrimp, lobsters, or crayfish (appendix section "Groups of organisms for use in laboratory toxicity testing," p 149) may be more appropriate than tests with these organisms for their entire life cycle. In addition, the endocrine system of these decapod crustaceans is reasonably well understood (Table 3-1). A case study for methoprene (p 134) demonstrates the utility of conducting partial life-cycle exposure to evaluate effects of EDCs on invertebrates. An approach for evaluating the sensitivity of other species, endpoints, or life stages to potential EDCs is described in the following section.

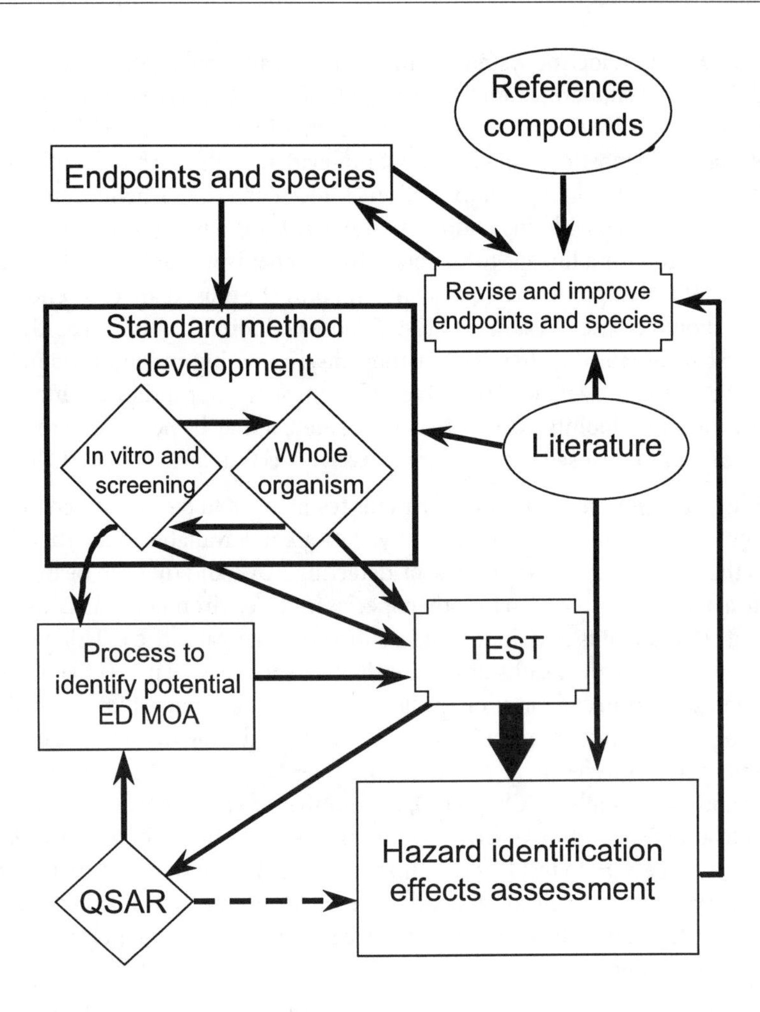

Figure 3-2 Standard method review and improvement process

Evaluation of endpoints and species for use in effects assessments

The endpoints used in toxicity tests are intended to define organism or population status (e.g., survival, growth). Endpoints currently used in acute and chronic toxicity testing typically include measures of survival, growth, and reproduction (Tables 3-1 and A-1). These endpoints may or may not be the most responsive endpoints for detecting chemicals that use endocrine-disrupting mode of action. To ensure that important effects are not missed due to a misalignment of the current, standard endpoints with endocrine-related effects, additional endpoints are suggested in Table A-1 for use in testing of suspected EDCs. These endpoints were selected based on knowledge of endocrine systems in invertebrates and the potential for effects caused by altered endocrine function. The endpoints selected are not necessarily exclusively responsive to an endocrine mechanism of action but should be used with more specific, diagnostic testing in a weight-of-evidence approach to identify ED as the mechanism of action (appendix section "Molecular and biochemical techniques," p 168). In considering endpoints, Table A-1 was developed to be as inclusive as possible so long as a linkage to an endocrine mechanism could be postulated. However, not all endpoints are equivalent, and only the most diagnostic endpoints for EDCs should be used in standard tests. Those endpoints cannot currently be identified because the data to select the most discriminating endpoints do not exist. Diagnostic endpoints will need to be identified through appropriate testing and comparisons of endpoints. As experience is gained, it may be possible to reduce the number of endpoints used in the future to assess effects associated with EDCs.

Selection of the most appropriate species for testing should be considered in the same vein as endpoint selection. Currently, data are not available to identify those species that have the greatest promise of detecting EDCs and that can be used to provide a degree of protection for other species in the environment. As with endpoint selection, it is envisioned that a subset of surrogate species will be identified for testing, with the results of tests with these species used in an effects assessment context to protect a range of species. There is no assurance that the goal of protecting invertebrate communities can be met by reducing the number of endpoints and or species. In fact, as more becomes known about invertebrate biology and endocrinology (Chapter 2), the number of endpoints and species needed to afford protection of invertebrate communities may increase. It will, however, be necessary to test this assumption to avoid redundancy of methods and unnecessary testing. On the other hand, data are not currently available to suggest that existing tests and effects assessment procedures do not adequately protect invertebrate communities.

There is a need to test a number of species and endpoints from those described in appendix section "Groups of organisms for use in laboratory toxicity testing" (p 149) with a diverse set of reference compounds suspected of having ED mode of action (Table 3-4). Based on the data obtained with these tests, the most appropriate endpoints and species would be selected for standardized testing (Figure 3-2). This is similar to the approach used historically to select the endpoints and species that currently are used in environmental effects assessments. In addition to sensitivity, endpoint selection factors might include

1) ease of measurement, integration (e.g., neonate production versus egg production and sperm viability),
2) linkage to an endocrine mode of action (e.g., histological identification of imposex versus time to molt),
3) utility in the effects assessment process (e.g., fecundity versus hormone titers), and statistical rigor ("Statistical analysis of data," p 119).

Table 3-4 Suggested reference chemicals for evaluating relative endpoint or species sensitivity to potential endocrine-disrupting compounds (see Figure 3-2)

Possible mode of action	Chemical
Juvenile hormone (JH) agonist	Methoprene
JH antagonist	Precocene
Ecdysone agonist	20–OH ecdysone
Ecdysone antagonist	Homobrassinolide, luteolin
Aromatase inhibitor	Fadrozole
Androgen agonist	Methyl testosterone
Androgen antagonist	Flutamide
Estrogen agonist (weak)	4-tertpentylphenol
Estrogen agonist (strong)	Ethynyloestradiol
Estrogen antagonist	ZM-189,154

These selection factors for endpoints are not absolute and should be considered with regard to the overall effects assessment process. For example, if a simple measure of reproductive impairment is not as sensitive as more complex reproductive endpoints, the simpler endpoint might be preferred in a standard method with appropriate uncertainty factors built into the effects assessment. Establishing the appropriate uncertainty factors for effects of EDCs will be a challenging task (i.e., is there a threshold of an effect?). Furthermore, the endpoints will also need to be indicative of biologically important responses (i.e., adverse effects on fitness).

Assuming each test species meets the selection factors outlined in "Factors to consider in selecting organisms to evaluate potential EDCs" (p 112), the only additional selection factors for a species would be the potential for exposure of the species in the field (Chapter 4) and the ability of that species or a group of species to serve as a surrogate for predicting effects across a wide range of species. Again, this approach and rationale has been applied to the current selection of test organisms for effects assessments. The endpoint and species selection process should be considered in the overall context of the effects assessment process. In these assessments, uncertainty or assessment factors of 10 to 50 are used with the lowest chronic toxicity value to extrapolate to a concentration broadly protective of invertebrate communities. There is considerable uncertainty regarding the utility of these uncertainty factors for assessing effects of EDCs. Alternatively, chronic toxicity results for a single species can be assumed to come from a known distribution of toxicity to many species. Based on this assumption, statistical approaches can be used to estimate a concentration protective of a high percentage of species (usually 95%). The applicability of these methods to EDCs is not well understood but should be considered in the endpoint selection process.

The process outlined in Figure 3-2 may lead to 1 of 2 possible outcomes:

1) it is possible to identify a subset of endpoints and species that are most sensitive to compounds acting by an ED mode of action and that can be used with appropriate uncertainty tools within the hazard identification and effects assessment process to protect a wide range of species or
2) the diversity of responses across endpoints and species is broad, occurring in a wide range of environmental concentrations such that it is not possible to identify a subset of endpoints and species that will confer protection on invertebrate communities.

At present, it is uncertain which of these outcomes is most likely. However, without a framework to address this issue, there will always be a high level of uncertainty.

Once a set of species and endpoints is identified, standard methods should be developed ("Further development of consensus-based standard methods," p 130). Revisions of existing standard methods or the development of new standards to address effects of EDCs should not be finalized until research has been conducted to evaluate relative endpoint and species sensitivity. While most of these methods are likely to be whole-organism tests, in vitro screening methods should be developed and compared with whole-organism testing to help in the screening process as well as to support mechanism of action studies (appendix section "Molecular and biochemical techniques," p 168). Once appropriate tests are identified and standardized, these methods could be used with other compounds suspected of having an ED mode of action. These data could be used to build quantitative structure-activity relationships to help identify and prioritize potential EDCs for future testing. The test data would also be used along with literature data in the hazard identification and effects assessment process. Finally, information gained during the hazard identification and effects assessment process would be used to refine further the selection of endpoints and species.

While research is being conducted on the process of endpoint and species identification, practical decisions will have to be made on compounds and their potential effects in the environment. Some decisions may need to be made before the most appropriate endpoints and species are identified. Unfortunately, it is difficult to propose an interim solution. One interim approach could be to use traditional hazard-and-effects assessment processes to make decisions. Another approach might be to include additional uncertainty in the assessment process. The goal of the process discussed above is to work toward developing a scientifically defensible dataset, using specific test compounds upon which future standard test species and decisions can be based.

Links from laboratory toxicity testing to the field

Results from laboratory toxicity tests with single species have been criticized as lacking in relevance for natural systems (e.g., Cairns 1983). Single-species tests may overestimate or underestimate toxicity because of the artificial exposure regime or the lack of normal intraspecific and interspecific processes. Single-species tests also may not reflect the range of sensitivities present in species of concern. One way in which natural exposure conditions have been assessed is through the in situ deployment of single species in cages. This has been done with amphipods (e.g., *Gammarus pulex* [Crane et al. 1995], *Hyalella azteca* [Chappie and Burton 1997]), cladocerans (Sasson-Brickson and Burton 1991), insects (e.g., chironomids; Chappie and Burton 1997; Sibley et al. 1999), mollusks (e.g., *Corbicula* and *Mytilus*; Widdows and Donkin 1991), and annelids (e.g., Monson et al. 1995; Olesen and Weeks 1998). Procedures for conducting in situ exposures are particularly useful for evaluating the effects of episodic events. For example, in situ effects with up to 9 species of freshwater invertebrates were evaluated with simulated acid-aluminium episodes (Ormerod et al. 1987) or with farm-waste episodes (McCahon et al. 1991). Almost any invertebrate used in laboratory toxicity tests can be caged in the field, but most of these in situ exposures are currently short term, with survival as the primary endpoint. Tests with amphipods and mysids have also measured growth and feeding rate as sublethal endpoints. Hence, critical life stages of organisms can be used to investigate effects in the field. The effects of metals on the feeding rate of amphipods (*Gammarus* spp.) have been well documented for both laboratory and in situ tests (e.g., Crane and Maltby 1991; Maltby and Crane 1994). Precopula guarding behavior in *Gammarus* spp. has also been evaluated using in situ exposures and might be a useful indicator of EDC effects on hormone control of mating behavior (Pascoe et al. 1994). Similarly, effects of EDCs on emergence of midge could be evaluated using in situ exposures (Schmude et al. 1999). These types of in situ exposures may help to establish links between the laboratory and the field when EDCs are expected to disrupt hormonal systems involved in growth, development, or feeding. However, considerable effort would need to be expended to develop in situ exposures into full life-cycle or reproduction tests for aquatic species because of problems in maintaining sufficient concentrations of food and avoiding fouling of cages. In contrast, full life-cycle in situ tests with terrestrial species such as earthworms, beetles, and springtails are feasible, particularly for those organisms with rapid life cycles (appendix section “Groups of organisms for use in laboratory toxicity testing,” p 149).

Mesocosms and microcosms are frequently presented as useful ways of increasing the realism of both exposure conditions and ecological interactions in toxicity tests (Crossland 1990). These methods may be useful systems for demonstrating chemical causality if the exposure route is mediated by environmental processes that are difficult to simulate in the laboratory. For example, methylated selenium is a reproductive toxicant with exposure primarily through the food chain, hence a mesocosm study is likely to be the most cost-effective and realistic experimental approach (Crane et al. 1992). However, experimenters and risk assessors should make sure that such studies are focused on specific taxa and endpoints and that the physical structure of the mesocosm is relevant for the natural systems in which exposure is likely to occur. If this is not the case, the results of mesocosm studies may either be difficult to interpret by decision-makers or have little predictive power (Crane 1997). Soil microcosms are available for investigating the effects of chemicals on soil faunal assemblages (e.g., Parmelee et al. 1993; Römbke et al. 1994).

Further development of consensus-based standard methods

"Evaluation of endpoints and species for use in effects assessments" (p 125) outlined an approach for identifying species, life stages, and endpoints that may be useful for identifying an ED mode of action in invertebrates (Figure 3-2, Table 3-4). Once there is an understanding of which endpoints, life stages, and species are sensitive to EDCs, consensus-based standard methods should be developed for these tests (Tattersfield et al. 1997). Such standard methods for addressing endocrine-specific effects of EDCs could be developed either as revisions to existing standards (e.g., for cladocerans, amphipods, midges, or oligochaetes) or as new standards (e.g., for crabs, shrimp, crayfish, lobsters). The following section outlines approaches for developing consensus-based standards within countries and harmonization of these methods between countries.

Numerous standard methods for conducting toxicity tests have been developed or are being developed at the national level (e.g., USEPA, Environment Canada, DIN in Germany, BS in the United Kingdom, and AFNOR in France) and at the international level (OECD, ISO, EU, ASTM). Use of standard methods can help to improve the quality of data, increase comparability of results among laboratories, and establish common criteria for evaluating the acceptability of a test. Results of tests conducted using standard methods can also be used in regulatory processes, including

1) chemical regulation (pre-manufacturing or pre-marketing notification of new chemicals, classification and labelling, risk assessment of existing chemicals, pesticides registration),
2) establishing environmental quality standards or objectives,
3) setting emission limits, or
4) assessing the toxicity of in-place contaminants.

In addition, the use of internationally harmonized methods is a prerequisite for the acceptance of data among countries.

The OECD (1997) and the Endocrine Disrupter Screening and Testing Advisory Committee (EDSTAC; Ankley et al. 1998) have reviewed methods for evaluating environmental effects of EDCs. OECD established a task force on endocrine disrupter testing and assessment (EDTA) to coordinate and prioritize method development. The task force agreed to start a validation project in which member countries and industry would cooperate and share the cost of the validation of new and existing standards for testing with mammals and nonmammals, including fish, birds, amphibians, and invertebrates. Similarly, EDSTAC established a framework in the United States for implementing a variety of screening assays and toxicity tests for evaluating potential effects of EDCs (Ankley et al. 1998). The screening assays recommended by EDSTAC included in vitro and in vivo assays with mammals, birds, reptiles, amphibians, or invertebrates and whole-organism reproductive tests with rats, birds, fish, and mysids.

None of the existing standard methods outlined in Tables 3-1 and A-1 for testing invertebrates were designed specifically to evaluate effects associated with potential EDCs. However, many of these standards describe approaches that can be used to measure endpoints that may be responsive to effects of EDCs (e.g., development, growth, reproduction). Figure 3-2, "Evaluation of endpoints and species for use in effects assessments" (p 125), and the appendix to this chapter outline how these existing standards or other published methods with invertebrates could be used to evaluate potential effects associated with EDCs. Results of tests conducted with these other organisms may provide functionally equivalent information, or they may provide unique information on effects of EDCs on invertebrates. Additional endpoints that could be measured in these tests and may provide information indicative of ED might include

1) sex ratio of adults or offspring,
2) mating success,
3) production of ephippia (resting eggs in cladocerans),
4) egg hatching success,
5) viability of the offspring,
6) morphological abnormalities,
7) molting frequency and timing,
8) time to emergence or pupation,
9) metamorphic or larval developmental success, or
10) pigmentation.

See Table 3-1 and the appendix to this chapter ("Groups of organisms for use in laboratory toxicity testing," p 149) for additional examples of endpoints that could be measured with modifications to existing standards or published methods.

The development of a harmonized method through OECD takes several years until final adoption. The 3 stages for developing a harmonized method include a description of a

1) a conceptual protocol (e.g., method described in the peer-reviewed literature but not described as a consensus-based standard),
2) a developmental protocol (e.g., method published as consensus-based standard but with limited data), and
3) validated protocol (e.g., consensus-based method supported by data on intralaboratory and interlaboratory variability [ring-testing]).

The approach suggested for validation of new methods is a necessary step in the development of these standard methods (see "Evaluation of endpoints and species for use in effects assessments," p 125, and Figure 3-2). Generating the data necessary to evaluate relative endpoint, life stage, and species sensitivity to EDCs will be time consuming and costly. Therefore, it is desirable for researchers developing these methods to be aware of the international requirements and time frame to validate and harmonize standard methods.

Case Studies

The following section outlines how methods described in "Factors to consider in selecting organisms to evaluate potential EDCs" (p 112) and in appendix section "Groups of organisms for use in laboratory toxicity testing" (p 149) have or could be used to evaluate potential EDCs. Three case studies are described:

1) effluent monitoring,
2) methoprene, and
3) tributyltin.

Detection of potential endocrine-disrupting chemicals in effluents

The testing of effluents with laboratory toxicity tests has been practiced in North America for several years and has recently been introduced in some European nations and other parts of the world. This section addresses the question of whether the effluent testing approaches currently in use are likely to detect EDCs. In the U.S. whole effluent toxicity (WET) testing of point source discharges is authorized under the Clean Water Act. Under this Act, all discharges into surface waters of the United States must have a National Pollutant Discharge Elimination System (NPDES) permit. These permits are based upon effluent guidelines, state water quality standards, and WET testing (Heber and Norberg-King 1996). Several acute mortality and short-term chronic toxicity tests have been developed for this program (USEPA 1993, 1994b, 1995; Table 3-5). It is clear that the chronic or sublethal tests have the potential to detect EDCs that affect endocrine systems in cladocerans or mysids. Hormonal mediation of sea urchin fertilization could also be detected in marine testing. However, other important groups of aquatic organisms, such as freshwater insects are protected only by surrogates such as the cladoceran *Ceriodaphnia dubia*.

In the United Kingdom, fewer test systems are used in direct toxicity assessment (DTA, a synonym for WET testing). Acute lethal 48-hour *Daphnia magna* tests are used to assess discharges to fresh water, and 24-hour oyster embryo larval (OEL) development tests with the Pacific oyster *Crassostrea gigas* are used to assess discharges to salt water (Wharfe 1996; Tinsley et al. 1996). This program is currently

Table 3-5 Toxicity test methods used to assess effluents in the U.S.

Toxicity test method	Type of organism	Scientific name	Test duration	Test endpoints
Acute lethality	Freshwater			
	Cladoceran	*Ceriodaphnia dubia*	24, 48; optional 96 h	Survival
		Daphnia magna	24, 48; optional 96 h	Survival
		Daphnia pulex	24, 48; optional 96 h	Survival
	Marine/estuarine			
	Mysid shrimp	*Mysidopsis bahia*	24, 48, 96 h	Survival
Short-term chronic or sublethal	Freshwater			
	Cladoceran	*Ceriodaphnia dubia*	7 d	Survival, reproduction
	Marine/estuarine			
	Mysid shrimp	*Mysidopsis bahia*	7 d	Survival, growth, and fecundity
	Sea urchin	*Arbacia punctulata*	1.5 h	Fertilization success

under development and has not yet been implemented in regulations in the UK. Clearly there is little potential for the detection of EDCs discharged to fresh water, as no developmental or reproductive endpoints are measured in test organisms. The OEL test has the potential to detect disruption of hormonal systems controlling post-fertilization development in bivalves up to the shelled larval stage. However, endocrine systems present in important saltwater invertebrates such as crustaceans are unlikely to be protected by tests with this bivalve mollusk.

The approaches used in WET testing do not directly consider the measuring ED in invertebrates. However, it is clear that those countries that have a more extensive tiered-testing approach that includes chronic testing of invertebrates are more likely to be able to detect reproductive and developmental effects of exposure to whole effluents using these tests (or modifications of these methods). European Economic Area member states that currently use chronic tests in their testing approaches include Denmark, France, Germany, The Netherlands, Norway, and Sweden. Current tests for the hazard assessment of effluents are at different stages of development in different countries. Discharges to fresh water and salt water containing chemicals that disrupt invertebrate endocrine systems are more likely to be detected in those countries that incorporate sublethal endpoints in WET testing (e.g., the U.S. and some European countries). While major groups of invertebrates remain potentially unprotected in countries that do not incorporate chronic testing of invertebrates, it is uncertain whether these tests can be used to detect effects of EDCs.

Methoprene

Advances in knowledge of the insect neuroendocrine system have led to the introduction of insect hormones and their analogs as "third generation pesticides" for use as biochemical biological control agents (Keeley et al. 1990). Those hormones involved in the regulation of reproductive and developmental processes in insects are of particular interest in strategies for the control of insect pests (Downer and Laufer 1983; Mulla 1995). In addition to natural insect juvenile hormones (JHs), several major types of synthetic insect growth regulators (IGRs) using this strategy have been studied to date.

Methoprene is a juvenile hormone analog (JHA) used in the production of a number of foods and used in aquatic areas to control mosquitoes and several types of flies, moths, beetles, and fleas. While rotifers (Schaefer et al. 1974), nematodes (Lucas 1978), mollusks (Schaefer et al. 1974), and fish (Takahashi and Miura 1975; Quistad et al. 1976; McKague and Pridmore 1978; Lee and Scott 1989) have been shown to survive exposure to low parts per million (ppm) and high parts per billion (ppb) concentrations of methoprene, methoprene exposures in the low ppb concentrations in the laboratory have been shown to

1) alter larval development in marine crabs and shrimp (Celestial and McKenney 1994),
2) affect both development and reproduction in freshwater cladocerans (Templeton and Laufer 1983; Chu et al. 1997), and
3) retard juvenile growth, delay the onset of reproduction, and reduce reproductive capacity in an estuarine mysid, *Mysidopsis bahia* (McKenney and Celestial 1996).

The fact that an IGR used in mosquito control is more toxic to aquatic crustaceans than to other biota is consistent with the close phylogenetic relationship between insects and crustaceans.

If the Altosid liquid larvicide formulation (used in mosquito control; containing 5% (S)-methoprene; 51 g/L) is applied at the maximum label rate (293 ml/hectare) to water 15-cm deep, the expected environmental concentration of (S)-methoprene would be 10 μg/L (= ppb). Methoprene concentrations in freshwater microcosms treated with sustained-release Altosid formulations detected the highest (S)-methoprene residue of 6 μg/L, while 85% of all samples contained residues of 1.0 μg/L (Ross et al. 1994). Residues of methoprene in natural waters were detected in 3 of 5 sites and ranged from 0.39 to 8.8 μg/L (Knuth 1989). Reproductive and developmental alterations in laboratory exposures of crustaceans to ppb concentrations of methoprene suggest that these physiological responses of crustaceans could serve as appropriate screening methods and biomarkers of the potential impact of these novel substances on the environment.

A suspected mode of action of JHAs is their disruption of the metamorphic process in targeted insects (Chapter 2, "Juvenile hormone-mediated processes," p 45). Given this mode of action, it is not surprising that insect larval stages (particularly final premetamorphic larval stages) are predominantly used during efficacy testing of these compounds as insecticides. Concentrations of methoprene that disrupt larval development in several species of decapod crustaceans (the grass shrimp *Palaemonetes pugio,* the mud crab *Rhithropanopeus harrisii,* and the blue crab *Callinectes sapidus*) were similar to concentrations that are lethal to a number of target insects, including salt marsh mosquitoes (Bookhout and Costlow 1975; Bookhout et al. 1976, 1980; Forward and Costlow 1978; Christiansen et al. 1977; Payen and Costlow 1977; Wilson and Costlow 1986; Clare and Costlow 1989; McKenney 1998). Similar concentrations of methoprene reduced reproductive success in entire life-cycle exposures of estuarine mysids (McKenney and Celestial 1996), but these tests required 4 weeks to conduct, as opposed to exposures for 2 to 3 weeks in partial life-cycle testing with larvae.

Similarly, a number of studies have demonstrated that compounds with JH activity, which interrupt developmental processes in insect larvae, also affect crustacean larval development. Cyprid larvae of the acorn barnacle *Balanus galeatus* metamorphosed prematurely when exposed to either the pesticide hydroprene, a JHA (Gomez et al. 1973), or synthetic juvenile hormone I (JHI; Ramenofsky et al. 1974). Treatment of cyprids from another barnacle, *Eliminus modestus,* with JHAs resulted in abnormal intermediates with both some cyprid and some nauplius characteristics (Tighe-Ford 1977), similar to larval–pupal intermediates reported for insects treated with JHAs (Downer and Laufer 1983; Sehnal 1983). JHAs produced morphological abnormalities in crab megalopae (Costlow 1977) and metamorphosed lobster larvae (Charmantier et al. 1988). Exposure to precocene, a group of IGRs known to inhibit JH production by the insect corpus allatum (CA), caused toxicity to early developmental stages of brine shrimp and barnacles (Landau and Rao 1980).

Tributyltin

The biocide tributyltin (TBT) may have subtle effects on organisms in the environment; these effects have not always been recognized. Recent work indicates an endocrine mode of action that was not fully appreciated in previous studies. Effects are not specifically looked for in laboratory toxicity studies and might not be recognized in the laboratory or field due to a poor understanding of the outward signs of endocrine-mediated effects (Chapter 4). Alternatively, it could be argued that the level at which environmental toxicology and effects assessments are currently practiced is sufficiently sophisticated that effects from all modes of action will be detected and incorporated into the effects assessment leading to an environment protected from the effects of endocrine modulators. A case study of TBT is useful in deciding which of these hypotheses is most likely to be correct (see Chapter 4 for additional case studies related to TBT in field assessments). TBT is one of the better known examples of a compound impacting the environment through an unsuspected mode of ED action. If any risk assessments were conducted on TBT before its use as an antifoulant, they appear to have provided an insufficient level of protection for many invertebrates. This raises the question whether routine toxicological testing and effects assessment procedures would identify a TBT concentration protective of marine gastropods. Toward that end, a brief summary of toxicity data on TBT is outlined below and used in a conventional effects-assessment procedure to establish a low risk concentration in the environment.

Organotin compounds have been used commercially in a variety of applications since the late 1930s (Fent 1996). The antifouling properties of TBT were recognized in the 1950s, leading to a marked increase in their production and use. As antifouling compounds, TBT and other organotins leach out of paints used to coat materials in contact with the freshwater and marine environments (i.e., docks and the hulls of ships). As the organotins leach out of the paint, their toxicity prevents organisms from attaching to the painted surface. This use and the limited degradation of TBT led to freshwater and marine concentrations that are associated with significant effects on gastropods through an endocrine mode of action (Chapter 4).

Chronic effects of triorganotin compounds (primarily TBT) on gastropods and marine bivalves have been reported at concentrations in the 0.1 to 10 ng/L range (Chapter 4, Table 4-3; Fent 1996). The effects observed included sterilization, imposex, intersex, and induction of spermatogenesis in ovarian tissue.

To determine whether toxicity test and effects assessments currently in use would have predicted adverse effects in the environment, the data in Fent (1996) were used to conduct an initial effects assessment. Acute and chronic toxicity data from conventional test species were selected and assessed using USEPA assessment factors (Table 3-6; Nabholz et al. 1993). In this type of evaluation, assessment factors are used in an attempt to extrapolate from acute and chronic test data on a limited number of species to field conditions (see Cowan et al. 1995 for more information). While this assessment approach has been criticized (Suter 1993), it has been successfully used to assess the risk of a large number and variety of materials (Nabholz et al. 1993; Cowan et al. 1995). This type of assessment is frequently applied to new compounds for which little toxicity data are available and should not be confused with more sophisticated and data-intensive assessment methods (see Cardwell et al. 1999 for a more complete TBT risk assessment). For this assessment, data selection was limited to those species that would conventionally be used in baseline toxicity testing. These tests are typically performed on a green alga, *Daphnia magna* (invertebrate), and *Oncorhynchus mykiss* (rainbow trout) in the freshwater environment and *Skeletonema costatum* (marine diatom), *Mysidopsis bahia* (mysid shrimp), and *Menidia* species (silversides) in the marine environment. All of the data used in this analysis came from typical biological endpoints for these tests rather than those that are specific for an endocrine mode of action. Where data for these species were missing, the lack of data was ignored.

Table 3-6 Toxicity data and assessment factors used to derive an initial effects assessment for tributyltin

Species	Test type	Test duration	Effect concentration (μg/L)	Assessment factor	Effect/AF[1] (μg/L)
Freshwater					
Daphnia magna	Acute LC50	48 h	4.4	100	0.044
Oncorhynchus mykiss	Acute LC50	96 h	3.5	100	0.035
Oncorhynchus mykiss	Chronic	10 d	0.2	10	0.020
Marine					
Marine diatoms	Acute EC50	72 h	0.35	100	0.0035
Mysid shrimp	Acute LC50	96 h	0.30	100	0.0030
Menidia beryllina	Acute LC50	96 h	3.0	100	0.030
Marine diatoms	Chronic		0.1	10	0.010
Menidia beryllina	Chronic growth	28 d	0.09	10	0.009

[1] Observed effect divided by the assessment factor. This is the value used in the effects assessment to establish a low level of concern.

In the effects assessment conducted for the TBT case study, acute and chronic effect concentrations ranged from approximately 0.2 to 4 μg/L in the freshwater environment and 0.1 to 3 μg/L in the marine environment. After the application of assessment factors, a concentration of low risk ranging from 20 to 44 ng/L in freshwaters and 3 to 73 ng/L in the marine environment was derived.

Effects on gastropods in the environment have been reported to occur at 0.1 to 10 ng/L (Chapter 4). A traditional screening effects assessment would have been able to protect against exposures down to approximately 3 ng/L but would not have conferred protection on all affected species. This analysis suggests that additional research into endocrine-mediated effects is needed to ensure that populations of sensitive species can be protected in the environment. In this one example, current effects-assessment procedures provide some degree of protection against effects mediated by an endocrine mode of action. However, we do not know if this one case represents the norm or is a special case. Hence, a deeper understanding of endocrine-mediated effects and the ability of various invertebrate groups to serve as surrogates for other organisms is warranted.

Conclusions and Recommendations

1) There is a need to define both ecologically relevant effects and a mechanism of endocrine system dysfunction when potential EDCs are evaluated. Adverse effects on whole-organism endpoints (e.g., development or reproduction) represent a greater level of potential ecological risk than does mechanistic evidence of endocrine activity *per se*. Ecologically relevant endpoints that may

be responsive to exposure to EDCs might include behavior, development, feeding rate, growth, reproduction, or long-term survival.

2) There are major uncertainties regarding the sensitivity of developmental stages and critical periods of endocrine function in invertebrates exposed to EDCs. Effects of EDCs are believed to be associated with certain sensitive life stages. But in the invertebrates, neither the most sensitive life stages nor the most sensitive endpoints have been identified in most groups. Therefore, it currently is difficult to identify specific life stages or endpoints that should be incorporated into laboratory testing methods for evaluating potential effects of EDCs. An alternative approach was considered for evaluating the utility of conducting full life-cycle tests with invertebrates. These full life-cycle tests could include embryolarval stages, periods of gonadal development, and reproduction as part of the assessment of EDCs. Exposures with some invertebrates can also be extended over more than one generation to include some or all of the key life stages in the next generation.

3) None of the standard methods for conducting toxicity tests with invertebrates were designed specifically to evaluate effects associated with EDCs. However, many of these methods describe approaches that can be used to measure endpoints that may be responsive to effects of EDCs (e.g., development, growth, reproduction) but that are not necessarily characteristic of them. Additional endpoints that could be measured in these laboratory tests and that may provide information indicative of ED include
   - sex ratio (of adults or offspring),
   - mating success,
   - production of ephippia (resting eggs in cladocerans),
   - egg hatching success,
   - viability of the offspring,
   - morphological abnormalities,
   - molting frequency and timing,
   - time to emergence or pupation,
   - metamorphic or larval developmental success, or
   - pigmentation.

4) If the mode of action of an EDC is known, testing of effects on invertebrates may be more expeditiously directed at those targeted processes. Partial life-cycle testing with particularly sensitive life stages of crabs, shrimp, lobsters, or crayfish may be more appropriate than tests with these organisms over their entire life cycle. In addition, the endocrine system of these decapod crustaceans is relatively well understood.

5) Factors to consider in selecting invertebrates for evaluating potential effects associated with EDCs might include mode of reproduction, laboratory culture and ability to conduct transgeneration exposures, generation time, size of organisms, knowledge of endocrinology, and the availability of consensus-based standards.
6) Sufficient information is outlined in publications to summarize methods for conducting toxicity tests with representatives of cnidarians, turbellarians, mollusks, annelids, mites, millipedes, centipedes, crustaceans (amphipods, cladocerans, isopods, fairy shrimp, copepods, mysids, mud crabs, grass shrimp, lobsters, crayfish, barnacles), insects (midges, other aquatic insects, and terrestrial insects), echinoderms, and tunicates. The Laboratory Testing Work Group was unaware of information that could be used to describe adequately methods for conducting toxicity tests with sponges, jellyfish, anemones, bryozoans, leeches, spiders, sea stars, brittle stars, or sea cucumbers.
7) A variety of surrogate organisms can be used with limited method development to conduct life-cycle or transgenerational exposures in the laboratory. However, there is limited knowledge of the endocrinology controlling molting, growth, reproduction, or sexual determination for most of these species (e.g., rotifers, nematodes, polychaetes, oligochaetes, amphipods, *Artemia* spp., cladocerans, copepods).
8) To the best of our knowledge, insects and crustaceans are the only groups of organisms for which surrogates with a relatively well-understood endocrine system are available for life-cycle testing. An important research need will be to determine whether surrogates of a particular group (e.g., *Chironomus* spp.) represent the range of responses in a larger group (e.g., insects).
9) To the best of our knowledge, the echinoderms are the only major group of organisms identified that would require moderate to extensive method development, but they are also important organisms for monitoring EDCs in the field.
10) There is a need to validate responses of a number of species and endpoints with a diverse set of reference compounds suspected of having an ED mode of action. Based on the data obtained with these tests, the most appropriate endpoints and species would be selected for standardized testing. Historically, a similar approach has been used to select the endpoints and species used in current environmental effects assessments.
11) Not all endpoints are equivalent, and only the most diagnostic endpoints for EDCs should be used in standard tests. Those endpoints cannot currently be identified because the data to select the most discriminating endpoints do not exist. Diagnostic endpoints will need to be identified through appropriate testing and comparisons of endpoints. As experience is gained, it may be

possible to reduce the number of endpoints used in the future to assess effects associated with EDCs.

12) Once appropriate tests and endpoints are identified, these methods could be used to test other compounds suspected of having an ED mode of action. These data could be used to build quantitative structure-activity relationships to help identify and prioritize potential EDCs for future testing. The test data also would be used along with literature data in the hazard identification and effects assessment process. Information gained during the hazard identification and effects assessment process would be used to refine further the selection of endpoints and species.
13) Many different types of discrete and continuous data may be produced from toxicity tests with EDCs. Dose-response data are best analyzed using regression or time-to-event techniques. These approaches are statistically and scientifically superior to calculation of the NOEC. Suitable techniques for analysis of nominal, ordinal, and multivariate data need to be considered in the development of any methods that will involve collection of this type of data. Because the focus of environmental toxicology is often at the level of populations, the use of survival and reproduction data to establish demographic parameters should be considered whenever possible.
14) Effluents containing chemicals that disrupt invertebrate endocrine systems are more likely to be detected in programs that incorporate sublethal endpoints in testing. While major groups of invertebrates remain potentially unprotected in countries that do not conduct chronic testing of invertebrates, it is uncertain whether these tests can be used to detect effects of EDCs.
15) Effects of TBT on gastropods in the environment have been reported to occur at 0.1 to 10 ng/L. A traditional screening effects assessment would have been able to protect against exposures down to approximately 3 ng/L but would not have conferred protection on all affected species. This analysis suggests that additional research into endocrine-mediated effects is needed to ensure that populations of sensitive species can be protected in the environment.
16) Almost any invertebrate used in laboratory toxicity tests can be caged in the field, but most of these in situ exposures are currently short term, with survival as the primary endpoint. Measurement of sublethal responses using in situ exposures may help to establish links between the laboratory and the field. Life-cycle testing of aquatic invertebrates in situ is not practical. However, life-cycle in situ tests with terrestrial species such as earthworms, beetles, and springtails are feasible, particularly for those organisms with rapid life cycles. In addition, periodic use of mesocosms and microcosms might increase the realism of both exposure conditions and ecological interactions in toxicity tests, although the objectives of these complex studies need to be clearly focused.

17) Once a set of species and endpoints is identified, standard methods should be developed. Revisions of existing standard methods or the development of new standards to address effects of EDCs should not be finalized until research has been conducted to evaluate relative endpoint and species sensitivity. Such standard methods for addressing endocrine-specific effects of EDCs could be developed either as revisions to existing standards (e.g., for cladocerans, amphipods, midges, or oligochaetes) or as new standards (e.g., for crabs, shrimp, crayfish, lobsters).

Appendix to Chapter 3

Endpoints for measuring potentially toxic effects of endocrine-disrupting compounds in laboratory tests

Introduction

The purpose of this appendix is to summarize endpoints that can be measured in laboratory toxicity tests with a variety of different invertebrates. The primary sources of information for this summary were consensus-based standard methods (ASTM, OECD, USEPA, Environment Canada, and APHA). Methods described in the peer-reviewed literature were also evaluated when standard methods either did not exist or were limited in scope relative to potential endpoints of interest.

The Laboratory Testing Work Group first developed a comprehensive list of endpoints that could be measured to evaluate potential effects of EDCs in laboratory toxicity tests. This list focuses on multiple endpoints for measuring effects on reproduction, growth, development, or behavior (Table A-1). The intent of this listing was to be more inclusive of potential endpoints compared to the limited number of endpoints described in standard methods or in other publications. For example, standard methods may outline only one primary endpoint for growth (e.g., biomass) or reproduction (e.g., number of young/female). However, Table A-1 lists other growth or reproductive endpoints that might also be measured with some additional effort (e.g., sexual maturation, sex ratio, growth rate, molting frequency). Additional endpoints that were not included in Table A-1 might be useful in evaluating the potential effects of EDCs, including regeneration of body parts (e.g., planarians, crustaceans; Chapter 2), gross morphology or histopathology (Chapter 4), plasticity (e.g., cladocerans), or biomarkers.

The list of potential endpoints in Table A-1 was used to summarize procedures described in standard methods or in the peer-reviewed literature (by major group [phylum, class, order] or by individual species). Additional endpoints that would require minor or major development are also tabulated in Table A-1 (i.e., endpoints not described in either standard methods or in other publications and requiring additional research before routine use). Endpoints that are either impractical or impossible to measure with a particular organism or group of organisms are also identified. Finally, the Laboratory Testing Work Group identified the need for additional research on methods for under-represented groups of organisms. There was sufficient information available to describe methods for laboratory toxicity tests with cnidarians, turbellarians, mollusks, annelids, mites, millipedes, centipedes, crustaceans (amphipods, cladocerans, fairy shrimp, isopods, copepods, mysids,

Table A-1 Potential endpoints for measuring endocrine-disrupting compounds in laboratory tests[1]

Group	Survival	Fecundity	Sexual maturation	Sex ratio	Mating behavior	Biomass	Growth rate	Molt time and success	Embryonic development	Larval development	Feeding behavior
Cnidarians											
Hydra spp.	P	P	P	D	NA	I	P	NA	I	NA	P
Marine hydrozoans	P	D	I	NA	NA	P	P	NA	NA	D	D
Anemones	P	D	I	NA	NA	D	D	NA	NA	D	D
Coral	P	D	I	NA	NA	P	P	NA	NA	P	NA
Turbellarians	P	I	I	NA	D	D	D	NA	I	I	P
Rotifers											
Brachionus spp.	S	P	P	P	P	P	P	NA	I	NA	P
Nematodes											
Caenorhabdites elegans	P	P	P	P	P	P	P	P	NA	NA	P
Mollusks											
Crassostrea gigas	S	P	P	D	P	P	P	NA	D	P	P
Crassostrea virginica	S	P	P	D	P	P	P	NA	P	P	P
Macoma baltica	S	D	D	D	D	D	D	NA	D	D	D
Macoma nausta	S	D	D	D	D	D	D	NA	D	D	D
Mercenaria mercenaria	S	P	P	D	P	P	P	NA	D	P	P
Mytilus edulis	S	D	P	D	P	P	P	NA	P	P	P
Voldia limatula	S	D	D	D	D	D	D	NA	D	D	D
Gastropods	D*	D	P	P	D	P	D*	NA	D*	D	P
Cephalopods	D*	D*	D*	D*	D*	D*	D*	NA	D*	D*	D

Table A-1 *continued*

Group	Survival	Fecundity	Sexual maturation	Sex ratio	Mating behavior	Biomass	Growth rate	Molt time and success	Embryonic development	Larval development	Feeding behavior
Annelids											
Polychaetes											
Capitella capitata	S	S	P	D	D	S	D	NA	D*	D*	D
Dinophilus gyrociliatus	S	S	P	NA	D	S	D	NA	D*	D*	D*
Neanathes spp.	S	S	P	P	P	S	S	NA	D*	D	P
Nereis diversicolor	D	D	D	D	D	D	D	NA	D*	D*	D*
Ophryotrocha diadema	P	P	P	NA	D	P	D	NA	D*	D*	D*
Oligochaetes											
Aporrectodea claiginosa	P	P	D	D	D	P	D	NA	D	I	I
Cognettia sphagnetorum	P	NA	NA	NA	D	P	D	NA	NA	NA	I
Eisenia spp.	S	S	D	D*	D	S	D	NA	D*	I	I
Lumbricus variegatus	D*	D*	I	I	D	S	D	NA	D*	I	I
Tubifex tubifex	S	S	P	D	D	D	D	NA	D	D	D
Mites	S	S	D	D	D	I	I	I	I	NA	I
Millipedes and centipedes											
Centipedes	P	D	D	D	P	P	D	D	D	D	P
Millipedes	P	P	D	D	P	D	D	D	D	P	P

Table A-1 *continued*

Group	Survival	Fecundity	Sexual maturation	Sex ratio	Mating behavior	Biomass	Growth rate	Molt time and success	Embryonic development	Larval development	Feeding behavior
Crustaceans											
Amphipods											
Ampelisca abdita	S	P	P	P	D*	P	P	P	D*	NA	D*
Amphiporeia virginiana	S	D	NA	NA	I	NA	NA	D*	NA	NA	I
Corophium spp.	S	P	P	P	D*	P	P	P	D	NA	I
Diporeia spp.	S	I	I	I	D*	D	D	D*	D*	NA	I
Euhaustorius spp.	S	NA	NA	NA	I	NA	NA	D*	NA	NA	I
Foxiphalus xixmeus	S	D	NA	NA	I	NA	NA	D*	NA	NA	I
Gammarus spp.	P	P	P	P	P	P	P	P	D	NA	P
Grandidierella japonica	S	P	P	P	I	P	P	P	D*	NA	I
Hyalella azteca	S	S	S	S	P	S	S	P	D*	NA	D*
Leptochirus spp.	S	S	P	P	D*	S	S	P	D*	NA	I
Rhepoxinius abronius	S	NA	NA	NA	I	NA	NA	D*	NA	NA	I
Isopods	S	S	D	D	D	S	D	D	D	NA	P
Cladocerans	S	S	P	P	P	P	P	P	I	P	P
Fairy shrimp											
Artemia spp.	P	P	P	D	P	P	P	P	I	P	P
Thamnocephalus spp.	P	P	P	D	P	P	P	P	I	P	P
Copepods	S	P	P	P	D	P	P	P	I	I	D
Mysid shrimp											
Holmesimysis costata	S	D	D	D	D	D	D	D	I	I	D
Mysidopsis bahia	S	S	S	P	D	S	P	D	I	I	D
Neomysis mercedis	S	D	D	D	D	D	D	D	I	I	D

Table A-1 *continued*

Group	Survival	Fecundity	Sexual maturation	Sex ratio	Mating behavior	Biomass	Growth rate	Molt time and success	Embryonic development	Larval development	Feeding behavior
Crustaceans *(continued)*											
Decapods											
Rhithropanopeus harrissii (mud crab)	P	P	P	P	D	P	P	P	D	P	D
Callinectes sapidus (blue crab)	P	P	P	P	P	P	P	P	P	P	P
Palaemonetes pugio (grass shrimp)	P	P	P	P	P	P	P	P	P	P	D
Homarus (lobster)	P	P	P	P	P	P	P	P	D	D	P
Procambarus (crayfish)	P	P	P	P	P	P	P	P	D	D	P
Balanus (barnacles)	D	P	D	NA	D	I	I	D	I	P	P
Insects											
Aquatic beetles	P	I	I	I	I	D	D	I	I	I	D
Aquatic bugs	P	I	I	I	I	D	D	D	I	D	D
Bees	S	I	I	I	NA	I	I	I	I	I	D
Blackflies	P	I	I	I	I	D	D	I	I	D	D
Caddisflies	P	I	I	I	I	D	D	I	I	D	D
Chironomus spp.	S	S	S	S	S	S	D	D	P	S	P
Dragonflies	P	I	I	I	I	D	D	D	I	D	D
Drosophila spp.	P	P	P	P	P	P	P	P	P	P	P
Hexagenia spp.	S	I	I	I	I	S	D	S	NA	S	D
Mosquitoes	P	P	P	P	P	P	P	P	P	P	P
Springtails	P	P	D	D	D	P	D	D	D	D	D
Staphylinid beetles	P	P	D	D	D	P	D	D	D	D	D
Stoneflies	P	I	I	I	I	D	D	D	NA	P	D

Table A-1 *continued*

Group	Survival	Fecundity	Sexual maturation	Sex ratio	Mating behavior	Biomass	Growth rate	Molt time and success	Embryonic development	Larval development	Feeding behavior
Echinoderms											
Echinoids	D*	D*	I	I	I	I	P	NA	S	S	D
Arbacia punctulata	D*	D*	I	I	I	I	D	NA	S	S	D
Dendraster excentricus	D*	D*	I	I	I	I	D	NA	S	S	D
Lytechnius pictus	D*	D*	I	I	I	I	D	NA	S	S	D
Strongylocentrotus spp.	D*	D*	I	I	I	P	P	NA	S	S	D
Sea stars	D*	D*	I	I	I	P	P	NA	P	D*	D
Brittle stars	D*	D*	I	I	I	P	P	NA	P	D*	D
Tunicates	D*	D	D*	D*	D*	D*	D*	NA	D*	D	D

[1] S = Consensus-based standard
P = Cited in peer-reviewed literature
D = Development of a new method required (* indicates major development needed)
I = Impractical
NA = Not applicable

crabs, shrimp, lobsters, crayfish, barnacles), insects (midges, other aquatic insects, and terrestrial insects), echinoderms, and tunicates. The Laboratory Testing Work Group was unaware of information that could be used to describe adequately methods for conducting toxicity tests with sponges, jellyfish, anemones, bryozoans, leeches, spiders, sea stars, brittle stars, or sea cucumbers. The need for additional research on methods for these under-represented groups of organisms is addressed in "Further development of methods and use of surrogate species" (p 123).

Groups of Organisms for Use in Laboratory Toxicity Testing

Cnidarians

There are no published standard methods for toxicity tests with either freshwater or marine cnidarians. However, a variety of methods have been developed using the freshwater hydrozoan *Hydra* spp. (Tables 3-1 and A-1) or sea anemones (Brooks et al. 1995). Tests currently available include measurements of mortality and feeding activity (Beach and Pascoe 1998) and growth rate (budding; Stebbing and Pomroy 1978; Adams and Haileselassie 1984). Inhibition of regeneration has been related to teratogenicity and could be a useful indicator of developmental toxicity (Wilby and Tesh 1990). The same authors also observed loss of osmoregulation and dilation of the body in *Hydra* spp. as potential endpoints that could be measured in toxicity tests. Although both sexual and asexual reproduction occur in *Hydra* spp., asexual budding tends to be predominant in the laboratory. Sexual reproduction is generally too erratic, and embryonic development too lengthy, to form the basis of a routine toxicity test (although both processes have been described in great detail; Grassi et al. 1995; Martin et al. 1997). With further development, tests based on changes in population structure of *Hydra* spp., in a predator–prey system with *Daphnia* spp. as the prey, could be developed (Taylor et al. 1995). Several species of *Hydra* spp. can easily be maintained in the laboratory using a specific culture medium (Lenhoff 1983), and techniques for identifying the various species are described by Campbell (1989). Being diploblastic, all of the cells in *Hydra* spp. may come into intimate contact with the surrounding medium, making these organisms potentially valuable indicators of water quality.

Several species of marine hydrozoans, including *Campanularia* spp. (Stebbing 1976), have also been used in laboratory toxicity tests. Inhibition of colony growth and hydranth degeneration are the primary endpoints measured in these tests. Spangenberg (1984) reported on the use of the schyphozoan *Aurelia* spp. in a test based upon metamorphosis of the polyp. This study revealed developmental delays and teratogenic effects resulting from exposure to hydrocarbons. Little work has been carried out with anthozoan anenomes, although Ormond and Caldwell (1982) did demonstrate reduced ova size in *Actinia equina* exposed to oil. Laboratory toxicity studies have also been carried out with marine corals; for example, Raimondi et al. (1997) examined the effects of drilling muds on adult brown cup corals, *Paracyathus stearnsi.* It is unlikely that this group of cnidarians has much potential for adoption as routine laboratory test organisms for EDCs because of the long generation time. However, marine hydrozoans have an important role for in situ monitoring (Chapter 4).

Turbellarians

Although no standard methods have been described for conducting toxicity tests with planarians, several toxicity test methods have been described since the early work of Jones (1937). Methods for measuring effects on survival of *Polycelis nipra* have been described in short-term tests with 4 compounds (Williams et al. 1984) and organochlorine compounds to *P. felina* (Kouyoumjian and Uglow 1974; Tables 3-1 and A-1). Methods for assessing contaminant effects on the predator–prey behavior and functional responses of *Dendrocoelum lacteum* feeding on an isopod (*Asellus aquaticus*) have also been described (Ham et al. 1995). However, it would not be realistic to develop these methods into standard tests. Freshwater triclad turbellaria are mostly hermaphrodite, but members of the group can reproduce asexually by transverse fission or sexually by laying eggs in cocoons. Cocoon production is variable, and developmental time is long. Therefore, routine life-cycle or transgenerational testing for EDCs with turbellarians is not practicable. Several triclad species are noted for their capacity to regenerate (Hoshina and Teshirogi 1991; Peter 1995), and with some method development, a regeneration endpoint could be used in toxicity testing. Triclad turbellarians can be maintained in laboratory cultures using a variety of methods and with living or dead food, depending on species. Laboratory studies on the feeding behavior of triclads have been described in relation to their interaction with other organisms, particularly leeches (Seaby et al. 1995, 1996).

Rotifers

Standard methods have been described for conducting toxicity tests with the rotifers *Brachionus calyciforus* and *B. plicatilis* (APHA 1999; ASTM 1999b). Endpoints measured in these standards are limited to survival, fecundity, or net population change over a few days (Tables 3-1 and A-1). The advantages of conducting standard toxicity tests with rotifers are the short test duration, the ease of culture, and a large ecological database. Rotifiers are common in the zooplankton in fresh- and brackish-water lakes and ponds. These small animals (about 250 μm) are occasionally dominant herbivores on algae and are probably important prey items for copepods and other small invertebrate predators and larval fish. Rotifers are easily cultured in the laboratory over many generations. *Brachionus* spp. are widely cultured in kg amounts for use in aquaculture to feed larval fish and small invertebrates. During exposure to vertebrate and invertebrate hormones, changes in the productivity and percent of mictic females were observed (Gallardo et al. 1997), suggesting that rotifers might be responsive to EDCs.

The standard methods outlined in ASTM (1999b) and APHA (1999) could be modified to produce additional endpoints that would signal the possibility of exposure to EDCs (Tables 3-1 and A-1). Like cladocerans, rotifers have a facultative sexual reproductive strategy (Snell and Carmona 1995). It is easy to distinguish males, parthenogenetically reproducing females, and sexually reproducing females by their distinctive eggs. Appearance of males or mictic females might be a signal that ED is occurring. Rotifers (such as *Brachionus* spp.) with spines and a lorica (hard shell) show phenotypic plasticity. Spine length is affected by both natural and artificial chemicals in the environment. Thus, measurements of spine length at the end of a long-term exposure could provide data indicating potential EDCs or developmental toxicity.

Female rotifers initiate reproduction within a few days after their birth. The length of this immature period might be increased in the presence of EDCs. Therefore, generation time could be a useful measurement of abnormal juvenile development. Use of the standard assay and assessment of net population growth (net production) would most likely detect this change in the immature period. Male fertility may be reduced by exposure to EDCs. An indicator of male fertility could be sperm count and normal sperm morphology. A more major modification of the standard test would allow for counting sperm. Sperm count requires that the males be fixed and observed at high magnification (Snell and Childress 1987). Rotifers, along with cladocera, have been used in mesocosms to test effects of pesticides (e.g., Fukunaga et al. 1992; Halbach 1984). The advantages of observing rotifers are their fast reproductive rates and large population sizes; disadvantages are their tendency to escape through all but the finest mesh enclosures and their insensitivity to many contaminants.

Nematodes

Caenorhabditis elegans is a small (about 1 mm long) transparent nematode with a short generation time (Table 3-1). This species has been thoroughly adapted to laboratory culture, it is easily cultured on artificial media, and its nutritional requirements are well understood (Wood 1988; Riddle 1997). However, relatively little is known about the ecology of these soil-dwelling nematodes. *Caenorhabditis elegans* is not routinely used in toxicity testing (Williams and Dusenberg 1990; Donkin and Williams 1995). However, this species is an excellent candidate for future research. The genetic characteristics of virtually every aspect of this nematode are well understood at both the gross and the molecular level (i.e., development, metabolism, life history, behavior), and its genetic code is completely described (Sequencing Consortium 1999). Although the endocrine system of these organisms is poorly understood, there is great potential for developing tests with *C. elegans* that are indicative of EDCs.

Mollusks

Standard methods are available for testing in several marine bivalves (ASTM 1999c, 1999d; USEPA 1997) but not for freshwater bivalves. Endpoints outlined in these standard tests include acute survival, development, and larval settlement assays (Tables 3-1 and A-1). The OECD Test Guidelines Program has given high priority to the development of an acute test on early life stages and a shell deposition test with marine mollusks (*Crassostrea, Mytilus, Mercenaria*; OECD 1998a). Methods are also available in the published literature for conducting long-term in situ exposures for marine bioaccumulation monitoring and physiological effects such as scope for growth (e.g., Salazar and Salazar 1991; Uhler et al. 1993). Bivalves are useful for evaluating bioconcentration and are able to integrate over time the contaminant load at a given site. However, little is known about the endocrinology of bivalves, which could make it difficult to diagnose an ED mechanism using these organisms. However, numerous studies have evaluated the effects of TBT on the endocrine system of mollusks (Chapters 2 and 4). Larval settlement is another aspect of the bivalve life-cycle that could be used to detect EDCs with only minor modifications of existing methods (Lawler and Aldrich 1987; Laughlin et al. 1988; Phelps and Warner 1990; Nell and Chvojka 1992; Ruiz et al. 1994; Ruiz, Bryan, Gibbs 1995; Ruiz, Bryan, Wigham, Gibbs 1995). An oyster spat settlement assay might also be useful in determining potential EDC effects on the settlement and early metamorphosis of this group of organisms. Conducting transgenerational exposures with bivalves may not be practical because of the lengthy generation time (months to years) for most species.

Methods have been published for freshwater gastropod mollusks to measure effects on survivorship, growth, and reproduction in short-term and long-term exposures (Tables 3-1 and A-1; Wells and Young 1965; DeRusha et al. 1989; Hanlon et al. 1989; Fiorito et al. 1998). However, these methods do not currently incorporate endpoints that could lead to an absolute determination of an ED mechanism (e.g., measuring hormone levels). Techniques have been developed to evaluate and rank levels of imposex and intersex within freshwater and marine species and across species (Fioroni et al. 1991; Deutsch and Brick 1993; Huet et al. 1995; Stroben et al. 1995). The most sensitive species currently used is *Nucella lapillus*, which is affected at 1 ng TBT/L in seawater. The imposex stages are also well defined, and this is a clear case of ED, making this an excellent species for comparison with other gastropods.

Another unique gastropod group that could be adapted to evaluate EDCs is the marine *Crepidula* spp. snails. Much is known about their serial hermaphroditic lifestyle (Joosse 1988), which could be incorporated into endpoints in a method for evaluating potential EDCs. However, life-cycle tests would be impractical due to generation time (months to years). The terrestrial mollusk *Helix pomatia* has been used to evaluate metal accumulation (Berger et al. 1993; Dallinger et al. 1993). The freshwater gastropod *Lymnaea peregra* can be cultured easily and requires only calcium-rich water and food such as lettuce (M. Crane, personal communication).

Cephalopods have been used extensively in neurobiology. Culture methods for octopus have been published but no information regarding testing of cephalopods with potential EDCs is available. However, some investigations into the reproductive system, including identification of steroid hormone receptors, may make studies of EDCs possible in the future (e.g., D'Aniello et al. 1996; Di Cosmo et al. 1998). Due to the highly intelligent nature of these animals, their long life-cycle, and the difficulty in confining them, cephalopods may not be a practical group for evaluation of effects associated with EDCs. The other major mollusk groups, Amphineura, Monoplacophora, and Scaphopoda, are not routinely cultured or used in toxicity testing. Use of these groups would require major development of techniques and advancement in knowledge of mollusk endocrinology.

Annelids

Methods for conducting chronic toxicity tests have been developed for marine and estuarine macrofaunal-sized polychaetes (including *Nereis (Neanthes) arenaceodentata, Capitella capitata,* and *Streblospio benedicti*) and for meiofaunal-sized species (*Dinophilus gyrociliatus, Ophryotrocha diadema,* and *Ctenodrilus serratus*; Reish 1977; Carr et al. 1989; Carr 1998; Moore and Dillon 1993; Bridges and Farrar 1997). Standard methods have been outlined for all of these species, except *S. benedicti* (APHA 1999; ASTM 1999e, 1999f). Environment Canada and the OECD Test Guidelines Program have given high priority to development of survival and growth tests with marine annelids (e.g., *Polydora* and *Arenicola;* OECD 1998a). All of the species listed above can be cultured in the laboratory. Therefore, full life-cycle exposures can be conducted with polychaetes (Carr et al. 1989; Bridges et al. 1993; Levin et al. 1996), and multigeneration exposures should be feasible.

Freshwater oligochaetes (including *Tubifex tubifex, Limnodrilus hoffmeisteri, Stylodrilus heringianus, Lumbriculus variegatus,* and *Branchiura sowerbyi*) also can be cultured in the laboratory and have been used to evaluate sublethal responses in toxicity tests. Endpoints measured in these tests include growth, reproduction, burrowing behavior, and change in morphology (McMurtry 1984; Keilty et al. 1988; Reynoldson et al. 1991; Casellato et al. 1992; Phipps et al. 1993). Standard methods for several of these species are described in ASTM (1999a, 1999g), USEPA (1999a), and APHA (1999; Table 3-1). Terrestrial oligochaetes (earthworms), including *Eisenia foetida,* can be cultured and have been used in chronic test exposures to measure effects of xenobiotics on growth, reproduction, and behavior (Van Gestel et al. 1989; Neuhauser and Callahan 1990). Standard methods for conducting toxicity tests with earthworms are described in OECD (1984) and ASTM (1999h).

Procedures used to culture and test annelids might be applicable to other species, provided generation times are suitably short (i.e., less than 6 months). Body sizes, generation times, and other ecological characteristics are quite diverse within the polychaete and oligochaete annelid classes. Therefore, numerous opportunities exist for developing additional toxicity tests to measure EDC-modulated systems.

Mites, millipedes, and centipedes

Standard methods have not been developed for conducting laboratory tests with mites, millipedes, and centipedes. However, toxicity tests have been developed for assessing the effects of pesticides on these nontarget arthropods. A document published by the European and Mediterranean Plant Protection Organization supports the use of combinations of tests for evaluating the side effects of a pesticide as part of the general program of biological evaluation of new products (Hassan 1985). International Organization for Biological and Integrated Control of Noxious Animals and Plants (IOBC/WPRS) work groups have agreed on common features for testing, including laboratory, controlled-field (i.e., mesocosm or field-plot experiment), and field testing (Hassan 1985). Regulatory testing procedures for pesticides with nontarget arthropods, including chelicerates are outlined in SETAC (1994; predatory mites *Typhlodromus pyri* and *Aphiodius ropalosiphy*).

Standard laboratory toxicity testing of representative chelicerates and other beneficial-use organisms is currently and routinely required for pesticide registration in European member states (Directive 91/414/EEC: Council Directive [15 July 1991] concerning plant protection products). Evaluation of effects in initial toxicity tests is carried out by measuring the reduction in beneficial capacity (i.e., parasitism) or performance (i.e., egg laying) compared to the control.

A number of tests with chelicerates include endpoints that may be of use in assessing the potential effects of EDCs. These species include the following:

1) *Lepthyphates tenuise* (Linyphidae, araneae). The linyphiids are a large group of spiders that are abundant in a wide range of crop ecosystems and are known to be predators of agricultural pests such as aphids. This species is found in a variety of habitats and can be readily collected from uncultivated areas.
2) *Phytoseiulus persimilis* (Phytoseiidae, Acari). This predatory mite lives on all stages of the spider mite *Tetranychus urticae*. It is used for biological control of spider mites in glasshouses in most European countries.
3) *Typhylodromus pyri* (Phytoseiidae, Acari). This is an important predator mite on *T. urticae* and *Panonychus ulmi* in orchards.
4) *Amblyseius potentillae* (Phytoseiidae, Acari). This phyoseiid mite is capable of regulating fruit-tree red spider mite populations.

Examples of laboratory tests with these species include the following:

1) *Lepthyphates tenuise.* Mortality is assessed after 48-hour exposure; for less toxic compounds (< 50%) survivors are transferred to sealed vessels in which *Drosophila* are added and feeding is assessed.
2) *Phytoseiulus persimilis.* Mortality is assessed by starting the test with animals < 48 hours old; after 13 days, exposure eggs are recorded and counted every second day.
3) *Amblyseius potentillae.* Mortality is assessed after one week, and the average egg production per female is assessed during the second week of exposure.

Field tests conducted under controlled conditions typically investigate the effects of single compounds. Field tests permit full investigation of long-term effects or the effects of repeated applications during a season. In this way, field effects may be observed. Nonstandard methods for testing mites, millipedes, and centipedes have also been described in the published literature. The following tests have been described in a handbook of soil invertebrate toxicity tests (Løkke and Van Gestel 1998) and were developed within the European Union-funded Sublethal Effects of Chemicals on Fauna Soil Ecosystem project (SECOFASE; Tables 3-1 and A-1). The mite test uses the oribatid mite *Platynothrus peltifer*. Tests have been described that measure reproduction and feeding as endpoints, with a test duration of 8 to 11 weeks (the life-cycle duration is from several months to more than a year). Tests on the centipede species *Lithobius mutabilis* (Chilopoda: Lithobiidae) have also been described (Løkke and Van Gestel 1998). Field-captured animals can be maintained in the laboratory for a number of months, the populations being periodically supplemented with new field-captured animals. The variables measured are mortality and growth rate. In more extensive experiments, respiration rate, locomotor, activity and biochemical biomarkers can also be measured. The duration of tests is 30 days (or 90 days for more persistent chemicals). Tests using the millipede *Brachydesmus superus* have also been described in Løkke and Van Gestel (1998). This species has a life cycle of one year, egg production can be assessed, and 7 postembryonic stages are distinguishable. Semiquantitative evaluation of feeding activity is possible by observation of the dark gut contents visible through the body tissues. In laboratory culture, high mortality of juvenile stadia during postembryonic development has been noted. Adults are only available for part of the year, and it is difficult to maintain permanent cultures of millipedes during the whole year.

The methods described above for conducting toxicity tests with mites, millipedes, or centipedes were not developed to assess potential effects of EDCs. However, measurement of sublethal endpoints such as reproduction or development in these tests may be indicative of direct or indirect effects on the endocrine system.

Crustaceans

Amphipods

Amphipods are one of the most toxicologically sensitive invertebrate taxa in benthic aquatic communities (Burton et al. 1992; Lamberson et al. 1992), and several amphipod species are used in standard 10-day toxicity tests (Environment Canada 1992a, 1997a; USEPA 1994a, 1994b, 1997; ASTM 1999a, 1999i; Table A-1). Chronic toxicity tests are available for some freshwater, marine, and estuarine amphipods (Lamberson et al. 1996; Emery et al. 1997; Ciarelli et al. 1998; Ingersoll et al. 1998). Standard methods will soon be published in the U.S. for the estuarine amphipod, *Leptocheirus plumulosus* (USEPA 1999b) and the freshwater amphipod, *Hyalella azteca* (USEPA 1999a). These standard tests measure survival, growth, and reproduction over one life cycle. With minor modification, these standard methods could be extended to include additional EDC-modulated endpoints, such as molting frequency, sex ratio, embryo development, or gross morphology. The OECD Test Guidelines Program has given high priority to development of survival and growth tests with marine and freshwater amphipods (e.g., *Corophium, Leptocheirus, Ampelisca,* and *Hyalella*; OECD 1998a).

The availability of simple laboratory culture methods for some amphipods (including those used in the standard methods) and short generation time makes it possible to measure transgenerational and population-level effects of EDCs with amphipods (DeWitt et al. 1997). For longer-lived species, methods are available to measure other sublethal endpoints, including growth, mating behavior, burrowing behavior, scope for growth, and population size-structure (Swartz et al. 1985; Pascoe et al. 1994). However, measurement of reproduction or transgenerational effects in these longer-lived species may be impractical. Amphipods have been used in short-term in situ exposures to measure survival, growth, fecundity, scope for growth, feeding rate, and mating behavior (Pascoe et al. 1994; Maltby 1995a; Ireland et al. 1996; DeWitt et al. 1999). Field studies have also demonstrated reduction in amphipod abundance and species diversity relative to the environmental concentration of toxic chemicals, including the distributions of genera or species used in standard tests (Burton and Ingersoll 1994; Swartz et al. 1994; Widbom and Oviatt 1994).

Cladocerans

Standard methods have been published in ASTM (1999a, 1999j, 1999k), USEPA (1994b, 1997), Environment Canada (1990a, 1990b, 1992b), and OECD (1998b) for conducting toxicity tests with 3 species of cladocerans (*Daphnia magna, D. pulex, Ceriodaphnia dubia*; Tables 3-1 and A-1). Draft guidance is also outlined in ISO (1998a) for conducting toxicity tests with *D. magna*. In addition, there is extensive information in the peer-reviewed literature on the use of daphnids in toxicity testing. Endpoints that can be measured in toxicity tests with cladocerans include effects on survival, growth, reproduction, and development in full life-cycle exposures (Tables 3-1 and A-1). The most widely used tests are short-term lethality tests. Longer-term tests (up to 7 days for *C. dubia* or 21 days for *Daphnia* spp.) are used to evaluate effects on survival, reproduction, and growth. These data can be used to calculate a population growth-rate coefficient (r; Meyer et al. 1986, 1987). The major advantages of testing *Daphnia* spp. or *C. dubia* are ecological relevance, an extensive ecological database, fast growth and maturation, ease of culture over many generations, and relative sensitivity to many contaminants. *Ceriodaphnia dubia* has a shorter generation time (relative to *Daphnia* spp.), but both genera typically have similar sensitivities to toxicants (Versteeg et al. 1997). *Ceriodaphnia dubia* is perhaps a little more difficult to culture, compared to *Daphnia* spp.

The standard tests described above can easily be modified to include additional endpoints that may be potential indicators of exposure to EDCs (Table A-1). For example, neonates are typically counted and discarded in these tests. If these organisms are observed using a low-magnification microscope, it is easy (especially using the larger *D. magna* or *D. pulex*) to distinguish male and female neonates. Cladocerans typically produce only female offspring in the standard tests. However, males can be produced in response to environmental signals such as exposure to specific chemicals (Dodson et al. 1999). The appearance of males in a toxicity test with cladocerans may indicate that ED or a change in metabolic capability of the organisms is occurring. When neonates are observed to determine gender, observations can also be made of gross morphological abnormalities, such as the short setae and curved tail spine that result from exposure to the suspected endocrine disrupter, nonylphenol (Baldwin et al. 1997; Shurin and Dodson 1997). Environmental signals also can induce adult females to produce ephippia (resting eggs). For example, toxic food can induce ephippia production in *D. magna* (Shurin and Dodson 1997). Ephippial eggs typically are not produced in the controls during 21-day tests, but development of ephippial eggs can be induced by environmental factors such as exposure to cyanobacteria (Shurin and Dodson 1997).

Molting and growth rate can be measured in the standard tests with cladocerans by observing when organisms are sexually mature (presence of eggs in the brood pouch) and by measuring body length of adults or neonates at the end of the test. Molting frequency can also be determined by counting the shed exoskeletons of the cladocerans (Ingersoll and Winner 1982). Molting, growth rate, and steroid metabolism are affected by suspected disrupters of the steroid system in *D. magna* (Baldwin et al. 1997). Effects of acetylcholinesterase inhibitors such as carbamate pesticides can be detected using exposures with *D. pulex*. These insecticides mimic natural chemicals (produced by predators) that induce the development of neck teeth in neonatal *D. pulex* (Hanazato and Dodson 1995). Cladocerans are ecologically important organisms whose abundance is often measured in mesocosms as an indicator of environmental quality (e.g., Graney et al. 1995).

Isopods

Standard methods have not been developed for isopods, but terrestrial isopods have been used extensively for testing the effects of chemicals. The woodlouse *Porcellio scaber* has been most studied (Drobne 1997), but other species such as *Oniscus asellus* and *Porcellio laevis* have also been used (Drobne and Hopkin 1994; Van Brummelen et al. 1996; Odendaal and Reineke 1999). Many different toxicological endpoints have been investigated in woodlice, including survival, growth, food consumption and assimilation, respiration, molt frequency, reproduction, induction of heat shock proteins, locomotory behavior, and ultrastructural changes in cell structures (Tables 3-1 and A-1; Drobne 1997; Sorensen et al. 1997). Woodlice have also been exposed to a wide range of different toxicants, including heavy metals (e.g., Odendaal and Reinecke 1999) and organic compounds (e.g., Brummelen et al. 1996; Zimmer 1999). Vitellogenin synthesis and ecdysteroid titers in the hemolymph can be measured in woodlice (Steel and Vafopoulou 1998), which could provide a useful method for confirming the endocrine effects of some EDCs. The effects of chemicals on field populations of woodlice has also been extensively investigated (e.g., Eckwert and Kohler 1998; Jones and Hopkin 1998), including effects on reproduction (Jones and Hopkin 1996). Aquatic isopods have been used less frequently in environmental toxicology. However, toxic effects on *Asellus aquaticus*, a common isopod found below sewage treatment works, have been investigated in both the laboratory and field (e.g., Whitehurst 1991; Maltby 1995b; Mulliss et al. 1996).

Fairy shrimp

Standard methods have not been developed for toxicity testing with fairy shrimp, but these organisms are widely used in acute tests and occasionally in chronic tests to measure effects of chemicals on survival, fecundity, and growth (Persoone and Wells 1987; Dumont and Munuswamy 1997; Tables 3-1 and A-1). Fairy shrimp are natives of ponds that lack fish (typically temporary ponds). For example, *Thamnocephalus* spp. occur in temporary freshwater ponds, and *Artemia* spp. occur in hypersaline salt ponds. Most species eat algae or bacteria-laden detritus. Fairy shrimp are larger than cladocerans or rotifers and have proportionally longer generation times. Fairy shrimp are obligate sexual organisms. The advantages of testing fairy shrimp are that they can be cultured for many generations in the laboratory, cultures are readily initiated from dry cysts, the adults are large enough to see, and there is an extensive ecological database, especially for *Artemia* (Artemia Reference Center 1999; Persoone and Wells 1987). *Artemia* spp. that are adapted to high temperature and high salinity may be too specialized to be a representative invertebrate in toxicity testing. Fairy shrimp are generally less sensitive to toxicants than organisms such as cladocerans, mysids, or grass shrimp.

Fairy shrimp exhibit variability in egg production and egg type (resting eggs or eggs that hatch immediately). It is possible that this variation in reproductive strategy could be linked with hormonal control and subsequently with ED. Male *Thamnocephalus* spp. have elaborate second antennae for grasping females during mating. It is possible that the growth and development of these complex structures could be affected by EDCs. *Artemia* spp. are specialists at living in saturated salt water and may be unique organisms for evaluating osmoregulation as an endpoint.

Copepods

Though standard methods have not been developed for conducting long-term toxicity tests with copepods, published information is available that could be used to develop standard methods (Chandler and Scott 1991; Chandler and Green 1996; Green et al. 1996). Methods have been described for measuring effects of contaminated sediments on survival, reproduction, and larval settlement of marine copepods (Chandler and Scott 1991). Contemporary laboratory studies are being developed to conduct entire life-cycle exposures (T. Hutchinson, personal communication). Toxicity of representative chemicals to specific life stages of copepods has been reported by Green et al. (1996). Relatively short reproductive tests (14-day) have been described for laboratory- and field-contaminated sediment (Chandler and

Green 1996). The OECD Test Guidelines Program has given high priority to the development of survival and reproduction tests with pelagic marine crustaceans, including copepods (e.g., *Acartia, Tisbe*; OECD 1998a). ISO (1998b) describes acute lethality tests for marine copepods.

Mysid shrimp

Standard guides for conducting life-cycle toxicity tests with saltwater mysids have been developed using the species *Mysidopsis bahia* as a model (USEPA 1995, 1997; ASTM 1999l). These guides provide a consensus-based method for determining the effects of waterborne chemicals on the survival, growth, and reproduction of *M. bahia* during exposure in the laboratory through an entire life cycle. While specifically derived from work with *M. bahia*, these standard methods can be modified for testing of other mysid species. Reviews of toxicity testing results with *M. bahia* demonstrated the sensitivity of this estuarine crustacean to a wide variety of contaminants and identified the utility of *M. bahia* in chronic toxicity studies in the marine environment (Nimmo and Hamaker 1982; McKenney 1998). For the majority of pesticides examined in life-cycle toxicity tests using *M. bahia*, sublethal reductions in growth and reproduction have proven to be the most sensitive endpoints of chronic effects. Modified procedures for focusing on alterations in growth and reproductive responses in this species are described by McKenney (1994). A standard guide for conducting static and flow-through acute toxicity tests with mysids from the west coast of the U.S. (*Holmesimysis costata* and *Neomysis mercedis*) is also available (ASTM 1999m). However, no standard methods have been developed for conducting life-cycle toxicity tests with these 2 west coast species.

The relatively short life-cycle of *M. bahia*, about 3 to 4 weeks, compared to about 3 months for *H. costata* and *N. mercedis*, makes this species more amenable for both life-cycle and transgenerational exposures. *Mysidopsis bahia* has been reared in the laboratory through to maturation of a second generation, allowing observations of transgenerational effects (Nimmo et al. 1980). This is an attractive attribute for examining the effects of EDCs, which would allow determination of the influence of a suspected EDC not only on sex ratios and sex determination of exposed juveniles but also on the unexposed offspring. Though never used in toxicity studies with mysids, both gross morphological and histopathological alterations associated with exposure to suspected EDCs may be observed during all life stages in the life history of these mysids. Though never quantified, mating behavior in this species can potentially be measured following both intermittent stage-specific exposure or continuous exposure starting with newly released juveniles. Feeding densities associated with optimal reproductive success have been determined empirically for *M. bahia* (McKenney 1987), suggesting that feeding behavior may be quantified. Osmoregulatory capabilities of juvenile *M. bahia* have been reported for this estuarine species (DeLisle and Roberts 1987).

Little information exists in the literature on the endocrinology of mysids. However, life-cycle exposure of this species to methoprene, a JHA used in mosquito control, produced mortality at concentrations similar to those preventing adult emergence of several mosquito species in the laboratory and in the field (McKenney and Celestial 1996). Unlike the inhibitory effects of traditional chemical pesticides, growth of *M. bahia* was reduced only at higher methoprene concentrations. In contrast, reproduction endpoints were quite sensitive to methoprene. Compounds with JH activity have been demonstrated to affect these same reproductive functions in crustaceans, and the responses of mysids to very low concentrations of methoprene resemble responses of insects to JH and JHA. These data suggest that responses of *M. bahia* to methoprene may be functioning through interference with an endogenous endocrine system in crustaceans, which utilizes JH-like compounds (McKenney and Celestial 1996).

Caged *M. bahia* have been successfully tested in the field to measure acute lethality associated with aerial application of pesticides (McKenney et al. 1985; Clark et al. 1986). Higher respiration rates in caged juvenile *M. bahia* following application of an organophosphate pesticide accompanied reduced growth of mysids (McKenney et al. 1985). These results in a field study confirmed earlier laboratory studies, indicating that short-term measurements of metabolic dysfunction in exposed mysids may be used to predict altered production rates in these dominant secondary producers.

Decapods

Standard methods have been described for conducting acute toxicity tests with aqueous effluents or single compounds using a variety of decapods (i.e., crayfish, shrimp, crabs, lobster; USEPA 1993, 1997; ASTM 1999n, 1999o). General approaches are also outlined for conducting toxicity tests with decapods, including various freshwater, marine, and estuarine species (APHA 1999). However, no specific standards have been developed for conducting chronic toxicity tests with decapod crustaceans. Nevertheless, several recent reviews suggest the utility of endocrine-mediated functions in decapod crustaceans as biomarkers of environmental contaminants acting as endocrine disrupters in partial life-cycle exposures (Fingerman et al. 1996, 1998). Use of developing decapod crustacean larvae as an endpoint to assess EDCs in the marine environment has been proposed (McKenney 1999). The following section outlines this information for representative crabs (the mud crab *Rhithropanopeus harrisii* and the blue crab *Callinectes sapidus*) shrimp (grass shrimp *Palaemonetes pugio*), lobsters, crayfish, and barnacles.

Mud crabs

The mud crab, *R. harrisii*, is a common estuarine crustacean with a wide distribution, and its biology has been relatively well studied (Williams 1984). Reproducing crabs have been maintained in the laboratory for up to 14 months. However, due to a relatively long life cycle (3 to 4 months), full life-cycle exposures have not been reported in the laboratory. Although mud crabs are seasonal breeders, their reproduction and development can be manipulated in the laboratory during the nonbreeding season (Bookhout et al. 1984; Goy et al. 1985).

Unlike mysids, for which larval development occurs within the confines of a marsupial pouch, *R. harrisii* is a suitable nontarget estuarine crustacean for developmental studies with EDCs intended to alter larval development in target insects. Larval development is completed in about 2 weeks (Costlow 1968), which would acommodate a rapid screening method. The duration of the larval molt cycle is generally 2 to 3 days under optimal salinity-temperature conditions (Costlow et al. 1966). In contrast, the molt cycles of adult decapods may last weeks or months.

The influence of EDCs on *R. harrisii* larvae have been reported in the literature. A series of studies demonstrated that several JHAs increase the time required for complete larval development of *R. harrisii* and limit metamorphic success at concentrations similar to those that are effective against developing larvae of various target insects. Toxic effects on *R. harrisii* were observed at 2 to 3 orders of magnitude lower than lethal concentrations for fish and other nontarget organisms (Bookhout and Costlow 1974; Christiansen et al. 1977; Costlow 1977; Celestial and McKenney 1994). An ecdysone mimic accelerated molting of zoeae of this crab (Clare, Rittschof, Costlow 1992). Exposure to TBT altered growth rates in larval *R. harrisii* (Sanders et al. 1985).

Other reports on physiological responses of *R. harrisii* pertinent to EDCs have been described in the literature. Methoprene exposure inhibited vitellogenesis in this crab (Payen and Costlow 1977). Baseline information on the process of autotomy-induced cheliped regeneration has been studied in *R. harrisii* megalopae, against which the effects of xenobiotics could be assessed (Clare, Costlow, Bedair 1992). An ultrastructural study of the larval cuticle of this crab described deformities in the endo- and exo-cuticle following exposure to the chitin-synthesis inhibitor diflubenzuron (Christiansen and Costlow 1982). The effects of the herbicide, alachlor, on respiration and osmoregulation of *R. harrisii* have also been studied (Diamond et al. 1989). Factors controlling egg hatching in *R. harrisii* have been reported by Forward and Lohmann (1983). Hormonal control of a variety of larval processes in *R. harrisii*, including molting, apolysis (detachment of the epidermis from the cuticle at the beginning of the premolt phase), and completion of metamorphosis, have been reviewed (Christiansen 1988). The ontogeny of osmoregulation and its hormonal control have been reported for this crab (Kalber and Costlow 1966).

Blue crab

The blue crab *Callinectes sapidus* is a commercially important decapod that has been extensively studied both ecologically and physiologically. Molting and claw regeneration assays, behavioral assays, reproductive studies, and biomarker studies have all been published for this species (Engel and Eggert 1974; Lewis and Haeffner 1976; McCarthy and Skinner 1979; Singer et al. 1980; Conner and Singer 1981; Ary et al.

1987; Schlenk and Brouwer 1991; Brouwer et al. 1992; Mangum 1992; Forward et al. 1995, 1996; Jop and Hoberg 1995; Wood 1995; Wood and Derby 1995; Wood et al. 1995; Oberdörster et al. 1998a). In addition, several studies on potential EDC compounds have been conducted with this species, with endpoints such as molting, developmental success (oocyte, embryonic, and larval), claw regeneration, lipovitellin production, and steroid metabolism (Bookhout and Costlow 1975; Bookhout et al. 1976; Mothershead and Hale 1992; Schlenk et al. 1993; Lee and Noone 1995; Lee et al. 1996; Oberdörster et al. 1998a). However, due to the tenacity of this species as well as their long life-cycle, multigenerational studies with *C. sapidus* are impractical.

Grass shrimp

The grass shrimp, *Palaemonetes pugio*, is a ubiquitous crustacean that maintains a dominant role in the energy cycles of tidal marsh estuaries. Its biology has been relatively well studied (Williams 1984). A life-cycle toxicity test with *P. pugio* exposed to endrin was conducted for 145 days to evaluate life-stage survival and growth, gonadal development, spawning, and metamorphic success and duration (Tyler-Schroeder 1979). Although grass shrimp are seasonal breeders, this species has been induced to reproduce out of season in the laboratory, and larvae have been successfully reared from these winter breeders (Little 1968).

Complete larval development of *P. pugio* within about 2 weeks can be used to evaluate effects of IGRs on nontarget crustaceans (McKenney 1998). Larvae of grass shrimp exhibited reduced rates of metamorphic success at concentrations of methoprene similar to those that are effective against developing larvae of various target insects. Toxic effects on *P. pugio* were observed at 2 to 3 orders of magnitude lower than lethal concentrations for fish or other nontarget organisms. Exposed larvae were more sensitive to toxicity during the early and final larval stages than during intermediate stages. Methoprene in these exposures retarded growth in early larval stages and postlarvae, but enhanced growth in premetamorphic larvae.

Other reports on physiological responses of *P. pugio* pertinent to effects of EDCs have been described in the literature. Exposure of embryos to diflubenzuron (a chitin synthesis inhibitor) did not effect embryonic development or hatching success of *P. pugio* at concentrations similar to concentrations altering metamorphic success during larval exposures alone (Cunningham et al. 1987). However, exposure to diflubenzuron resulted in malformed larvae and reduced larval viability (Wilson et al. 1995). Viability of pelagic *P. pugio* larvae was unaffected by the pyrethroid insecticide fenvalerate bound to sediment until near completion of metamorphosis, with significant mortality occurring among postlarvae settling on to the contaminated sediment (Weber et al. 1996). Fenvalerate-laden sediment altered the growth and metabolism of developing *P. pugio*, depending on fenvalerate concentration, age of shrimp, and whether shrimp were premetamorphic or postmetamorphic in development (McKenney et al. 1998). Osmoregulatory patterns in *P. pugio* are modified by PCBs (Roesijadi et al. 1976). Several pesticides inhibited limb regeneration in this shrimp (Conklin and Rao 1982) and resulted in additional physiological responses coupled with histopathological deformities (Rao et al. 1982).

Additional aspects of the biology of *P. pugio* applicable to studies of potential EDCs have been reported in the literature. Reproductive behavior of this shrimp has been described (Berg 1979). The molt cycle of *P. pugio* and its hormonal control have been studied (Freeman and Bartell 1975). Histological examination of the embryonic coat of *P. pugio* has been reported (Glas et al. 1997). The influence of diet on larval development of *P. pugio* have also been reported (Broad 1957). Seasonal distribution, grazing patterns, growth rates, and reproductive potential have been described from field studies of *P. pugio* (Wood 1967; Morgan 1980; Kneib 1987).

Crayfish

Standard methods for using crayfish have been described only for acute toxicity testing (ASTM 1999n). The red swamp crayfish *Procambarus clarkii* is native to the southeastern U.S., but has been introduced into dozens of countries as a hardy aquaculture species and is available to many researchers worldwide. Crayfish, unlike the crabs and shrimps, developed directly from egg to adult without a series of metamorphic stages. The lack of metamorphic events may make crayfish less useful as sentinels of exposure to EDCs, since stage-specific toxicity of EDCs in other crustaceans is usually most evident at metamorphosis. The reproductive biology of *P. clarkii* has been documented (Penn 1943; Daniels et al. 1994; Tuberty 1998). Fingerman (1995) reviewed the endocrine mechanisms of crayfish, emphasizing reproduction and neurotransmitter regulation of hormone release. Use of crayfish as biological indicators of contaminants is reviewed by Alikhan et al. (1990) and by Fingerman et al. (1996, 1998).

Exposure of *P. clarkii* to cadmium causes induced hyperglycemia, which is thought to be mediated by crustacean hyperglycemic hormone (Reddy et al. 1994). Cadmium arrested ovarian maturation of *P. clarkii* by inhibiting release of gonad-stimulating hormone (Reddy et al. 1997) and decreased fecundity and hatching success (Naqvi and Howell 1993). Reddy et al. (1997) also reported arrested ovarian maturation by mercury exposure due to inhibition of the stimulatory effects of serotonin. Ovarian atresia was induced by naphthalene, which was reported to block release of gonad-stimulating hormone by brains and thoracic ganglia (Sarojini et al. 1994, 1995). Devi and Fingerman (1995) reported decreased acetylcholinesterase (AChE) activity in *P. clarkii* from exposure to heavy metals. Anderson et al. (1997) reported increased pH and swollen and vacuolated R cells in the hepatopancreas of *P. clarkii* placed for a week in a petroleum-contaminated site. The organophosphate pesticide, fenitrothion, impeded AChE activity and induced the cytochrome P450 pathway in laboratory and field studies of *P. clarkii* (Porte and Escartin 1998). Juvenile *P. clarkii* exposed to the herbicide terbufos also reported inhibition of AChE activity, along with loss of motor control and equilibrium (Fornstrom et al. 1997).

Cherax quadricarinatus is a species of Australian crayfish that is commonly used in aquaculture because of its ability to reproduce several times a year (Barki et al. 1997). Recent descriptions of this intersex crayfish, from the Department of Life Sciences at Ben-Gurion University of the Negev in Israel, make it particularly interesting for the study of androgen hormone mimicking and antiandrogenic compounds (Sagi et al. 1996, 1999; Khalaila et al. 1999). This species has a naturally occurring subpopulation that contains both male and female sex organs. The androgenic gland, if present, will make these crayfish functionally male, while a nonfunctioning ovary is present on the other half of the animal. Upon removal of the androgen gland the crayfish will develop a functional ovary, and the testes become atretic. This crayfish species may be used for a novel assay of the masculinization or feminization effects of xenobiotics on population sex ratios.

Lobster

Lobsters (especially *Homarus americanus*) have been extensively studied because of their commercial value and use in aquaculture. Because of this large body of information, assays could be developed to detect EDCs in lobster. Antibodies for lobster vitellogenin have also been developed (Byard and Aiken 1984). In addition, much research has been done to identify key steroid-metabolizing monooxygenases and the effects of xenobiotics on these enzymes (Burns et al. 1984; James 1984, 1989; Jame and Little 1984; James and Shiverick 1984; Boyle and James 1996; James et al. 1992, 1996; Snyder 1998). Recently, 2 cytochrome P450 isozymes have been cloned from lobster (CYP 2L and CYP 45; Boyle and James 1996; Snyder 1998). Although much of lobster endocrinology is known, they are impractical models for routine laboratory testing due to their size and longevity.

Barnacles

Barnacles are important members of the fouling community. The unique metamorphosis of barnacles from free-swimming nauplii to cypris larvae to settled pinheads has been used extensively in the antifouling research arena as an indicator of toxicity (Rittschof et al. 1994; Rittschof and Holm 1997). The larval settlement assays have been well developed and are easily reproducible. The exact hormonal regulation of metamorphosis has not yet been identified; however, the ecdysteroid system is involved (Chapter 2). These assays have recently been adapted to investigate the action of environmental estrogens (Billinghurst et al. 1998a, 1998b).

Insects

Midges

Standard methods have been described for conducting toxicity tests with 2 species of midges: *Chironomus tentans* or *C. riparius* (Biologischen Bundesanstalt fur Land- und Forstwirtschaft [BBA] 1995; Environment Canada 1997b; ASTM 1999a; USEPA 1999a; Tables 3-1 and A-1). Draft standard methods are also being developed by OECD (1998c) for conducting toxicity tests with midges, and numerous publications describe the use of midges in laboratory toxicity testing (e.g., Williams et al. 1986, 1987; Pascoe et al. 1989; Ingersoll et al. 1990; McCahon and Pascoe 1991; Taylor et al. 1991, 1993; Sibley et al. 1996; Watts and Pascoe 1996, 1998; Benoit et al. 1997). Endpoints that can be measured routinely in toxicity tests with *Chironomus* spp. include effects on survival, growth, reproduction, and development in full life-cycle exposures (Table A-1). Any of the additional endpoints listed in Table A-1 could also be applied to toxicity tests with midges (i.e., morphology, behavior, biomarkers). Methods outlined in standard toxicity tests with midges could be adapted with minor modification to conduct transgenerational studies. Toxic effects of potential EDCs on *Chironomus* spp. could be conducted using water, sediment, or dietary routes of exposure. While midges have a relatively rapid life-cycle (about 3 to 4 weeks at 23°C), the amount of tissue may be a limiting factor for use in in-vitro testing (individual 4th instar larvae are about 0.5 to 1.0 mg ash-free dry weight).

Midge tests have been evaluated using in situ field exposures (i.e., Sibley et al. 1999). Comparisons of responses of midges in companion laboratory-to-field comparisons have also been reported (Giesy et al. 1988; Maund et al. 1992; Taylor et al. 1994; Canfield et al. 1996, 1998; Conrad et al. 1999). West et al. (1997) evaluated the toxicity and bioaccumulation of 2,3,7,8-tetrachlorodibenzo-*p*-dioxin (TCDD) in the diet of *Chironomus tentans* and the oligochaete *Lumbriculus variegatus.* No toxic effects of this suspected EDC were observed in full life-cycle tests with either species at accumulated tissue concentrations up to 9533 ng TCDD per gram of lipid. The authors concluded that the lack of sensitivity of the 2 species to TCDD was consistent with the presumed absence of the aryl hydrocarbon receptor in aquatic invertebrates. However, they also concluded that the ability of these invertebrates to accumulate relatively high concentrations of TCDD in the absence of toxic effects may be relevant to the transfer of contaminants through aquatic food webs to potentially sensitive vertebrates. Similarly, Kahl et al. (1997) exposed *C. tentans* to 4-nonylphenol in a life-cycle test and observed little effect that could be attributed to ED.

Other aquatic insects

Standard methods for conducting 21-day sediment toxicity tests with burrowing mayflies of the genus *Hexagenia* are described in ASTM (1999a) and in Tables 3-1 and A-1. Endpoints measured in this standard method include survival, growth, molting and behavior. General guidance for conducting life-cycle tests (90 to 120 days) with mayflies has also been described (APHA 1999). Exposures with mayflies may be started with either eggs or newly hatched larvae. General procedures for conducting toxicity tests with *Hexagenia* spp. are described by Friesen (1979), Fremling and Mauck (1980), and Ort et al. (1995).

In addition to *Hexagenia* spp., other representatives of the Ephemeroptera, as well as stoneflies (Plecoptera), dragonflies (Odonata), caddisflies (Trichoptera), beetles (Coleoptera), and bugs (Hemiptera), play important roles within the freshwater ecosystem and have been used in toxicity testing. However, there are no standard methods developed for these other insects. Nevertheless, representatives from these groups have been used in survival studies (e.g., Brown and Pascoe 1988), and some have also been used in short-term sublethal tests, particularly using behavioral responses such as net spinning in the trichopteran *Hydropsyche angustipennis* (Petersen and Petersen 1984) and case building in *Agapetus fuscipes* (McCahon et al. 1989). Dragonfly larvae (*Aeshna* spp.) are also valuable in predator–prey studies of functional responses (Woin and Larsson 1987).

Several species of mayfly, stonefly, and caddisfly have been tested using in situ exposures (Ormerod et al. 1987) to examine the effects of acid and aluminium pulse episodes. In view of their importance as vectors of disease, mosquitoes (Culicidae) and blackflies (Simuliidae) have been investigated intensively, and there are a variety of methods for testing the efficacy of insecticides targeted against the aquatic larval stages of both of these dipteran groups (e.g., Muirhead-Thomson 1973). With relatively little modification, such tests could be used for laboratory testing of potential EDCs. However, with the exception of the midges, *Hexagenia* spp., mosquitoes, and blackflies, other aquatic insects provide little immediate opportunity for incorporation into laboratory exposures involving life cycle or reproductive studies. This largely reflects the difficulties of maintaining insects in the laboratory for significant and lengthy phases of the life cycle, which is a prerequisite for examining effects on reproduction.

Terrestrial insects

Methods have been published in the European Community for several terrestrial insects (Løkke and van Gestel 1998). However, these methods have not yet been developed as consensus-based standards. Tests with 3 species of springtails and a species of staphylinid beetle are described in Løkke and van Gestel (1998). The rove beetle *Aleochara bilineata*, a staphylinid beetle, has also been used extensively (Samsøe-Peterson 1987, 1993, 1995a, 1995b; Samsøe-Petersen and Moreth 1994), and the springtail *Folsomia candida* has been used in toxicity tests for many years (e.g., Thompson and Gore 1972). Approaches for evaluating effects on survival of honeybees in acute tests are described in standard methods (Organisation européenne et méditerranéenne pour la protection des plantes/European and Mediterranean Plant Protection Organization [OEPP/EPPO] 1992; ISO 1994; USEPA 1997; OECD 1998d, 1998e). However, the life-history strategy of honeybees makes these insects unsuitable for the development of generally relevant larval or reproductive tests. Draft ISO standards have also been developed for evaluating survival of springtails (ISO 1994). Survival, growth, development, and reproduction can be routinely measured in tests designed to investigate contaminated soils with springtails (Wiles and Krogh 1998) and staphylinid beetles (Metge and Heimbach 1998). Additional endpoints that could be included with minor modification to these methods might include behavior (e.g., Fábián and Petersen 1994; Sørensen et al. 1995) or morphology. Short-term, controlled-field deployment systems have been used with beetles (Metge and Heimbach 1998) and longer-term exposure systems that include reproduction could be developed with both beetles and springtails. The moth *Lymantria dispar* has also been used in laboratory toxicity testing (Ortel et al. 1993). Survival, growth, and reproduction can be measured in this species. Field effects have been observed on soil insect populations after the application of endocrine-disrupting pesticides (e.g., Edwards et al. 1967; Frampton 1997) and in areas contaminated by heavy metals (e.g., Smit and van Gestel 1996). Tests with these species could probably be developed into transgenerational studies with relative ease.

Several other terrestrial species that have not been used previously in environmental toxicity studies could be of use because of extensive knowledge of their biology. Different species of flies (Chapter 2), grasshoppers, locusts, crickets, cockroaches, termites, aphids, scale bugs, moths, butterflies, two-winged flies, bees, ants, wasps, and beetles are all widely used in pesticide efficacy, neuroendocrine, or genetic studies and are bred in the laboratory for these purposes (Hopkin 1989). Transgenerational toxicity tests with these terrestrial insects could be developed easily. For example, *Drosophila melanogaster*, the laboratory fruit fly, is easily cultured and is used frequently in toxicity testing to evaluate genetic change (Hopkins 1989; Hoffman and Parsons 1991). Although there is no consensus-based standard method for toxicity testing, this small fruit fly has been extensively tested by geneticists for nearly a century. A great deal is known about the genetics, development, metabolism, physiology, life history, morphology, behavior and ecology of fruit flies. *D. melanogaster* can be cultured easily in large numbers through multiple generations, and its nutritional requirements are well understood. The *D. melanogaster* endocrine system is relatively well known. On balance, *Drosophila* would be an excellent candidate for further development for assessing EDC activity.

Echinoderms

Echinoderms are sensitive to environmental contaminants, and toxicity tests have been developed using partial life-cycle exposures with gametes, larvae, or adults (ASTM 1999p). The fertilization success and larval development tests for sea urchins and sand dollars (Bay et al. 1993) have been standardized and are used in regulatory water quality testing programs (Environment Canada 1992c; USEPA 1995; ASTM 1999p). However, it is doubtful that fertilization or early embryological development in echinoderms is hormonally controlled, or that these tests have utility for measuring effects caused by EDCs.

Whole-organism toxicity tests with echinoderms are less common and have not been developed as a standard method. However, limb regeneration in sea stars, crinoids, and ophiuroids (Walsh et al. 1986; D'Andrea et al. 1996; Candida Carnevali et al. 1998), growth in sand dollars (Casillas et al. 1992), and gonad development in echinoids (Bay and Greenstein 1994) might be useful tests for evaluating EDC impacts on developmental and reproductive systems. We are unaware of any toxicological studies conducted with holothurians (sea cucumbers).

Full life-cycle toxicity tests have not been reported for any echinoderms. Although methods exist to culture sea urchins and brittle stars (Yamashita 1985; Blin 1997), full life-cycle or transgenerational exposures are probably impractical because of the long generation time (several weeks to months) for most species. Echinoids have been exposed in cages to test the in situ toxicity (mortality) of contaminated sediments. It may be possible to include sublethal endpoints such as gonad growth (Bay and Greenstein 1994) in these in situ exposures. Conceivably, in situ exposure of asteroids or ophiuroids to measure limb regeneration could also be conducted. Field studies have also demonstrated reductions in the abundance and species diversity of echinoderms (especially ophiuroids and echinoids) relative to increasing concentrations of sediment-associated contaminants (Swartz et al. 1986).

Tunicates

Tunicates are an important invertebrate group because of their place in phylogeny between invertebrates and vertebrates. Tunicates are in the phylum Chordata and are therefore the most closely linked to the vertebrates. Standard methods have not been developed for conducting toxicity tests with tunicates. However, tunicates have been used to conduct cadmium toxicity tests (Liebrich et al. 1995) and microcosm recolonization studies (Flemer et al. 1995). Tunicates can be easily cultured in the laboratory and have been used to study general life-cycle endpoints. Tunicates have some unique reproductive strategies that could be used to evaluate potential EDCs. Specifically, the asexual budding could possibly be used in short-term testing. Larval settlement studies have also been performed as a part of general life-cycle studies. However, considerable development would need to be done to use either of these methods for routine testing of potential EDCs. Settlement assays similar to what has been described for barnacles, tunicates, and anemones have also been developed for other members of the fouling community, including bryozoans (Bryan et al. 1997).

Molecular and Biochemical Techniques

To determine whether a potential EDC is acting on the endocrine system, it is essential to determine the mechanism of action (see "Approach," p 109; Table A-2). Several biochemical and molecular techniques have been developed for both in vivo and in vitro testing, which provide data to support an endocrine mechanism of toxic action. On a cautionary note, these techniques have not yet been standardized to allow them to be used routinely as screening tools. Biochemical techniques involving primary cell cultures and tissue explants are currently being developed for several phyla and could be used in determining an EDC mechanism. Several stable insect cell lines have also been developed. Immunoassays for specific P450s and lipovitellin are also published, as are whole animal steroid metabolism assays.

Ecdysone (20 HE) responsive cell lines have been developed from insects to study ecdysteroid-dependent responses, including growth and differentiation (Peel and

Table A-2 Summary of endocrine-disrupting mechanisms and in vitro test methods

Endocrine mechanism	Molecular or biochemical endpoints
Agonist or antagonist action on hormone-responsive cell lines of tissues	• Change in proliferation of hormone-responsive cell lines • Change in lipovitellin production (either cell lines or whole-organism studies)
Changes in hormone titers	• Radioimmunoassay (RIA) for juvenile hormone (JH; isolated glands) • RIA for ecdysone (isolated glands or whole-organism studies)
Altered metabolism or biosynthesis of hormones	• Induction of cytochrome P450 (tissue extracts or whole-organism studies) • Induction of phase II enzymes (tissue extracts or whole-organism studies) • RIA for JH and ecdysone
Receptor xenobiotic interactions at the molecular level	• Receptor binding assays • Reporter gene assays

Milner 1992; Cottam and Milner 1997; Lan and Riddiford 1997; Lan et al. 1997). Imaginal disc cells from *Drosophila*, Cl 8+, differentiate in response to ecdysteroid agonists (Peel and Milner 1992; Cottam and Milner 1997). This cell line has been used to investigate potential EDCs as well the ability of phytochemicals to hinder the normal development of these cells (Oberdörster, Wilmon et al. 1999). Some of these cell lines demonstrate modified responses to ecdysteroids in the presence of JHs (Cherbas et al. 1989; Cottam and Milner 1998). More work is needed to define specific responses to JHs. Primary cultures of *Manduca* tobacco hornworm epidermis have been used to evaluate both ecdysteroid and JH action on various ecdysteroid-induced genes (Zhou et al. 1998). Two of these genes, E75A and the Broad Complex (BRC), respond differently to 20E in the presence and absence of JH (Zhou et al. 1998). Such a system could be used to evaluate both juvenile hormone agonists and antagonists.

Recently, there have been additional activities to develop cell lines in invertebrates other than insects. Primary cultures that may have hormonal responses would be the most useful for detecting an EDC mechanism (i.e., hepatopancreas, Y-organ, and ovarian cultures). Primary cultures of hepatopancreas or digestive glands and ovaries would be useful to study the effects of xenobiotics on the production and distribution of lipovitellin in crustaceans. A meeting was held in October 1998 in Brest, France to summarize current knowledge on primary cell culturing of invertebrate tissues (Table A-3). Culturing techniques for several different groups of organisms have been developed, with successful passaging (splitting and growing cells) for up to several months. The representative taxa include crustaceans (Deca-

Table A-3 Invertebrate primary cell cultures

Common name	Scientific name	Tissue type[1]	Select references
Poriferans			
Sponge	*Geodia cydonium Jameson*		Batel et al. 1993
Sponge		Mesenchymal cell	Bond and Harris 1988
Sponge	*Latrunculia magnifica*		Contini 1994
Sponge		Mesenchymal cell	Leys 1997
Cnidarians			
Colonial cnidarians			Frank et al. 1994
			Mialhe et al. 1988
Mollusks			
Abalone	Haliotis rufescens	Larvae	Naganuma et al. 1994
Clam	Mya arenaria Linnae	Heart	Kleinschuster et al. 1996
Clam	*Meretrix lusoria*	Heart	Wen et al. 1993
Clam	*Mercenaria mercenaria*	Brown cells	Zaroogain and Anderson 1995
Butterfish clam	*Ruditapes decussatus*	Gill	Auzoux et al. 1993
Freshwater mussel		Gill	Gardiner et al. 1991
Mussel	*Mytilus edulis*	Digestive gland	Birmelin et al. 1997
Oyster	*Ostrea edulis*	Hemocytes	Mourton et al. 1992
Oyster	*Crassostrea gigas*		Mourton et al. 1992; Chen et al. 1993
Oyster	*Crassostrea virginica*	Amoebocytes, heart, mantle; other tissues	Perkins and Menzel 1964; Tripp et al. 1966; Stephens and Hetrick 1979a, 1979b; Ellis 1987; Ellis and Bishop 1989
Japanese scallop	*Mizuchopecten yessoensis*	Embryo	Odinstova and Khomenko 1991
Arthropods			
Crustaceans			Brody and Chang 1989; Chang and Brody 1989
Crab	*Carcinus maenas*	Y-organ	Toullec and Dauphin-Villemant 1994
Red swamp crayfish	*Procambarus clarkii*		Birmelin et al. 1999a, 1999b
Lobster		Lymphatic organ	Fisher-Piette 1931
Shrimp	*Penaeus* spp.	Hepatopancreas, hemocytes, embryo, hemolymph	Ellender et al. 1988; Ellender et al. 1992; Najafabadi et al. 1992
Shrimp	*Penaeus indicus*		Toullec et al. 1996
Shrimp	*Penaeus monodon* Fabrisus	Lymphoid organ	Hsu et al. 1995

Table A-3 *continued*

Common name	Scientific name	Tissue type[1]	Select references
Crustaceans *(cont'd)*			
Shrimp	*Penaeus orientalis* Kishinouye	Hepatopancreas	Hu 1990; Ke et al. 1990
Shrimp	*Penaeus penicillatus*		Chen and Kou 1989
Shrimp	*Penaeus stylirostris*	Lymphoid organ, nerves, ovaries	Leudman and Lightner 1992; Nadala et al. 1993; Taypay et al. 1995
Shrimp	*Penaeus vannamei*	Lymphoid organ, nerves, ovaries	Leudman and Lightner 1992; Ndala et al. 1993; Toullec et al. 1996
Hemichordatas			
Encrusting tunicate	*Botryllus schlosseri*	Embryo, hemocytes	Rinkevich and Rabinowitz 1993, 1994
Tunicate	*Styela clava*		Raftos et al. 1990; Raftos, Cooper et al. 1991; Raftos, Stillman, Cooper 1991

[1] Blank cell indicates whole-organism preparations

pods), mollusks (Bivalvia), sponges (Porifera), hydra (Cnidaria), and tunicates (Hemichordata).

Tissue and organ explants are another useful method to investigate mechanisms of action of potential EDCs. In insects, radiochemical assays for JH biosynthesis (Feyereisen 1985) and radioimmunoassays for ecdysone biosynthesis (King et al. 1974) by isolated glands are regularly used. King et al. (1974) first reported ecdysone biosynthesis by prothoracic glands in vitro and these methods have been modified for many different insects. Ovarian explants and Y-organ explants have been developed for crustaceans, echinoderms, and mollusks to assess the steroid-metabolizing pathways in these organs (deLongcamp et al. 1974; Chang and O'Connor 1978; Schoenmakers and Voogt 1980; James and Little 1984; Quackenbush 1989). Similarly, microsomes or tissue extracts can be used to evaluate the hydroxylase activity in previously exposed animals (James and Shiverick 1984; Jewell et al. 1997; Oberdörster et al. 1998b).

Lipovitellin production has been fairly well studied in insects and crustaceans. In insects, lipovitellin production can be controlled by either ecdysteroids or juvenile hormones. Antibodies against lipovitellin of several species of insects, crabs, shrimp, crayfish, and barnacles have been produced to study the reproductive cycle (Shafir et al. 1992; Hasegawa et al. 1993; Lee and Watson 1994; E. Oberdörster unpublished data). ELISAs have been developed for some crustaceans (Lee and Watson 1994). However, little is known about the utility of immunoassays to determine EDCs. Reproductive toxicants could modulate lipovitellin production in the blue crab (Lee and Noone 1995).

In both insects and crustaceans, hemolymph titers of ecdysteroids are measured by radioimmunoassay (Warren and Gilbert 1988). Titers of JHs are measured by coupled gas chromatography–mass spectrometry or by radioimmunoassay (Granger and Goodman 1988; Goodman et al. 1995). Both radioimmunoassays and ELISAs are used to measure insect and crustacean neuropeptides such as proctolin, crustacean hyperglycemic, and black pigment-dispersing hormones (Keller 1988) and the allatostatic neuropeptides (Audsley et al. 1999).

Enzymes that metabolize steroid hormones can be used specifically to determine effects on distribution, uptake, and elimination of steroids. For example, CYP 45 in lobster is an ecdysone hydroxylase, which can be modulated by phenobarbitol, heptachlor, and naphtoflavone (Snyder 1998). Development of an antibody for immunoassays would help in elucidating the effects of xenobiotics on this P450, although northern blotting of the induced RNA may be just as useful. CYP 2L in lobster has also been identified, but its role in ecdysone metabolism is limited (James et al. 1996). Antibodies to CYP 2L have been produced, although the application of an immunoassay for this P450 may be limited in evaluation of EDCs (Boyle and James 1996).

Whole-organism testosterone metabolism assays have been described for cladocerans and gastropod mollusks (Baldwin and LeBlanc 1994; Baldwin et al. 1997; Oberdörster et al. 1998b). These assays have the utility of analyzing several enzyme systems that metabolize steroids concurrently, including the transferases, hydroxylases, and reductases. Both the cladoceran and gastropod assays have been used to detect EDCs. Use of testosterone as a tool in assessing steroid metabolism has utility, although the role of vertebrate sex steroids is not well understood in invertebrates (Chapter 2).

Molecular techniques have been developed largely for the arthropods, primarily the insects. These include ecdysone receptor (EcR) binding assays, EcR reporter gene assays, ecdysteroid-responsive and JH-responsive cell lines, and tissue explants. Differential gene expression, which has been used to detect EDCs in vertebrates, only now is being developed for some invertebrate groups.

EcR binding assays are labor intensive but could be adapted to determine EDC mechanisms of toxic action in arthropods. Binding affinity of a xenobiotic to EcR/USP (ultraspiracle) can be tested using radiolabeled ligands ([3H]- or [125I]-ponasterone A). Either protein extracts from 20E-responsive tissues or insect cells or proteins induced in vitro from cloned cDNAs encoding the EcR and USP (Cherbas et al. 1988) can be used. Although they have not yet been used to identify EDCs *per se*, these assays have been used in the study of insect and plant chemical interactions, as well as in understanding insect physiology (Cherbas et al. 1988; Dinan et al. 1997). No et al. (1996) have developed a stable Chinese Hamster Ovary cell line that expresses the EcR/RXR (retinoic acid receptor) construct. This cell line is commercially available and has potential as a screen for EDCs. Studies on PAHs, PCBs, and phytosterols have been conducted using this system (Oberdörster, Clay, McLachlan 1999; Oberdörster, Wilmon et al. 1999). In addition, reporter gene assays in ecdysone-responsive insect cell lines have been used to investigate the effects of plant compounds (Cottam and Milner 1997, 1998; Dinan et al. 1997).

Acknowledgements: The Laboratory Testing Work Group would like to thank the following individuals for their thoughtful review of this chapter: Gary Ankley of the U.S. Environmental Protection Agency, Tracy Andacht of the University of Georgia, Richard Campbell of the University of California, Peter Campbell and Lisa Tattersfield of Zeneca Agrochemicals, Joel Coates of Iowa State University, Paul Sibley of the University of Guelph, J.O. Young of the University of Liverpool, and Peter deFur of Virginia Commonwealth University.

References

Adams JA, Haileselassie M. 1984. The effects of polychlorinated biphenols (Arochlors 1016 and 1254) on mortality, reproduction and regeneration in *Hydra oligactis*. *Arch Environ Contam Toxicol* 13:493–499.

Alikhan MA, Bagatto G, Zia S. 1990. The crayfish as a "biological indicator" of aquatic contamination by heavy metals. *Water Research* 24:1069–1076.

Anderson MB, Reddy P, Preslan JE, Fingerman M, Bollinger J, Jolibois L, Maheshwarudu G, George WJ. 1997. Metal accumulation in crayfish, *Procambarus clarkii*, exposed to a petroleum-contaminated bayou in Louisiana. *Ecotoxicol Environ Safety* 37:267–272.

Ankley G, Mihaich E, Stahl R, Tillitt D, Colborn T, McMaster S, Miller R, Bantle J, Campbell P, Denslow N, Dickerson R, Folmar L, Fry M, Giesy J, Gray LE, Guiney P, Hutchinson T, Kennedy S, Kramer V, LeBlanc G, Mayes M, Nimrod A, Patino R, Peterson R, Purdy R, Ringer R, Thomas P, Touart L, Van Der Kraak G, Zacharewski T. 1998. Overview of a workshop on screening methods for detecting potential (anti-) estrogenic/androgenic chemicals in wildlife. *Environ Toxicol Chem* 17:68–87.

[APHA] American Public Health Association, American Water Works Association, and Water Pollution Control Federation Standard. 1999. Methods for the examination of water and wastewater, 20th ed. Washington DC: APHA.

Artemia Reference Center. 1999. http://allserv.rug.ac.be/~booghe/index.html.

Ary RD Jr, Bartell CM, Poirrier MA. 1987. The effects of chelotomy on molting in the blue crab, Callinectes sapidus. *J Shellfish Res* 6:103–108.

[ASTM] American Society for Testing and Materials. 1999a. Standard test methods for measuring the toxicity of sediment-associated contaminants with freshwater invertebrates. In: Annual book of standards, Volume 11.05. Philadelphia PA: ASTM. E1706-95b.

[ASTM] American Society for Testing and Materials. 1999b. Guide for acute toxicity test with the rotifer *Brachionus*. In: Annual book of standards, Volume 11.05. Philadelphia PA: ASTM. E1440-91.

[ASTM] American Society for Testing and Materials. 1999c. Guide for conducting static acute toxicity tests starting with embryos of four species of saltwater bivalve mollusks. In: Annual book of standards, Volume 11.05. Philadelphia PA: ASTM. E724-94.

[ASTM] American Society for Testing and Materials. 1999d. Practice for conducting bioconcentration tests with fishes and saltwater bivalve mollusks. In: Annual book of standards, Volume 11.05. Philadelphia PA: ASTM. E1022-94.

[ASTM] American Society for Testing and Materials. 1999e. Standard guide for conducting sediment toxicity tests with marine and estuarine polychaetous annelids. In: Annual book of standards, Volume 11.05. Philadelphia PA: ASTM. E1611-94.

[ASTM] American Society for Testing and Materials. 1999f. Standard guide for conducting acute, chronic, and life-cycle aquatic toxicity tests with polychaetous annelids. In: Annual book of standards, Volume 11.05. Philadelphia PA: ASTM. E1562-94.

[ASTM] American Society for Testing and Materials. 1999g. Standard guide for determination of bioaccumulation of sediment-associated contaminants by benthic invertebrates. In: Annual book of standards, Volume 11.05. Philadelphia PA: ASTM. E1688-97a.

[ASTM] American Society for Testing and Materials. 1999h. Standard guide for conducting a laboratory soil toxicity tests with the lumbricid earthworm *Eisenia foetida*. In: Annual book of standards, Volume 11.05. Philadelphia PA: ASTM. E1676-97.

[ASTM] American Society for Testing and Materials. 1999i. Standard guide for conducting static sediment toxicity tests with marine and estuarine amphipods. In: Annual book of standards, Volume 11.05. Philadelphia PA: ASTM. E1367-92.

[ASTM] American Society for Testing and Materials. 1999j. Guide for conducting *Daphnia magna* life-cycle toxicity tests. In: Annual book of standards, Volume 11.05. Philadelphia PA: ASTM. E1193-97.

[ASTM] American Society for Testing and Materials. 1999k. Guide for conducting three-brood, renewal toxicity tests with *Ceriodaphnia dubia*. In: Annual book of standards, Volume 11.05. Philadelphia PA: ASTM. E1295-89 (1995).

[ASTM] American Society for Testing and Materials. 1999l. Guide for conducting life-cycle toxicity tests with saltwater mysids. In: Annual book of standards, Volume 11.05. Philadelphia PA: ASTM. E1191-97.

[ASTM] American Society for Testing and Materials. 1999m. Guide for conducting static and flow-through acute toxicity tests with mysids from the west coast of the United States. In: Annual book of standards, Volume 11.05. Philadelphia PA: ASTM. E1463-92.

[ASTM] American Society for Testing and Materials. 1999n. Guide for conducting acute toxicity tests on test materials with fishes, macroinvertebrates, and amphibians. In: Annual book of standards, Volume 11.05. Philadelphia PA: ASTM. E729-96.

[ASTM] American Society for Testing and Materials. 1999o. Guide for conducting acute toxicity tests on aqueous ambient samples and effluents with fishes, macroinvertebrates, and amphibians. In: Annual book of standards, Volume 11.05. Philadelphia PA: ASTM. E1192-97.

[ASTM] American Society for Testing and Materials. 1999p. Standard guide for conducting static acute toxicity tests with echinoid embryos. In: Annual book of standards, Volume 11.05. Philadelphia PA: ASTM. E1563-95.

Audsley N, Weaver RJ, Edwards JP. 1999. Enzyme-linked immunosorbent assay of *Manduca sexta* allatostatin (MAS-AS), isolation and measurement of MAS-AS immunoreactive peptide in *Lacanobia oleracea*. *Insect Biochem Mol Biol*: In press.

Auzoux S, Domart-Coulon I, Doumenc D. 1993. Gill cell cultures of the butterfish clam *Ruditapes decussatus*. *J Mar Biotechnol* 1:79–81.

Baldwin WS, Graham SE, Shea D, LeBlanc GA. 1997. Metabolic androgenization of female *Daphnia magna* by the xenoestrogen 4-nonylphenol. *Environ Toxicol Chem* 16:1905–1911.

Baldwin WS, Grahm SE, Shea D, LeBlanc GA. 1998. Altered metabolic elimination of testosterone and associated toxicity following exposure of *Daphnia magna* to nonylphenol polyethoxylate. *Ecotox Env Safety* 39:104–111.

Baldwin WS, LeBlanc GA. 1994. Identification of multiple steroid hydroxylases in *Daphnia magna* and their modulation by xenobiotics. *Environ Toxicol Chem* 13:1013–1021.

Barki A, Levi T, Hulata G, Karplus I. 1997. Annual cycle of spawning and molting in the red-claw crayfish, *Cherax quadricarinatus*, under laboratory conditions. *Aquaculture* 157:239–249.

Batel R, Bihari N, Rinkevich B, Dapper J, Schaecke H, Schroeder HC, Mueller WEG. 1993. Modulation of organotin-induced apoptosis by the water pollutant methyl mercury in a human lymphoblastoid tumor cell line and a marine sponge. *Mar Ecol Prog Ser* 93:245–251.

Bay SM, Burgess DR, Nacci D. 1993. Status and application of echinoid (Phylum Echinodermata) toxicity test methods. In: Landis WG, Hughes JJ, Lewis MA, editors. Environmental toxicology and risk assessment. Philadelphia PA: ASTM. STP 1179. p 281–302.

Bay SM, Greenstein D. 1994. Sediment toxicity test methods for the brittle star, *Amphipodia urtica*. In: Cross JN, Francisco C, Hallock D, editors. Southern California coastal waters research project, Annual Report 1992–1993. Westminster CA: Southern California Coastal Waters Research Project. p 130–135.

Beach MJ, Pascoe, D. 1998. The role of *Hydra vulgaris* (Pallas) in assessing the toxicity of freshwater pollutants. *Water Research* 32:101–106.

Benoit DA, Sibley PK, Jeunemann JJ, Ankley GT. 1997. *Chironomus tentans* life-cycle test: Design and evaluation for use in assessing toxicity of contaminated sediments. *Environ Toxicol Chem* 16:1165–1176.

Berg ABV. 1979. Reproductive behavior of *Palaemonetes pugio* Holthuis (Decapoda, Caridea) [M.S. Thesis]. Charleston SC: The College of Charleston.

Berger B, Dallinger R, Felder E, Moser J. 1993. Budgeting the flow of cadmium and zinc through the terrestrial gastropod, *Helix pomatia* L. In: Dallinger R, Rainbow PS, editors. Ecotoxicology of metals in invertebrates. Boca Raton FL: Lewis. p 291–313.

Berkson, J. 1944. Application of the logistic function to bioassay. *J American Stat Soc* 39:357–365.

Billinghurst Z, Clare AS, Fileman T, Mcevoy J, Readman J, Depledge MH. 1998a. Inhibition of barnacle settlement by the environmental oestrogen 4-nonylphenol and the natural oestrogen 17b-oestradiol. *Mar Poll Bull* 36:833–839.

Billinghurst Z, Clare AS, Fileman T, McEvoy J, Readman J, Depledge MH. 1998b. Settlement of cypris larvae of *Balanus amphitrite* is inhibited by the environmental oestrogen, 4-nonylphenol and the naturally occurring oestrogen 17ß-oestradiol. *J Mar Pollut*: In press.

[BBA] Biologischen Bundesanstalt fur Land- und Forstwirtschaft. 1995. Long-term toxicity test with *Chironomus riparius*: Development and validation of a new test system. Streloke M, Kopp H, editors. Berlin, Germany: Blackwell WissenschaftsVerlag Gmbh.

Birmelin C, Escartin E, Goldfarb PS, Livingstone DR, Porte C. 1999a. Enzyme effects and metabolism of fenitrothion in primary cell culture of the Red Swamp Crayfish; *Procambarus clarkii. Mar Env Res*: In press.

Birmelin C, Escartin E, Goldfarb PS, Livingstone DR, Porte C. 1999b. Enzyme effects and metabolism of fenitrothion in primary cell culture of the Red Swamp Crayfish; *Procambarus clarkii. Mar Env Res*: In press.

Birmelin C, Mitchelmore CL, Goldfarb PS, Livingstone DR. 1997. Characterisation of biotransformation enzyme activities and DNA integrity in isolated cells of digestive gland of the common mussel *Mytilus edulis* L. *Comp Biochem Physiol:* In press.

Blin JC. 1997. Culturing the purple sea urchin, *Paracentrotus lividus*, in a recirculation system. In: Parson GT, editor. Proceedings of the Sea Urchin Culture Workshop; 1997 March. *Bull Aquacult Assoc Can.* Sachville, NB, Canada: Tribune Pr. p 8–13.

Bliss CI. 1935. The calculation of the dosage-mortality curve. *Ann Appl Biol* 22:134–167.

Bond C, Harris AK. 1988. Locomotion of sponges and its physical mechanism. *J Exp Zool* 246:271–284.

Bookhout CG, Costlow JD Jr, Monroe R. 1976. Effects of methoxychlor on larval development of the mud crab and blue crab. *Water Air Soil Poll* 5:349–365.

Bookhout CG, Costlow JD Jr, Monroe R. 1980. Kepone effects on larval development of mud-crab and blue-crab. *Water Air Soil Poll* 13:57–77.

Bookhout CG, Costlow JD Jr. 1975. Effects of Mirex on the larval development of blue crab. *Water Air Soil* Poll 4:113–126.

Bookhout CG, Costlow DJ Jr. 1974. Crab development and effects of pollutants. *Thalassia Jugoslavica* 10:77–87.

Bookhout CG, Monroe RJ, Forward RB Jr, Costlow JD Jr. 1984. Effects of hexavalent chromium on development of crabs, *Rhithropanopeus harrisii* and *Callinectes sapidus*. *Water Air Soil Pollut* 21:199–216.

Boyle SM, James MO. 1996. Cross-reactivity of an antibody to spiny lobster P450 2L with microsomes from other species. *Mar Env Res* 42:1–6.

Bridges TS, Dillon TM, Moore DW. 1993. The use of demographic modeling to assess sediment toxicity with the polychaete *Neanthes arenaceodentata* (Abstract). Society of Environmental Toxicology and Chemistry (SETAC) 14th Annual Meeting; Houston TX. Pensacola FL: SETAC.

Bridges TS, Farrar JD. 1997. The influence of worm age, duration of exposure and endpoint selection on bioassay sensitivity for *Neanthes arenaceodentata* (Annelida: Polychaeta). *Environ Toxicol Chem* 16:1650–1658.

Broad AC. 1957. The relationship between diet and larval development of *Palaemonetes*. *Biol Bull* 112:162–170.

Brody MD, Chang ES. 1989. Development and utilization of crustacean long-term primary cell cultures: Ecdysteroid effects in vitro. *Invertebrate Repro Dev* 16:141–147.

Brooks WR, Ceperly L, Rittschof D. 1995. Disturbance and reattachment behavior of sea anemones *Calliactis tricolor* (Le Sueur): Temporal, textural and chemical mediation. *J Chem Ecol* 21:1–12.

Brouwer M, Schlenk D, Ringwood AH, Brouwer-Hoexum T. 1992. Metal-specific induction of metallothionein isoforms in the blue crab *Callinectes sapidus* in response to single- and mixed-metal exposure. *Arch Biochem Biophys* 294:461–468.

Brown AF, Pascoe D. 1988. Studies on the acute toxicity of pollutants to freshwater macroinvertebrates. The acute toxicity of cadmium to twelve species of predatory macroinvertebrate. *Arch. Hydrobiol* 114:311–319.

Bruce RD Versteeg DJ. 1992. A statistical procedure for modeling continuous toxicity data. *Environ Toxicol Chem* 11:1485–1494.

Bryan PJ, Rittschof D, Qian YP. 1997. Settlement inhibition of bryozoan larvae by bacterial films and aqueous leachates. *Bull Mar Sci* 61:849–857.

Burns BG, Sanagalang GB, Freeman HC, McMenemy M. 1984. Isolation and identification of testosterone from serum and testes of the American lobster (*Homarus americanus*). *Gen Comp Endocrinol* 54:429–436.

Burton GA, Ingersoll CG. 1994. Evaluating the toxicity of sediments. Chapter 7. In: Fox RA, editor. Assessment of contaminated Great Lakes sediment. Chicago IL: USEPA. EPA/905-B94/002.

Burton GA Jr, Nelson MK, Ingersoll CG. 1992. Freshwater benthic toxicity tests. In: Burton GA Jr, editor. Sediment toxicity assessment. Boca Raton FL: Lewis. p 213–240.

Byard EH, Aiken DE. 1984. The relationship between molting, reproduction, and hemolymph female-specific protein in lobster *Homarus americanus*. *Comp Biochem Physiol* 77A:749–757.

Cairns J. 1983. Are single species toxicity tests alone adequate for estimating environmental hazard? *Hydrobiologia* 100:47–57.

Campbell RD. 1989. Taxonomy of the European *Hydra* (Cnidaria: Hydrozoa): A re-examination of its history with emphasis on the species *H. vulgaris* Pallas, *H. attenuata* Pallas and *H. circumcincta* Schultz. *Zool J Lin Soc* 95:219–244.

Candida Carnevali MD, Bonasaro F, Patruno, Thorndyke MC. 1998. Cellular and molecular mechanisms of arm regeneration in crinoid echinoderms: The potential of arm explants. *Dev Genes Evol* 208:421–430.

Canfield TJ, Dwyer FJ, Fairchild JF, Haverland PS, Ingersoll CG, Kemble NE, Mount DR, La Point TW, Burton GA, Swift MC. 1996. Assessing contamination in Great Lakes sediments using benthic invertebrate communities and the sediment quality triad approach. *J Great Lakes Res* 22:565–583.

Canfield TJ, Brunson EL, Dwyer FJ, Ingersoll CG, Kemble NE. 1998. Assessing sediments from the upper Mississippi river navigational pools using a benthic invertebrate community evaluation and the sediment quality triad approach. *Arch Environ Contam Toxicol* 35:202–212.

Cardwell RD, Brancato MS, Toll T, DeForest D, Tear L. 1999. Aquatic ecological risks posed by tributyltin in United States surface waters: Pre-1989 to 1996 Data. *Environ Toxicol Chem* 18:567–577.

Carr RS. 1998. Marine and estuarine porewater toxicity testing. In: Wells PG, Lee K, Blaise C, editors. Microscale aquatic toxicology - Advances, techniques and practice. Boca Raton FL: CRC Lewis. p 523-538.

Carr RS, Williams JW, Fragata CTB. 1989. Development and evaluation of a novel marine sediment pore water toxicity test with the polychaete *Dinophilus diadema*. *Environ Toxicol Chem* 8:533–543.

Casellato S, Aiello R, Negrisolo PA, Seno M. 1992. Long-term experiment on *Branchiura sowerbyi* Beddard (Oligocheata, Tubificidae) using sediment treated with LAS (linear alkyl benzene sulphonate). *Hydrobiologia* 232:169–173.

Casillas E, Weber D, Haley C, Sol S. 1992. Comparison of growth and mortality in juvenile sand dollars (*Dendraster excentricus*) as indicators of contaminated marine sediments. *Environ Toxicol Chem* 8:559–569.

Caswell H. 1996. Demography meets ecotoxicology: Integrating the population level effects of toxic substances. In: Newman MC, Jagoe CH, editors. Ecotoxicology: A hierarchical treatment. Boca Raton FL: Lewis.

Celestial DM, McKenney CL Jr. 1994. The influence of an insect growth regulator on the larval development of the mud crab, *Rhithropanopeus harrisii*. *Environ Pollut* 85:169–173.

Chandler GT, Green AS. 1996. A 14-day harpaticoid copepod reproduction bioassay for laboratory and field contaminated muddy sediment. In: Ostrander GK, editor. Techniques in aquatic toxicology. Boca Raton FL: CRC. p 23–39.

Chandler GT, Scott GI. 1991. Effects of sediment-bound endosulfan on survival, reproduction and larval settlement of meiobenthic polychaetes and copepods. *Environ Toxicol Chem* 10:375–382.

Chang ES, Brody MD. 1989. Crustacean organ culture. *Advances in Cell Culture* 7:19–86.

Chang ES, O'Connor JD. 1978. In vitro secretion and hydroxylation of a-ecdysone as a function of the crustacean molt cycle. *Gen Comp Endocrinol* 36:151–160.

Chapman PF, Crane M, Wiles J, Noppert F, McIndoe E. 1996. Improving the quality of statistics in regulatory ecotoxicity tests. *Ecotoxicology* 5:169–186.

Chappie DJ, Burton GA Jr. 1997. Optimization of in situ bioassays with *Hyalella azteca* and *Chironomus tentans*. *Environ Toxicol Chem* 16:559–564.

Charmantier G, Charmantier-Daures M, Aiken DE. 1988. Larval development and metamorphosis of the American Lobster *Homarus americanus* (Crustacea, Decapoda): Effect of eyestalk ablation and juvenile hormone injection. *General Comparative Endocrinology* 70:319–333.

Chen F, Schüürmann G. 1997. Quantitative structure -activity relationships in environmental science - VII. Pensacola FL: Society of Environmental Toxicology and Chemistry (SETAC).

Chen SN, Kou GH. 1989. Infection of cultured cells from the lymphoid organ of *Penaeus monodon* Fabriscus by monodon-type baculovirus (MBV). *J Fish Dis* 12:73–76.

Chen SN, Wen CM, Kou GH. 1993. Establishment of cell lines of the pacific oyster. *In vitro Cell Dev Biol* 29A:901–903.

Cherbas P, Cherbas L, Lee SS, Nakanishi K. 1988. 26-[125I]Iodoponasterone A is a potent ecdysone and a sensitive radioligand for ecdysone receptors. *Proceedings of the National Academy of Science USA* 85:2096–2100.

Cherbas L, Koehler MM, Cherbas P. 1989. Effects of juvenile hormone on the ecdysone response of *Drosophila* Kc cells. *Dev Gen* 10:177–188

Christiansen ME, Costlow JD Jr, Monroe R. 1977. Effects of juvenile hormone mimic ZR-515 (Altosid) on larval development of the mud crab *Rhithropanopeus harrisii* in various salinities and cyclic temperatures. *Mar Biol* 39:269–279.

Christiansen ME. 1988. Hormonal processes in decapod crustacean larvae. *Zoological Symposium* 59:47–68.

Christiansen ME, Costlow JD Jr. 1982. Ultrastructural study of the exoskeleton of the estuarine crab *Rhithropanopeus harrisii*: Effect of the insect growth regulator Dimilin® (Diflubenzuron) on the formation of the larval cuticle. *Mar Biol* 66:217–226.

Chu KH, Wong CK, Chiu, KC. 1997. Effects of the insect growth regulator (S)-methoprene on survival and reproduction of the freshwater cladoceran *Moina macrocopa*. *Environ Pollut* 96:173–178.

Ciarelli S, Vonck WAPMA, Van Straalen NM, Stronkhorst J. 1998. Ecotoxicity assessment of contaminated dredged material with the marine amphipod *Corophium volutator*. *Arch Environ Contam Toxicol* 34:350–356.

Clare AS, Costlow JD Jr. 1989. Effects of the Insecticide methomyl on development and regeneration in the megalopa and juveniles of the mud crab, *Rhitropanopeus harrisii* (Gould). Proceedings of a National Research Conference, Virginia Water Resources Research Center, Virginia Polytechnic Institute and State University.

Clare AS, Costlow JD, Bedair HM. 1992. Assessment of crab limb regeneration as an assay for developmental toxicity. *Can J Fish Aquat Sci* 49:1268–1273.

Clare AS, Rittschof D, Costlow JD Jr. 1992. Effects of the nonsteroidal ecdysone mimic RH 5849 on larval crustaceans. *J Exp Zool* 262:436–440.

Clark JR, Goodman LR, Borthwick PW, Patrick JM, Moore JC, Lores EM. 1986. Field and laboratory toxicity tests with shrimp, mysids, and sheepshead minnows exposed to fenthion. In: Poston TM, Purdy R, editors. Volume 9, Aquatic toxicology and environmental fate. Philadelphia PA: ASTM. STP 921. p 161–176.

Clarke KR, Warwick RM. 1994. Change in marine communities: An approach to statistical analysis and interpretation. Plymouth UK: Plymouth Marine Laboratory.

Conklin PJ, Rao RK. 1982. Effects of two dithicarbamates on the grass shrimp, *Palaemonetes pugio*: Molt-related toxicity and inhibition of limb regeneration. *Arch Environ Contam Toxicol* 11:431–435.

Conner JW, Singer SC. 1981. Purification scheme for cytochrome P450 of blue crab *Callinectes sapidus* Rathburn. *Aquat Toxicol* 1:271–278.

Conrad A, Fleming R, Crane M. 1999. Laboratory and field response of *Chironomus riparius* to a pyrethroid insecticide. *Wat Res* 33:1603–1610.

Contini H. 1994. Growth rate and propagation of the Red Sea sponge *Latrunculia magnifica*: A study in situ and in vitro. *31st Meeting of the Zoo Soc of Israel* 41:82–83.

Costlow JD Jr. 1968. Metamorphosis in crustaceans. In: Etkin W, Gilbert LI, editors. Metamorphosis: A problem in developmental biology. New York: Meredith Corp. p 3–41.

Costlow JD Jr. 1977. The effect of juvenile hormone mimics on development of the mud crab, *Rhithropanopeus harrisii* (Gould). In: Vernberg FJ, Calabrese A, Thurberg FP, Vernberg WB, editors. Physiological responses of marine biota to pollutants. New York: Academic Pr. p 439–457.

Costlow JD Jr, Bookhout CG, Monroe R. 1966. Studies on the larval development of the crab, *Rhithropanopeus harrisii* (Gould). I. The effect of salinity and temperature on larval development. *Physiological Zool* 39:81–100.

Cottam DM, Milner MJ. 1997. The effects of several ecdysteroids and ecdysteroid agonists on two Drosophila imaginal disc cell lines. *Cellular and Molecular Life Sci* 53:600–603.

Cottam DM, Milner MJ. 1998. The effect of juvenile hormone on the response of the *Drosophila* imaginal disc cell line C18+ to molting hormone. *J Insect Physiol* 44:1137–1144.

Cowan CE, Versteeg DJ, Larson RJ, Kloepper-Sams P. 1995. Integrated approach for environmental assessment of new and existing substances. *Regulatory Toxicol Pharmac* 21:3–31.

Crane M. 1997. Research needs for predictive multispecies tests in aquatic toxicology. *Hydrobiologia* 346:149–155.

Crane M, Delaney P, Parker P, Walker C, Watson S. 1995. The effect of Malathion 60 on *Gammarus pulex* (L.) below watercress beds. *Environ Toxicol Chem* 14:1181–1188.

Crane M, Flower T, Holmes D, Watson S. 1992. The toxicity of selenium in experimental freshwater ponds. *Arch Environ Contam Toxicol* 48:63–69.

Crane M, Maltby L. 1991. The lethal and sublethal responses of *Gammarus pulex* to stress: Sensitivity and variation in an in situ bioassay. *Environ Toxicol Chem* 10:1331–1339.

Crossland NO. 1990. The role of mesocosm studies in pesticide registration. Proceedings of the Brighton Crop Protection Conference 1990:499–508.

Cunningham PA, Wilson JH, Evans DW, Costlow JD Jr. 1987. Effects of sediment on the persistence and toxicity of diflubenzuron (Dimilin) in estuarine waters: A laboratory evaluation using larvae of two estuarine crustaceans. In: Vernberg WB, Calabrese A, Thurberg FP, Vernberg FJ, editors. Pollution physiology of estuarine organisms. Columbia SC: Univ of South Carolina Pr. p 299–331.

Dallinger R, Berger B, Graber A. 1993. Quantitative aspects of zinc and cadmium binding in *Helix pomatia:* Differences between essential and nonessential trace elements. In: Dallinger R, Rainbow PS, editors. Ecotoxicology of metals in invertebrates. Boca Raton FL: Lewis. p 315–332.

D'Aniello A, Cosmo AD, Di Cristo C, Assisi L, Botte V, Di Fiore MM. 1996. Occurrence of sex steroid hormones and their binding proteins in *Octopus vulgaris* lam. *Biochem Biophys Res Commun* 227:782–8 (published erratum in *Biochem Biophys Res Commun* 229:361).

D'Andrea AF, Stancyk SE, Chandler GT. 1996. Sublethal effects of cadmium on arm regeneration in the burrowing brittlestar, *Microphiopholis gracillima*. *Ecotoxicology* 5:115–133.

Daniels WH, Abramo LR, Graves KF. 1994. Ovarian development of female red swamp crayfish (*Procambarus clarkii*) as influenced by temperature and photoperiod. *J Crustacean Biol* 14:530–537.

De Lisle PF, Roberts MH Jr. 1987. Osmoregulation in the estuarine mysid, *Mysidopsis bahia* Molenock: comparison with other mysid species. *Comp Biochem Physiol* 88A:369–372.

deLongcamp D, Lubet P, Drosdowsky M. 1974. The in vitro biosynthesis of steroids by the gonad of the mussel (*Mytilus edulis*). *Gen Comp Endocrinol.* 22:116–127.

DeRusha RH, Forsythe JW, DiMarco FP, Hanlon RT. 1989. Alternative diets for maintaining and rearing cephalopods in captivity. *Lab Anim Sci* 4:306–312.

Deutsch U, Brick M. 1993. Morphological effects of tributyltin (TBT) in vitro on the genital system of the mesogastropoda *Littorina littorea* (Prosobranchia). *Helgol Meeresunters* 47:49–60.

Devi M, Fingerman M. 1995. Inhibition of acetylcholinesterase activity in the central nervous system of the red swamp crayfish, *Procambarus clarkii*, by mercury, cadmium, and lead. *Bull Environ Contam Toxicol* 55:746–750.

DeWitt TH, Glover SA, Emlen JM. 1997. Using multi-generation exposure experiments with amphipods to test predictions of chronic toxicity tests and population models (abstract). Society of Environmental Toxicology and Chemistry (SETAC) 18th Annual Meeting; 1997. San Francisco CA. Pensacola FL: SETAC.

DeWitt TH, Hickey CW, Morrisey DJ, Nipper MG, Roper DS, Williamson RB, Van Dam L, Williams EK. 1999. Do amphipods have the same concentration-response to contaminated sediments in situ as in vitro? *Environ Toxicol Chem* 18:1026–1039.

Di Cosmo A, Paolucci M, Di Cristo C, Botte V, Ciarcia G. 1998. Progesterone receptor in the reproductive system of the female of *Octopus vulgaris*: Characterization and immunolocalization. *Mol Reprod Dev* 50:451–60.

Diamond DW, Scott LK, Forward RB Jr, Kirby-Smith W. 1989. Respiration and osmoregulation of the estuarine crab, *Rhithropanopeus harrisii* (Gould): Effects of the herbicide, alachlor. *Comp Biochem Physiol* 93A:313–3118.

Dinan L, Whiting P, Girault JP, Lafont R, Dhadialla TS, Cress DE, Mugat B, Antoniewski C, Lepesant JA. 1997. Cucurbitacins are insect steroid hormone agonists acting at the ecdysteroid receptor. *Biochem J* 327:643–650.

Dodson SI, Merritt CM, Shannahan JP, Shults CM. 1999. Low exposure concentrations of atrazine increase male production in *Daphnia pulicaria*. *Environ Toxicol Chem*: In press.

Donkin SG, Williams PL. 1995. Influence of developmental stage, salts, and food presence on various endpoints using *Caenorhabditis elgans* for aquatic toxicity testing. *Environ Toxicol Chem* 14:2139–2147.

Downer RGH, Laufer H. 1983. Endocrinology of insects. New York: Alan R. Liss.

Drobne D. 1997. Terrestrial isopods - A good choice for toxicity testing of pollutants in the terrestrial environment. *Environ Toxicol Chem* 16:1159–1164.

Drobne D, Hopkin SP.1994. Ecotoxicological laboratory test for assessing the effects of chemicals on terrestrial isopods. *Bull Environ Contam Toxicol* 53:390–397.

Dumont HJ, Munuswamy N. 1997. The potential of freshwater Anostraca for technical applications *Hydrobiologia* 358:193–197.

Eckwert H, Kohler HR. 1997. The indicative value of the hsp70 stress response as a marker for metal effects in *Oniscus asellus* (Isopoda) field populations: Variability between populations from metal-polluted and uncontaminated sites. *Appl Soil Ecol* 6:275–282.

Edwards CA, Dennis EB, Empson DW. 1967. Pesticides and the soil fauna: Effects of aldrin and DDT in an arable field. *Annals Appl Biol* 60:11–22.

Ellender RD, Middlebrooks BL, McGuire SL. 1988. Evaluation of various growth enhancement factors, media formulation, and support matrices for the development of primary and established cell lines from *Penaeus hepatopancreas*. U.S. shrimp farming program progress report, Volume 1. Washington DC: U.S. Department of Agriculture.

Ellender RD, Najafabadi AK, Middlebrooks BL. 1992. Observations on the primary cell culture of hemocytes of *Penaeus*. *J Crustacean Biol* 12:178–185.

Ellis LL. 1987. Mitotic activity of oyster cell cultures supplemented with embryo extract and lipososmes. *In Vitro Cell Dev Biol* 23:39A.

Ellis LL, Bishop SH. 1989. Isolation of cell lines with limited growth potential from marine bivalves. In: Mitsuhashi J, editor. Invertebrate cell system applications. Boca Raton FL: CRC. p 244–251.

Emery VL Jr, Moore DW, Gray BR, Duke BM, Gibson AB, Wright RB, Farrar JD. 1997. Development of a chronic sublethal sediment bioassay using the estuarine amphipod *Leptocheirus plumulosus* (Shoemaker). *Environ Toxicol Chem* 16:1912–1920.

Engel DW, Eggert LD. 1974. The effect of salinity and sex on the respiration rates of excised gills of the blue crab, *Callinectes sapidus*. *Comp Biochem Physiol* 47A:1005–1011.

Environment Canada. 1990a. Biological test method: Reference method for determining acute lethality of effluents to *Daphnia magna*. Ottawa, Ontario: Environment Canada. Technical report EPS 1/RM/14.

Environment Canada. 1990b. Biological test method: Acute lethality test using *Daphnia* spp. Ottawa, Ontario: Environment Canada. Technical report EPS 1/RM/11.

Environment Canada. 1992a. Biological test method: Acute test for sediment toxicity testing using marine or estuarine amphipods. Ottawa, Ontario: Environment Canada. Technical report EPS 1/RM/26.

Environment Canada. 1992b. Biological test method: Test of reproduction and survival using the cladoceran *Ceriodaphnia dubia* Ottawa, Ontario: Environment Canada. Technical report EPS 1/RM/21.

Environment Canada. 1992c. Biological test method: Fertilisation assay using echinoids (sea urchins and sand dollars). Ottawa, Ontario: Environment Canada. Technical report EPS 1/RM/27.

Environment Canada. 1997a. Biological test method: Test for growth and survival in sediment using the freshwater amphipod *Hyalella azteca*. Ottawa, Ontario: Environment Canada. Technical report EPS 1/RM/33.

Environment Canada. 1997b. Biological test method: Test for growth and survival in sediment using larvae of freshwater midges (*Chironomus tentans* or *Chironomus riparius*). Ottawa, Ontario: Environment Canada. Technical report EPS 1/RM/32.

Fábián M, Petersen H. 1994. Short-term effects of the insecticide dimethoate on the activity and spatial distribution of a soil inhabiting collembolan *Folsomia fimetaria* L. (Collembola: Isotomidae). *Pedobiologia* 38:289–302.

Fent K. 1996. Ecotoxicology of organotin compounds. *Crit Rev Toxicol* 26:1–117.

Feyereisen R. 1985. Radiochemical assay for juvenile hormone III biosynthesis in vitro. *Meth Enzomol* 111B:530–539.

Fingerman M. 1995. Endocrine mechanisms in crayfish, with emphasis on reproduction and neurotransmitter regulation of hormone release. *American Zoologist* 35:68–78.

Fingerman M, Devi M, Reddy PS, Katyayani R. 1996. Impact of heavy metal exposure on the nervous system and endocrine-mediated processes in crustaceans. *Zoological Studies* 35:1–8.

Fingerman M, Jackson NC, Nagabhushanam R. 1998. Hormonally-regulated functions in crustaceans as biomarkers of environmental pollution. *Comp Biochem Physiol* 120:343–350.

Fiorito G, Agnisola C, d'Addio M, Valanzano A, Calamandrei G. 1998. Scopolamine impairs memory recall in *Octopus vulgaris*. *Neurosci Lett* 253:87–90.

Fioroni P, Oehlmann J, and Stroben, E. 1991. The pseudohermaphroditism of Prosobranchs; Morphological aspects. *Zool Anz* 226:1–26.

Fischer-Piette E. 1931. Culture de tissus de crustaces: la gland lymphatique du homard. *Archives de zoologie Experimentale et Generale* 74:33–52.

Flemer DA, Stanley RS, Ruth BF, Bundrick CM, Moody PH, Moore JC. 1995. Recolonization of estuarine organisms: Effects of microcosm size and pesticides. *Hydrobiologia* 304:85–101.

Fornstrom CB, Landrum PF, Weisskopf CP, La Point TW. 1997. Effects of terbufos on juvenile red swamp crayfish (*Procambarus clarkii*): Differential routes of exposure. *Environ Toxicol Chem* 16:2514–2520.

Forward RB Jr, Costlow JD Jr. 1978. Sublethal effects of insect growth regulators upon crab larval behavior. *Wat Air Soil Pollut* 9:227–238.

Forward RB Jr, Costlow JD. 1978. Sublethal effects of insect growth regulators upon crab larval behavior. *Wat Air Soil Poll* 9:227–238.

Forward RB, Devries MC, Rittschof D, Frankel DAZ, Bischoff JP, Fisher CM, Welch JM. 1996. Effects of environmental cues on metamorphosis of the blue crab *Callinectes sapidus*. *Marine Ecology-Progress Series* 131:165–177.

Forward RB Jr, Lohmann KJ. 1983. Control of egg hatching in the crab *Rhithropanopeus harrisii* (Gould). *Biol Bull* 165:154–166.

Forward RB, Tankersley RA, Devries MC, Rittschof D. 1995. Sensory physiology and behavior of blue crab (*Callinectes sapidus*) postlarvae during horizontal transport. *Marine Behavior Physiology* 26:233–248.

Frampton GK. 1997. The potential of Collembola as indicators of pesticide usage: Evidence and methods from the UK arable ecosystem. *Pedobiologia* 41:179–184.

Frank U, Rabinowitz C, Rinkevich B. 1994. In vitro establishment of continuous cell cultures and cell lines from ten colonial cnidarians. *Mar Biol* 120:491–499.

Freeman JA. 1983. Spine regeneration in larvae of the crab, *Rhithropanopeus harrisii*. *J Exp Zool* 225:443–448.

Freeman JA, Bartell CL. 1975. Characterization of the molt cycle and its hormonal control in *Palaemonetes pugio* (Decapoda, Caridea). *Gen Comp Endocrin* 25:517–528.

Fremling CR, Mauck WL. 1980. Methods for using nymphs of burrowing mayflies (Ephemeroptera, *Hexagenia*) as toxicity organisms. In: Buikema AL, Cairns J Jr, editors. Aquatic invertebrate bioassays. Philadelphia PA: ASTM. STP 715. p 81–87.

Friesen MK. 1979. Use of eggs of the burrowing mayfly *Hexagenia rigida* in toxicity testing. In: Scherer E, editor. Toxicity tests for freshwater organisms. *Can Spec Publ Fish Aquat Sci* 44:27.

Fukunaga I, Takamizawa K, Inoue Z, Hasebe T, Konae EM, Hatano K, Mori S. 1992. Appearance of plankton and its correlation with water quality in the stabilization ponds at a sea-based dredged sludge disposal site. *Environ Technol* 13:449–460.

Gallardo WG, Hagiwara A, Tomita Y, Soyano K, Snell TW. 1997. Effect of some vertebrate and invertebrate hormones on the population growth, mictic female production, and body size of the marine rotifer *Brachionus plicatilis* Müller. *Hydrobiologia* 358:113–120.

Gardiner DB, Turner FS, Myers JM, Dietz TH, Silverman H. 1991. Long term culture of fresh water mussel gill strips: Use of serotonine to affect aseptic conditions. *Biol Bull* 181:175–180.

Giesy JP, Graney RL, Newsted JL, Rosiu CJ, Benda A, Kreis RG, Horvath FJ. 1988. Comparison of three sediment bioassay methods using Detroit River sediments. *Environ Toxicol Chem* 7:483–498.

Glas PS, Courtney LS, Rayburn JR, Fisher WS. 1997. Embryonic coat of the grass shrimp *Palaemonetes pugio*. *Biol Bull* 192:231–242.

Glennon JP. 1982. Surrogate species workshop report. Washington DC: USEPA TR-507-36B.

Gomez ED, Faulkner DJ, Newman WA, Ireland C. 1973. Juvenile hormone mimics: Effect on cirriped crustacean metamorphosis. *Science* 179:813–814.

Goodman WG, Orth AP, Toong YC, Ebersohl R, Hiruma K, Granger NA. 1995. *Arch Insect Biochem Physiol* 30:295–306.

Goy JW, Morgan SG, Costlow JD Jr. 1985. Studies on the reproductive biology of the mud crab, *Rhithropanopeus harrisii* (Gould): Induction of spawning during the non-breeding season (Decapoda, Brachyura). *Crustaceana* 49:83–87.

Graney RL, Giesy JP, Clark JR. 1995. Chapter 9: Field Studies. In: Rand GM, editor. Fundamentals of aquatic toxicology, 2nd ed. Washington DC: Taylor and Francis.

Granger NA, Goodman WG. 1988. Radioimmunoassay: Juvenile hormones. In: Gilbert LI, Miller TA, editors. Immunological techniques in insect biology. New York: Springer-Verlag. p 215–251.

Grassi M, Tardent R, Tardent P. 1995. Quantitative data about gametogenesis and embryonic development in *Hydra vulgaris* Pall. (Cnidaria, Hydrozoa). *Invertebrate Reproduction and Development* 27:219–232.

Green AS, Chandler GT, Piegorsch WW. 1996. Life-stage–specific toxicity of sediment-associated chlorpyrifos to a marine, infaunal copepod. *Environ Toxicol Chem* 15:1182–1188.

Halbach U. 1984. Population dynamics of rotifers and its consequences for ecotoxicology. *Hydrobiologia* 109:79–96.

Ham L, Quinn R, Pascoe D. 1995. Effects of cadmium on the predator–prey interaction between the turbellarian *Dendrocoelum lacteum* (Muller, 1774) and the isopod crustacean *Asellus aquaticus* (L.). *Arch Environ Contam Toxicol* 29:358–365.

Hanazato T, Dodson SI. 1995. Synergistic effects of low oxygen concentration, predator kairomone, and a pesticide on the cladoceran *Daphnia pulex*. *Limnol Oceanogr* 40:700–709.

Hanlon RT, Bidwell JP, Tait R. 1989. Strontium is required for statolith development and thus normal swimming behavior of hatchling cephalopods. *J Exp Biol* 141:187–95.

Hasegawa Y, Hirose E, Katakura Y. 1993. Hormonal control of sexual differentiation and reproduction in Crustacea. *Amer Zool* 33:403–411.

Hassan SA. 1985. Standard methods to test the side-effects of pesticides on natural enemies of insects and mites developed by the IOBC/WPRS Working Group Pesticides and Beneficial Organisms. *EPPO Bulletin* 15:214-255.

Heber MA, Norberg-King TJ. 1996. United States Environmental Protection Agency's water quality based approach to toxics control. In: Tapp JF, Hunt SM, Wharfe JR, editors. Toxic impacts of wastes on the aquatic environment. Cambridge, UK: Royal Society of Chemistry. p 175–187.

Hoffman AA, Parsons PA. 1991. Evolutionary genetics and environmental stress. Oxford UK: Oxford Pr.

Holmes P, Harrison P, Bergman A, Brandt I, Brouwer B, Keiding N, Randall G, Sharpe R, Skakkebaek N, Ashby J, Barlow S, Dickerson R, Humfrey C, Smith LM. 1997. European workshop on the impact of endocrine disrupters on human health and wildlife. Proceedings of a workshop; 2–4 December 1996. Weybridge UK: MRC Institute for Environment and Health. Report No. EUR 17549.

Hopkin SP. 1989. Ecophysiology of metals in terrestrial invertebrates. London UK: Elsevier.

Hoshina T, Teshirogi W. 1991. Formation of malformed pharynx and neoplasia in the planarian *Bdellocephala brunnea* following treatment with a carcinogen. *Hydrobiologia* 227:61–70.

Hsu YL, Yang YH, C, Tung MC, Wu JL, Engelking MH, Leong JC. 1995. Development of an in vitro subculture system for the oka organ (lymphoid tissue) of *Penaeus monodon*. *Aquaculture* 136:43–55.

Hu K. 1990. Studies on a cell culture from the hepatopancreas of the oriental shrimp *Penaeus orientalis* Kishinouye. *Asian Fish Science* 3:299–307.

Huet M, Fioroni P, Oehlmann J, Stroben E. 1995. Comparison of imposex response in three Prosobranch species. *Hydrobiologia* 309:29–35.

Ingersoll CG, Brunson EL, Dwyer FJ, Hardesty DK, Kemble NE. 1998. Use of sublethal endpoints in sediment toxicity tests with the amphipod *Hyalella azteca*. *Environ Toxicol Chem* 17:1885–1894.

Ingersoll CG, Dwyer FJ, May TW. 1990. Toxicity of inorganic and organic selenium to *Daphnia magna* (Cladocera) and *Chironomus riparius* (Diptera). *Environ Toxicol Chem* 9:1171–1181.

Ingersoll CG, Winner RW. 1982. Effects on *Daphnia pulex* (De Geer) of daily pulse exposures to copper or cadmium. *Environ Toxicol Chem* 1:321–327.

Ireland DS, Burton GA Jr, Hess GG. 1996. In situ toxicity evaluations of turbidity and photoinduction of polycyclic aromatic hydrocarbons. *Environ Toxicol Chem* 15:574–581.

[ISO] International Organization for Standardization. 1994. Soil quality - Effects of soil pollutants on Collembola (*Folsomia candida*): Method for the determination of effects on reproduction. Geneva Switzerland: ISO. ISO/DIS 11267 (draft).

[ISO] International Organization for Standardization. 1998a. Water quality - Determination of long term toxicity of substances to *Daphnia magna* Straus. Geneva Switzerland: ISO. ISO/DIS 10706 (draft).

[ISO] International Organization for Standardization. 1998b. Water quality - Determination of acute lethal toxicity to marine copepods (Copepoda, Crustacea). Geneva Switzerland: ISO. ISO/FDIS 14669 (draft).

James MO. 1984. Catalytic properties of cytochrome P-450 in hepatopancreas of the spiny lobster, *Panulirus argus*. *Mar Environ Res* 14:1–11.

James MO. 1989. Cytochrome P450 monooxygenases in crustaceans. *Xenobiotica* 19:1063–1076.

James MO, Altman AH, Li CJ, Boyle SM. 1992. Dose- and time-dependent formation of benzo[a]pyrene metabolite DNA adducts in the spiny lobster, *Panulirus argus*. *Mar Env Res* 34:299–302.

James MO, Boyle SM, Trapido-Rosenthal HG, Smith WC, Greenberg RM, Shiverick KT. 1996. cDNA and protein sequence of a major form of P450, CYP2L, in the hepatopancreas of the spiny lobster, *Panulirus argus*. *Arch Biochem Biophy* 329:31–38.

James MO, Little PJ. 1984. 3-Methylcholanthrene does not induce in vitro xenobiotic metabolism in spiny lobster hepatopancreas, or affect in vivo disposition of benzo[a]pyrene. *Comp Biochem Physiol* 78C:241–245.

James MO, Shiverick K. 1984. Cytochrome P450-dependent oxidation of progesterone, testosterone, and ecdysone in the spiny lobster, *Panulirus argus*. *Arch Biochem Biophy* 233:1–9.

Jewell CSE, Mayeaux MH, Winston GW. 1997. Benzo[a]pyrene metabolism by the hepatopancreas and green gland of the red swamp crayfish, *Procambarus clarkii*, *in vitro*. *Comp Biochem Physiol* 118C:369–374.

Jobling S, Sumpter JP. 1993. Detergent components in sewage effluent are weakly oestrogenic to fish: An in vitro study using rainbow trout (*Oncorhynchus mykiss*) hepatocytes. *Aquat Toxicol* 27:361–372.

Jones DT, Hopkin SP. 1996. Reproductive allocation in the terrestrial isopods *Porcellio scaber* and *Oniscus asellus* in a metal-polluted environment. *Funct Ecol* 10:741–750.

Jones DT, Hopkin SP. 1998. Reduced survival and body size in the terrestrial isopod *Porcellio scaber* from a metal-polluted environment. *Environ Pollut* 99:215–223.

Jones JRE. 1937. The toxicity of dissolved metallic salts to *Polycelis nigra* (Muller) and *Gammarus pulex* (L.). *J Exp Biol* 14:351–363.

Joosse J. 1988. The hormones of mollusks. Endocrinology of selected invertebrate types. New York: Alan R Liss p 89–140.

Jop KM, Hoberg JR. 1995. Concentration of metal and organic compounds in blue crabs *Callinectes sapidus* from the Lower Quinnipiac and Connecticut River estuaries. *J Environ Sci Health Part A: Environmental Science Engineering Toxic Hazardous Substance Control* 30:1835–1842.

Kahl MD, Makynen EA, Kosian PA, Ankley GT. 1997. Toxicity of 4-nonylphenol in a life-cycle test with the midge *Chironomus tentans*. *Ecotoxicol Environ Saf* 38:155–160.

Kalber FA Jr, Costlow JD Jr. 1966. Ontogeny of osmoregulation and its neurosecretory control in the decapod crustacean, *Rhithropanopeus harrisii* (Gould). *Zoologist* 6:221–229.

Kaplan EL, Meier P. 1958. Nonparametric estimation from incomplete observations. *J Amer Stat Assn* 53:457–481.

Ke H, Liping D, Yumei D, Shuji Z. 1990. Studies on a cell culture from the hepatopancreas of the oriental shrimp *Penaeus orientalis* Kishinouye. *Asian Fisheries Sci* 3:299–307.

Keeley LL, Hayes TK, Bradfield JY. 1990. Insect neuroendocrinology: Its past; its present; future opportunities. In: Borkovec AB, Masler EP. Insect neurochemistry and neurophysiology. Clifton NJ: Humana Pr. p 163–203.

Keilty TJ, White DS, Landrum PF. 1988. Sublethal responses to endrin in sediment by *Limnodrilus hoffmeisteri* (Tubificidae) and in mixed-culture with *Stylodrilus heringianus* (Lumbriculidae). *Aquat Toxicol* 13:227-250.

Keller R. 1988. Radioimmunoassays and ELISAS: Peptides. In: Gilbert LI and Miller TA, editors. Immunological techniques in insect biology. New York: Springer-Verlag. p 253–272.

Kerr DR, Meador JP. 1996. Modeling dose response using generalized linear models. *Environ Toxicol Chem* 15:395–401.

Khalaila I, Weil S, Sagi A. 1999. Endocrine balance between male and female components of the reproductive system in intersex *Cherax quadricarinatus* (Decapods: Parastacidae) *J Exp Zool* 283:286–294.

King DS, Bollenbacher WE, Borst DW, Vedeckis WV, O'Connor JD, Ittycheriah PI, Gilbert LI. 1974. The secretion of alpha ecdysone by the prothoracic glands of *Manduca sexta in vitro*. *Proc Nat Acad Sci USA* 71:793–796.

Kleinschuster SJ, Parent J, Walker CW, Farley CA. 1996. A cardiac cell line from *Mya arenaria* (Linnaeus 1759). *J Shellfish Res* 15:695–707.

Kneib RT. 1987. Seasonal abundance, distribution and growth of postlarval and juvenile grass shrimp (*Palaemonetes pugio*) in a Georgia, USA, salt marsh. *Mar Biol* 96:215–223.

Knuth ML. 1989. Determination of the insect growth regulator methoprene in natural waters by capillary gas–liquid chromatography. *Chemosphere* 18:2275–2281.

Kouyoumjian HH, Uglow RF. 1974. Some aspects of the toxicity of DDT, DDE and DDD to the freshwater planarian *Polycelis felina* (Tricladida). *Environ Pollut* 7:103–109.

Lamberson JL, DeWitt TH, Swartz RC. 1992. Assessment of sediment toxicity to marine benthos. In: Burton GA Jr, editor. Sediment toxicity assessment. Boca Raton FL: Lewis. p 183–211.

Lamberson JL, Swartz RC, Ozretich RJ, Jones JKP, Sewall JE, Vance PM. 1996. Field validation of acute and chronic marine amphipod sediment toxicity tests with *Grandidierella japonica* (abstract). Society of Environmental Toxicology and Chemistry (SETAC) 17th Annual Meeting; Washington DC. Pensacola FL: SETAC.

Lan Q, Riddiford LM. 1997. DNA transfection in the ecdysteroid-responsive GV1 cell line from the tobacco hornworm, *Manduca sexta*. *In vitro Cell Dev Biol* 33:615–621.

Lan Q, Wu Z, Riddiford LM. 1997. Regulation of the ecdysone receptor, USP, E75 and MHR3 mRNAs by 20-hydroxyecdysone in the GV1 cell line of the tobacco hornworm, *Manduca sexta*. *Insect Molecular Biol* 6:3–10.

Landau M, Rao KR. 1980. Toxic and sublethal effects of precocene II on the early developmental stages of the brine shrimp *Artemia salina* (L.) and the barnacle *Balanus eburneus* Gould. *Crustaceana* 39:218–221.

Laughlin RB Jr, Gustafson R, Pendoley P. 1988. Chronic embryo–larval toxicity of tributyltin (TBT) to the hard shell clam *Mercenaria mercenaria*. *Mar Ecol Prog Ser* 48:29–36.

Lawler IF, Aldrich JC. 1987. Sublethal effects of bis(tri-n-butyltin)oxide on *Crassostrea gigas* spat. *Mar Pollut Bull* 18:274–248.

Lee BM, Scott GI. 1989. Acute toxicity of temephos, fenoxycarb, diflubenzuron, and methoprene and *Bacillus thuringiensis* var. *israelensis* to the mummichog (*Fundulus heteroclitus*). *Bull Environ Contam Toxicol* 43:827–832.

Lee C, Watson D. 1994. Development of a quantitative enzyme-linked immunosorbent assay for vitellin and vitellogenin of the blue crab *Callinectes sapidus*. *J Crustacean Biol*14:617–626.

Lee RF, Noone T. 1995. Effect of reproductive toxicants on lipovitellin in female blue crabs, *Callinectes sapidus*. *Mar Env Res* 39:151–154.

Lee RF, O'Malley K, Oshima Y. 1996. Effects of toxicants on developing oocytes and embryos of the blue crab, *Callinectes sapidus*. *Mar Env Res* 42:125–128.

Lenhoff HM. 1983. Culturing large numbers of *Hydra*. In: Lenhoff HM, editor. *Hydra*: Research methods. New York: Plenum Pr. p 53–62.

Leudman R, Lightner DV. 1992. Development of an in vitro primary cell culture from the penaeid shrimp *Penaeus stylirostris* and *Penaeus vannamei*. *Aquaculture* 101:205–211.

Levin L, Caswell H, Bridges T, DiBacco C, Cabena D, Plaia G. 1996. Demographic responses of estuarine polychaetes to pollutants: Life table response experiments. *Ecological Applications* 6:1295–1313.

Lewis EG, Haeffner PA Jr.1976. Oxygen consumption of the blue crab, *Callinectes sapidus* Rathburn, from proecdysis to postecdysis. *Comp Biochem Physiol* 54A:55–60.

Leys SP. 1997. Sponge cell culture: a comparative evaluation of adhesion to a native tissue extract and other culture substrates. *Tissue and Cell* 29:77–87.

Liebrich W, Brown AC, Botes DP. 1995. Cadmium-binding proteins from a tunicate, *Pyura stolonifera*. *Comparative Biochemistry and Physiology C. Pharmacology Toxicology Endocrinology* 112:35–42.

Little G. 1968. Induced winter breeding and larval development in the shrimp, *Palaemonetes pugio* Holthuis (Caridea, Palaemonidae). *Crustaceana Suppl* 1:19–26.

Løkke H, Van Gestel CAM, editors. 1998. Handbook of soil invertebrate toxicity tests. Chichester UK: J Wiley.

Lucas RE. 1978. Altosid - A unique tool for use in an integrated mosquito control program. *Proc Calif Mosq Control Assoc* 46:118–119.

Maltby L. 1995a. Predicting life-history adaptations to pollutants. Abstract. Society of Environmental Toxicology and Chemistry (SETAC) 2nd World Congress; 12–16 Nov 1995; Vancouver BC, Canada. Pensacola FL: SETAC.

Maltby L. 1995b. Sensitivity of the crustaceans *Gammarus pulex* (L) and *Asellus aquaticus* (L) to short-term exposure to hypoxia and unionized ammonia - observations and possible mechanisms. *Wat Res* 29:781–787.

Maltby L, Crane M. 1994. Responses of *Gammarus pulex* (Amphipoda, Crustacea) to metalliferous effluents: Identification of toxic components and the importance of interpopulation variation. *Environ Pollut* 84:45–52.

Mangum C. 1992. Physiological aspects of molting in the blue crab, *Callinectes sapidus*. *American Zoologist* 32:459–469.

Martin VJ, Littlefield CL, Archer WE, Bode HR. 1997. Embyogenesis in *Hydra*. *Biol Bull* 192:345–363.

Matthiessen P, Gibbs PE. 1998. Critical appraisal of the evidence for tributyltin-mediated endocrine disruption in mollusks. *Environ Toxicol Chem* 17:44–48.

Maund SJ, Peither A, Taylor EJ, Juettner I, Beyerle-Pfnuer R, Lay JP, Pascoe D. 1992. Toxicity of lindane to freshwater insect larvae in compartments of an experimental pond. *Ecotox Env Saf* 23:76–88.

McCahon CP, Pascoe D. 1991. Brief exposure of first and fourth instar *Chironomus riparius* larvae to equivalent assumed doses of cadmium: effects on adult emergence. *Water Air Soil Pollut* 60:395–403.

McCahon CP, Poulton MJ, Thomas PC, Xu Q, Pascoe D, Turner C. 1991. Lethal and sublethal toxicity of field simulated farm waste episodes to several freshwater invertebrate species. *Water Res* 25:661–671.

McCahon CP, Whiles AJ, Pascoe D. 1989. The toxicity of cadmium to different larval instars of the trichopteran larvae *Agapetus fuscipes* Curtis and the importance of life cycle information to the design of toxicity tests. *Hydrobiologia* 185:153–162.

McCarthy JF, Skinner DM. 1979. Changes in ecdysteroids during embryogenesis of the blue crab, *Callinectes sapidus* Rathburn. *Dev Biol* 69:627–633.

McKague AB, Pridmore RB. 1978. Toxicity of altosid and dimilin to juvenile rainbow trout and coho salmon. *Bull Environ Contam Toxicol* 20:167–169.

McKenney CL Jr, Celestial DM. 1996. Modified survival, growth, and reproduction in an estuarine mysid (*Mysidopsis bahia*) exposed to a juvenile hormone analog through a complete life cycle. *Aquat Toxicol* 35:11–20.

McKenney CL Jr, Matthews E, Lawrence DA, Shirley MA. 1985. Effects of ground ULV application of fenthion on estuarine biota. IV. Lethal and sublethal responses of an estuarine mysid. *J Fl Anti-mosq Assoc* 56:72–75.

McKenney CL Jr, Weber DE, Celestial DM MacGregor MA. 1998. Altered growth and metabolism of an estuarine shrimp (*Palaemonetes pugio*) during and after metamorphosis onto fenvalerate-laden sediment. *Arch Environ Contam Toxicol* 35:464–471.

McKenney CL Jr. 1987. Optimization of environmental factors during the life cycle of *Mysidopsis bahia*. Cincinnati OH: USEPA. EPA/600/M-87/004. 6 p.

McKenney CL Jr. 1994. Alterations in growth, reproduction, and energy metabolism of estuarine crustaceans as indicators of pollutant stress. In: Salanki J, Jeffery D, Hughes GM, editors. Biological monitoring of the environment: A manual of methods. Paris, France: International Union of Biological Sciences, CAB International. p 111–115.

McKenney CL Jr. 1998. Physiological dysfunction in estuarine mysids and larval decapods with chronic pesticide exposure. In: Wells PG, Lee K, Blaise C, editors. Microscale testing in aquatic toxicology: Advances, techniques, and practice. Boca Raton FL: CRC Pr. p 465–476.

McKenney CL Jr. 1999. Hormonal processes in decapod crustacean larvae as biomarkers of endocrine disrupting chemicals in the marine environment. In: Henshel DS, Black MC, Harrass, editors. Environmental Toxicology and Risk Assessment, 8th Volume. West Conshohocken PA: ASTM. STP 1364: In press.

McMurtry MJ. 1984. Avoidance of sublethal doses of copper and zinc by tubificid oligochaetes. *J Great Lakes Res* 10:267-272.

Metge K, Heimbach U. 1998. Tests on the staphylinid *Philonthus cognatus* Steph. 1832. In: Løkke H, van Gestel CAM, editors. Handbook of soil invertebrate toxicity tests. Chichester UK: J Wiley. p 131–156.

Meyer JS, Ingersoll CG, McDonald LL, Boyce MS. 1986. Estimating uncertainty in population growth rates: Jackknife vs. bootstrap techniques. *Ecology* 67:1156–1166.

Meyer JS, Ingersoll CG, McDonald LL, Boyce MS. 1987. Sensitivity analysis of population growth rates estimated from cladoceran chronic toxicity tests. *Environ Toxicol Chem* 6:115–126.

Mialhe E, Boulo V, Grizel H. 1988. Bivalve mollusk cell culture. *Am Fish Soc Sp Publ* 18:311–315.

Monson PD, Ankley GT, Kosian PA.1995. Phototoxic response of *Lumbriculus variegatus* to sediments contaminated by polycyclic aromatic hydrocarbons. *Environ Toxicol Chem* 14: 891–894.

Moore CG, Stevenson JM. 1991. The occurrence of intersexuality in harpaticoid copepods and its relationship with pollution. *Mar Pollut Bull* 22:72–74.

Moore CG, Stevenson JM. 1994. Intersexuality in benthic harpacticoid copepods in the firth of Forth, Scotland. *J Natural History* 28:1213–1230.

Moore DRJ, Caux P-Y. 1997. Estimating low toxic effects. *Environ Toxicol Chem* 16:794–801.

Moore DW, Dillon TM. 1993. The relationship between growth and reproduction in the marine polychaete *Nereis (Neanthes) arenaceodentata* (Moore): Implications for chronic sublethal sediment bioassays. *J Exp Mar Biol Ecol* 173:231–246.

Morgan MD. 1980. Grazing and predation of the grass shrimp *Palaemonetes pugio. Limnol Oceanogr* 25:896–902.

Mothershead RF, Hale RC. 1992. Influence of ecdysis on the accumulation of polycyclic aromatic hydrocarbons in field exposed blue crabs (*Callinectes sapidus). Mar Environ Res* 33:145–156.

Mourton C, Boulo V, Chagot D. 1992. Interactions between *Bonamia ostraea* (Protozoa: Ascetospora) and hemocytes of *Ostrea edulis* and *Crassostrea gigas* (Molluska: Bivalvia): In vitro system establishment. *J Invert Pathol* 59:235–240.

Muenchow G. 1986. Ecological use of failure time analysis. *Ecology* 67:246–250.

Muirhead-Thomson RC. 1973. Laboratory evaluation of pesticide impact on stream invertebrates. *Freshwater Biol* 3:479–498.

Mulla MS. 1995. The future of insect growth regulators in vector control. *J Amer Mosq Control Assoc* 11:269–273.

Nabholz JV, Miller P, Zeeman M. 1993. Environmental risk assessment of new chemicals under the Toxic Substances Control Act (TSCA) Section Five. In Landis WG, Hughes JS, Lewis MA, editors. Environmental Toxicology and Risk Assessment. Philadelphia PA: ASTM. STP 1179. p 40–55.

Nadala EC, Loh PC, Lu Y. 1993. Primary cell culture of lymphoid nerve and ovary cells from *Penaeus stylirostris* and *Penaeus vannamei. In Vitro Cell and Dev Biol* 29A:620–622.

Naganuma T, Degnan MB, Horikoshi K, Morse DE. 1994. Myogenesis in primary cell cultures from larvae of the abalone *Haliotis rufescens. Molecular Mar Biol Biotechnol* 3:131–140.

Najafabadi AK, Ellender RD, Middlebrooks BL. 1992. Analysis of shrimp hemolymph and ionic modification of a *Penaeus* cell culture formulation. *J Aqua Animal Health* 4:143–148.

Naqvi SM, Howell RD. 1993. Toxicity of cadmium and lead to juvenile red swamp crayfish, *Procambarus clarkii*, and effects on fecundity of adults. *Bull Environ Contam Toxicol* 51:303–308.

Nell JA, Chvojka R. 1992. The effect of bis-tributyltin oxide (TBTO) and copper on the growth of juvenile Sydney rock oysters *Saccostrea commercialis* (Iredale and Roughley) and Pacific oysters *Crassostrea gigas* Thunberg. In: Batley GE, editor. *Trace Metals in the Aquatic Environment* 125:193–201.

Neuhauser EF, Callahan CA. 1990. Growth and reproduction of the earthworm *Eisenia fetida* exposed to sublethal concentrations of organic chemicals. *Soil Biol Biochem* 22:175-179.

Newman MC. 1995. Quantitative methods in aquatic ecotoxicology. Boca Raton FL: Lewis.

Newman MC, Keklak MM, Doggett MS. 1994. Quantifying animal size effects on toxicity: A general approach. *Aquat Toxicol* 28:1–12.

Nimmo DR, Hamaker TL. 1982. Mysids in toxicity testing - a review. *Hydrobiologia* 93:171–178.

Nimmo DR, Hamaker TL, Moore JC, Moore RA. 1980. Acute and chronic effects of dimilin on survival and reproduction of *Mysidopsis bahia*. In: Eaton JG, Parrish RP, Hendricks AC, editors. Aquatic toxicology. Philadelphia PA: ASTM. STP 707. p 366–376.

No D, Yao TP, Evans RM. 1996. Ecdysone-inducible gene expression in mammalian cells and transgenic mice. *Proc Natl Acad Sci* 93:3346–3351.

Oberdörster E, Clay MA, McLachlan JA. 1999. Common flavonoid phytoestrogens inhibit ecdysone-dependent gene transcription: Evolutionary significance. *Endocrinology*: In review.

Oberdörster E, Rittschof D, McClellan-Green P. 1998a. Induction of cytochrome P450 3A and heat shock protein by tributyltin in blue crab, *Callinectes sapidus*. *Aquat Toxicol* 41:83–100.

Oberdörster E, Rittschof D, McClellan-Green P. 1998b. Testosterone metabolism in imposex and normal *Ilyanassa obsoleta*: A comparison of field and TBT Cl-induced imposex. *Mar Pollut Bull* 36:144–151.

Oberdörster E, Wilmon F, Cottam D, Milner M, McLachlan JA. 1999. Interaction of PAHs and PCBs with the arthropod ecdysone receptor. *Toxicol Appl Pharmacol*: In review.

Odendaal JP, Reinecke AJ. 1999. The toxicity of sublethal lead concentrations for the woodlouse, *Porcellio laevis* (Crustacea, Isopoda). *Biology and Fertility of Soils* 29:146–151.

Odinstova NA, Khomenko AV. 1991. Primary cell culture from embryos of the Japanese scallop *Mizuchopecten yessoensis* (Bivalvia). *Cytotechnology* 6:49–54.

[OECD] Organization for Economic Cooperation and Development. 1984. Guidelines for the Testing of Chemicals. No. 207. Earthworm acute toxicity test. Paris, France: OECD.

[OECD] Organization for Economic Cooperation and Development. 1997. Detailed review paper on appraisal of test methods for sex-hormone disrupting chemicals. Draft dated 1997. Environmental Health and Safety Publications. Series on Testing and Assessment. Paris, France: OECD.

[OECD] Organization for Economic Cooperation and Development. 1998a. Detailed review paper on aquatic testing methods for pesticides and industrial chemicals. Part 1: Report - Part 2: Annexes. No 11. Paris, France: OECD.

[OECD] Organization for Economic Cooperation and Development. 1998b. Guidelines for the Testing of Chemicals. No. 211. *Daphnia magna* reproduction test. Paris, France: OECD.

[OECD] Organization for Economic Cooperation and Development. 1998c. Guidelines for the testing of chemicals. Draft new guideline on chironomid toxicity test using spiked water and a draft new guideline on chironomid toxicity testing using spiked sediment. Draft documents dated May 1998. Paris, France: OECD.

[OECD] Organization for Economic Cooperation and Development. 1998d. Guidelines for the Testing of Chemicals. No. 213: Honeybees, acute oral toxicity test. Paris, France: OECD.

[OECD] Organization for Economic Cooperation and Development. 1998e. Guidelines for the Testing of Chemicals. No. 214: Honeybee, acute contact toxicity test. Paris, France: OECD.

[OECD] Organization for Economic Cooperation and Development. 1998f. Report of the OECD Workshop on Statistical Analysis of Aquatic Toxicity Data; 15–17 October 1996; Braunschweig, Germany. No. 10. Paris, France: OECD.

[OEPP/EPPO] Organisation européenne et méditerranéenne pour la protection des plantes/European and Mediterranean Plant Protection Organization. 1992. Guideline on test methods for evaluating the side-effects of plant protection products on honeybees. *OEPP/EPPO Bull* 22:203–215.

Olesen TM, Weeks JM. 1998. Use of a model ecosystem for ecotoxicology studies with earthworms. In: Sheppard S, Bembridge J, Holmstrap M, Posthuma L, editors. Advances in earthworm ecotoxicology. Pensacola FL: Society of Environmental Toxicology and Chemistry (SETAC). p 381–386.

Ormerod SJ, Boole P, McCahon P, Weatherly N, Pascoe D, Edwards R. 1987. Short term experimental acidification of a Welsh stream: Biological effects of hydrogen ions and aluminium. *Freshwater Biol* 17:341–356.

Ormond RF, Caldwell S. 1982. The effect of oil pollution on the reproduction and feeding behavior of the sea anenome *Actinia equina*. *Mar Poll Bull* 13:118–122.

Ort MP, Finger SE, Jones JR. 1995. Toxicity of crude oil to the mayfly, *Hexagenia bilineata* (Ephemeroptera, Ephemeridae). *Environ Pollut* 90:105–110.

Ortel J, Gintenreiter S, Nopp H. 1993. Metal bioaccumulation in a host insect (*Lymantria dispar* L. Lepidoptera) during development - Ecotoxicology implications. In: Dallinger R, Rainbow PS, editors. Ecotoxicology of metals in invertebrates. Boca Raton FL: Lewis. p 404–425.

Parmelee RW, Wentsel RS, Phillips CT, Simini M, Checkai RT. 1993. Soil microcosm for testing the effects of chemical pollutants on soil fauna communities and trophic structure. *Environ Toxicol Chem* 12:1477–1486.

Pascoe D, Kedwards TJ, Maund SJ, Muthi E, Taylor EJ. 1994. Laboratory and field evaluation of a behavioral bioassay - the *Gammarus pulex* (L.) precopula separation (GaPPS) test. *Wat Res* 28:369–372.

Pascoe D, Williams KA, Green DWJ. 1989. Chronic toxicity of cadmium to *Chironomus riparius* (Meigen)-effects upon larval development and adult emergence. *Hydrobiologia* 175:109–115.

Payen GG, Costlow JD Jr. 1977. Effects of a juvenile hormone mimic on male and female gametogenesis of the mud-crab *Rhithropanopeus harrisii* (Gould) (Brachyura: Xanthidae). *Biological Bull* 152:199–208.

Peel DJ, Milner MJ. 1992. The response of *Drosophila* imaginal disc cell lines to ecdysteroids. *Roux's Arch Developmental Biol* 202:23–35

Penn GH. 1943. A study of the life history of the Louisiana red-crawfish, *Cambarus clarkii* Girard. *Ecology* 24:1–18.

Perkins FO, Menzel RW. 1964. Maintenance of oyster cells in vitro. *Nature* 304:1106–1107.

Persoone G, Wells PJ. 1987. Artemia in aquatic toxicology: A review. *The Brine Shrimp Artemia* 1:259–275.

Peter R. 1995. Regenerative and reproductive capacities of the fissiparous planarian *Dugesia tahitiensis*. *Hydrobiologia* 305:261.

Petersen LBM, Petersen RC. 1984. Effect of Kraft pulp mill effluent and 4,5,6 trichloroguaiacol on the net spinning behavior of *Hydropsyche angustipennis* (Trichoptera). *Ecological Bull* 36:68–74.

Phelps H, Warner K. 1990. Estuarine sediment bioassay with oyster pediveliger larvae (*Crassostrea gigas*). *Bull Environ Contam Toxicol* 44:197–204.

Phipps GL, Ankley GT, Benoit DA, Mattson VR. 1993. Use of the aquatic oligochaete *Lumbriculus variegatus* for assessing the toxicity and bioaccumulation of sediment-associated contaminants. *Environ Toxicol Chem* 12:269-274.

Pinder LCV, Pottinger TG. 1998. Draft final report to the Environment Agency and Endocrine Modulators Steering Group. Endocrine function in aquatic invertebrates and evidence for disruption by environmental pollutants. Ambleside Cumbria UK: Centre for Ecology and Hydrology.

Porte C, Escartin E. 1998. Cytochrome P450 system in the hepatopancreas of the red swamp crayfish *Procambarus clarkii*: A field study. *Comp Bioch Physiol* 121C:333–338.

Quackenbush LS. 1989. Vitellogenesis in the shrimp, *Penaeus vannamei*: In vitro studies of the isolated hepatopancreas and ovary. *Comp Biochem Physiol* 94B:253–261.

Quistad GB, Schooley DA, Staiger LE, Bergot BJ, Sleight BH, Macek KJ. 1976. Environmental degradation of the insect growth regulator methoprene. IX. Metabolism by bluegill fish. *Pestic Biochem Physiol* 6:523–529.

Raftos DA, Cooper EL, Habitch GS, Beck G. 1991. Invertebrate cytokines: tunicate cell proliferation stimulated by an interleukin-1 like molecule. *Proc Natl Acad Sci USA* 88:9518–9522.

Raftos DA, Stillman DL, Cooper EL. 1990. In vitro culture of tissue from the tunicate *Styela clava*. *In vitro Cell Dev Biol* 26:962–970.

Raftos DA, Stillman DL, Cooper EL. 1991. Interleukin-2 and phytohaemagglutinin stimulate the proliferation of tunicate cells. *Immunol Cell Biol* 69:225–234.

Raimondi PT, Barnett AM, Krause PR. 1997. The effects of drilling muds on marine invertebrate larvae and adults. *Environ Toxicol Chem* 16:1218–1228.

Ramenofsky M, Faulkner DJ, Ireland C. 1974. Effect of juvenile hormone on cirriped metamorphosis. *Biochem Biophys Res Commun* 60:172–178.

Rao KR, Doughtie DG, Conklin PJ. 1982. Physiological and histopathological evaluation of dithiocarbamate toxicity to the grass shrimp, *Palaemonetes pugio*. In: Vernberg WB, Calabrese A, Thurberg FP, Vernberg FJ, editors. Physiological mechanisms of marine pollutant toxicity. New York: Academic Pr. p. 413–445.

Reddy PS, Devi M, Sarojini R, Nagabhushanam R, Fingerman M. 1994. Cadmium chloride induced hyperglycemia in the red swamp crayfish, *Procambarus clarkii*: Possible role of crustacean hyperglycemic hormone. *Comp Biochem Physiol* 107C:57–61.

Reddy PS, Tuberty SR, Fingerman M. 1997. Effects of cadmium and mercury on ovarian maturation in the red swamp crayfish, *Procambarus clarkii*. *Ecotoxicol Environ Safety* 37:62–65.

Reish DJ. 1977. Effects of chromium on the life history of *Capitella capitata* (Annelida: Polychaeta). In: Physiological responses of marine biota to pollutants. New York: Academic Pr. p 199–207.

Reynoldson TB, Thompson SP, Bamsey JL. 1991. A sediment bioassay using the tubificid oligochaete worm *Tubifex tubifex*. *Environ Toxicol Chem* 10:1061-1072.

Riddle DA. 1997. *Caenorhabditis elegans* II. Cold Spring Harbor monograph series 33. Plainview NY: Cold Spring Harbor Laboratory Pr.

Rinkevich B, Rabinowittz C. 1993. In vitro culture of blood cells from the colonial protochordate *Botryllus schlosseri*. *In Vitro Cell Dev Biol* 29A:79–85.

Rinkevich B, Rabinowitz C. 1994. Acquiring embryo-derived cell cultures and aseptic metamorphosis of larvae from the colonial protochordate *Botryllus schlosseri*. *In Vitro Cell Dev Biol* 29A:79–85.

Rittschof D, Holm ER. 1997. Antifouling and foul-release: A primer. In: Fingerman R, Nagabhushanam R, Thompson MF, editors. Recent advances in marine biotechnology. Volume I, Endocrinology and reproduction. New Delhi, India: Oxford & IBH. p 497–512.

Rittschof D, Sasikumar N, Murlless D, Clare AS, Gerhart DJ Bonaventura J. 1994. Mixture interactions of lactones, furans, and a commercial biocide: toxicity and antibarnacle settlement activity. In: Thompson MF, Nagabhushanam R, Sarojini R, Fingerman M, editors. Recent developments in biofouling control. New Delhi, India: Oxford and IBH. p. 269–274.

Roesijadi G, Anderson JW, Petrocelli SR, Giam CS. 1976. Osmoregulation of the grass shrimp *Palaemonetes pugio* exposed to polychlorinated biphenyls (PCBs). I. Effect on chloride and osmotic concentrations and chloride and water-exchange kinetics. *Mar Biol* 27:343–355.

Römbke J, Knacker T, Förster B, Maciorowski A. 1994. Comparison of effects of two pesticides on soil organisms in laboratory tests, microcosms and in the field. In: Donker M, Eijsackers H, Heimbach F, editors. Ecotoxicology of soil organisms. Boca Raton FL: Lewis. p 229–240.

Ross DH, Judy D, Jacobson B, Howell R. 1994. Methoprene concentrations in freshwater microcosms treated with sustained-release Altosid7 formulations. *J Amer Mosq Control Assoc* 10:202–210.

Ruiz JM, Bryan GW, Gibbs PE. 1994. Bioassaying the toxicity of tributyltin-(TBT)-polluted sediment to spat of the bivalve *Scrobicularia plana*. *Mar Ecol Prog Ser* 113:1–2.

Ruiz JM, Bryan GW, Gibbs PE. 1995. Effects of tributyltin (TBT) exposure on the veliger larvae development of the bivalve *Scrobicularia plana* (da Costa). *J Exp Mar Biol Ecol* 186:53–63.

Ruiz JM, Bryan GW, Wigham GD, Gibbs PE. 1995. Effects Of tributyltin (TBT) exposure on the reproduction and embryonic development of the bivalve *Scrobicularia plana. Mar Environ Res* 40:363–379.

Sagi A Khalaila I, Abdu U, Shoukrun R, Weil S. 1999. A newly established ELISA showing the effect of the androgenic gland on secondary vitellogenic specific protein in the hemolymph of the crayfish Cherax quadricarinatus. Gen Comp Endocrinol 115:37–45.

Sagi A, Khalaila I, Barki I, Hulata G, Karplus I. 1996. Intersex red claw crayfish, *Cherax quadricarinatus* (von Martens): Functional males with pre-vitellogenic ovaries. *Biol Bull* 190:16–23.

Salazar MH, Salazar SM. 1991. Assessing site-specific effects of TBT contamination with mussel growth rates. *Mar Environ Res* 32:131–150.

Samsøe-Petersen L. 1987. Laboratory method for testing side effects of pesticides on the rove beetle *Aleochara bilineata* - adults. *Entomophaga* 32:73–81.

Samsøe-Petersen L. 1993. Effects of 45 insecticides, acaricides and molluskicides on the rove beetle *Aleochara bilineata* (Col. Staphylinidae) in the laboratory. *Entomophaga* 38:371–382.

Samsøe-Petersen L. 1995a. Effects of 67 herbicides and plant growth regulators on the rove beetle *Aleochara bilineata* (Col. Staphylinidae) in the laboratory. *Entomophaga* 40:95–104.

Samsøe-Petersen L. 1995b. Effects of 37 fungicides on the rove beetle *Aleochara bilineata* (Col. Staphylinidae) in the laboratory. *Entomophaga* 40:145–152.

Samsøe-Petersen L, Moreth L. 1994. Initial and extended laboratory tests for the rove beetle *Aleochara bilineata* Gyll. *Bull Intl Organization for Biological and Integrated Control of Noxious Animals and Plants*, West Palaearctic Regional Section (IOBC/WPRS) 17:89–97.

Sanders B, Laughlin RB Jr, Costlow JD Jr. 1985. Growth regulation in larvae of the mud crab *Rhithropanopeus harrisii*. In: Wenner AM, editor. Crustacean issues Volume 2. Larval growth. Rotterdam, The Netherlands: Balkema. p 155–161.

Sarojini R, Nagabhushanam R, Fingerman M. 1994. A possible neurotransmitter-neuroendocrine mechanism in naphthalene-induced atresia of the ovary of the red swamp crayfish, *Procambarus clarkii. Comp Biochem Physiol* 108C:33–38.

Sarojini R, Nagabhushanam R, Fingerman M. 1995. Naphthalene-induced atresia in the ovary of the crayfish, *Procambarus clarkii. Ecotoxicol Environ Safety* 31:76–83.

Sasson-Brickson G, Burton GA Jr. 1991. In situ and laboratory sediment toxicity testing with *Ceriodaphnia dubia*. *Environ Toxicol Chem* 10:201–207.

Schaefer CH, Miura T, Mulligan FS III, Dupras EF Jr. 1974. Insect developmental inhibitors: Formulation research on Altosid. *Proc Calif Mosq Control Assoc* 42:140–145.

Schlenk D, Brouwer M. 1991. Isolation of three copper metallothionein isoforms from the blue crab (*Callinectes sapidus*). *Aquat Toxicol* 20:25–34.

Schlenk D, Ringwood AH, Brouwer-Hoexum T, Brouwer M. 1993. Crustaceans as models for metal metabolism: II. Induction and characterization of metallothionein isoforms from the blue crab (*Callinectes sapidus*). *Mar Environ Res* 35:7–11.

Schmude KL, Liber K, Corry TD, Stay FS. 1999. Effects of 4-nonylphenol on benthic macroinvertebrates and insect emergence in littoral enclosures. *Environ Toxicol Chem* 18:386–393.

Schoenmakers HJN, Voogt PA. 1980. In vitro biosynthesis of steroids from progesterone by the ovaries and pyloric ceca of the starfish *Asterias rubens. Gen Comp Endocrinol* 41:408–416.

Seaby RMH, Martin AJ, Young JO. 1995. The reaction time of leech and triclad species to crushed prey and the significance of this for their coexistence in British Lakes. *Freshwater Biol* 34:21–28.

Seaby RMH, Martin AJ, Young JO. 1996. Food partitioning by lake-dwelling triclads and glossiphonid leeches: Field and laboratory experiments. *Oecologia* 106:544–550.

Sehnal F. 1983. Juvenile hormone analogs. In: Downer RGH, Laufer H, editors. Endocrinology of insects. New York: Alan R. Liss. p 657–672.

Sequencing Consortium. 1999. Genome sequence of the nematode *C. elegans*: A platform for investigating biology. *Science* 282:2012–2017.

[SETAC] Society of Environmental Toxicology and Chemistry. 1994. Guidance document on regulatory testing procedures for pesticides with non-target arthropods. The Workshop on European Standard Characteristics of Beneficial Regulatory Testing (ESCORT); 28–30 March 1994. The Netherlands: SETAC.

Shafir S, Tom M, Ovadia M, Lubzens E. 1992. Protein, vitellogenin, and vitellin levels in the hemolymph and ovaries during ovarian development in *Penaeus semisulcatus* (de Haan). *Biol Bull* 183:394–400.

Shurin J, Dodson SI. 1997. Sublethal toxic effects of cyanobacteria and nonlyphenol on environmental sex determination and development in *Daphnia*. *Environ Toxicol Chem* 16:1269–1276.

Sibley PK, Ankley GT, Cotter AM, Leonard EN. 1996. Predicting chronic toxicity of sediments spiked with zinc: An evaluation of the acid volatile sulfide (AVS) model using a life-cycle test with the midge *Chironomus tentans*. *Environ Toxicol Chem* 15:2102-2112.

Sibley PK, Benoit DA, Balcer MD, Phipps GL, West CW, Hoke RA, Ankley GT. 1999. An in situ bioassay chamber for the assessment of sediment toxicity and bioaccumulation using benthic invertebrates. *Environ Toxicol Chem*: In press.

Singer SC, March PE Jr, Gonsoulin F, Lee RF. 1980. Mixed function oxygenase activity in the blue crab, *Callinectes sapidus*: Characterization of enzyme activity from stomach tissue. *Comp Biochem Physiol* 65C:129–134.

Smit CE, van Gestel CAM. 1996. Comparison of the toxicity of zinc for the springtail *Folsomia candida* in artificially contaminated and polluted field soils. *Appl Soil Ecol* 3:127–136.

Snell TW, Carmona MJ. 1995. Comparative toxicant sensitivity of sexual and asexual reproduction in the rotifer *Brachionus calyciflorus*. *Environ Toxicol Chem* 14:415–420.

Snell TW, Childress M. 1987. Aging and loss of fertility in male and female *Brachionus plicatilis* (Rotifera). *Intl J Invertebrate Reproduction and Development* 12:103–110.

Snyder MJ. 1998. Identification of a new cytochrome P450 family, CYP45, from the lobster, *Homarus americanus*, and expression following hormone and xenobiotic exposures. *Arch Biochem Biophys* 358:271–276.

Sørensen FF, Bayley M, Baatrup E. 1995. The effects of sublethal dimethoate exposure on the locomotor behavior of the collembolan *Folsomia candida* (Isotomidae). *Environ Toxicol Chem* 14:1587–1590.

Spangenberg DB. 1984. Use of the *Aurelia* metamorphosis test system to detect subtle effects of hydrocarbons and petroleum oil. *Mar Environ Res* 14:281–303.

Stebbing ARD, Pomroy AJ. 1978. A sublethal technique for assessing the effects of contaminants using *Hydra littoralis*. *Wat Res* 12:631–635.

Stebbing ARD. 1976. The effects of low metal levels on a clonal hydroid. *J Mar Biolog Assn UK* 56:977–994.

Steel CGH, Vafopoulou X. 1998. Ecdysteroid titres in haemolymph and other tissues during molting and reproduction in the terrestrial isopod, *Oniscus asellus* (L.). *Invert Reprod Dev* 34:187–194.

Stephens EB, Hetrick FM. 1979a. Decontamination of the American oyster tissues for cell and organ culture. *TCA Man* 21:1029.

Stephens EB, Hetrick FM. 1979b. Preparation of primary cell cultures from the American oyster; *Crassastrea virginica*. *In Vitro Cell Dev Biol* 15:198.

Stroben E, Schulte Oehlmann U, Fioroni P, Oehlmann J. 1995. A comparative method for easy assessment of coastal TBT pollution by the degree of imposex in Prosobranch species. *Haliotis* 24:1–12.

Sun K, Krause GF, Mayer FL Jr, Ellersieck MR, Basu AP. 1995. Predicting chronic lethality of chemicals to fishes from acute toxicity test data: Theory of accelerated life testing. *Environ Toxicol Chem* 14:1745–1752.

Suter GW. 1993. Ecological risk assessment. Chelsea MI: Lewis.

Swartz RC, Cole FA, Lamberson JO, Ferraro SP, Schults DW, DeBen WA, Lee H II, Ozretich RJ. 1994. Sediment toxicity, contamination and amphipod abundance at a DDT- and dieldrin-contaminated site in San Francisco Bay. *Environ Toxicol Chem* 13:949–962.

Swartz RC, Cole FA, Schults DW, DeBen WA. 1986. Ecological changes on the Palos Verdes Shelf near a large sewage outfall: 1980–1983. *Mar Ecol Prog Ser* 31:1–13.

Swartz RC, DeBen WA, Jones JKP, Lamberson JO, Cole FA. 1985. Phoxocephalid amphipod bioassay for marine sediment toxicity. In: Caldwell RD, Purdy R, Bahner RC, editors. Aquatic toxicology and hazard assessment: Seventh Symposium. Philadelphia PA: ASTM. STP 854. p 284–307.

Takahashi RM, Miura T. 1975. Insect developmental inhibitors- multiple applications of Dimilin and Altosid to *Gambusia affinis*. *Proc Calif Mosq Control Assoc* 43:85–87.

Tattersfield L, Matthiessen P, Campbell P, Grandy N, Länge R. 1997. SETAC-Europe/OECD/EC expert workshop on endocrine modulation in wildlife: Assessment and testing (EMWAT). Brussels, Belgium: Society of Environmental Toxicology and Chemistry (SETAC)-Europe.

Taylor EJ, Blockwell SJ, Maund SJ, Pascoe D. 1993. Effects of lindane on the life-cycle of the freshwater macroinvertebrate *Chironomus riparius* Meigen (Insecta: Diptera). *Arch Environ Contam Toxicol* 24:145–150.

Taylor EJ, Maund SJ, Bennett D, Pascoe D. 1994. Effects of 3-4 dichloroaniline on the growth of two freshwater macroinvertebrates in a stream mesocosm. *Ecotox Environ Safety* 29:80–85.

Taylor EJ, Maund SJ, Pascoe D. 1991. Evaluation of a chronic toxicity test using growth of the insect *Chironomus riparius* Meigen. In: Jeffrey DW, Madden B, editors. Bio-indicators and environmental management. Proc. 6th International Bio-indicators Conference. Dublin, Ireland: Academic Pr. p 345–352.

Taylor EJ, Morrison JE, Blockwell SJ, Tarr A, Pascoe D. 1995. Effects of lindane on the predator-prey interaction between *Hydra oligactis* Pallas and *Daphnia magna* Strauss. *Arch Environ Contam Toxicol* 29:291–296.

Taypay LM, Lu Y, Brock JA, Nadala EC, Loh PC. 1995. Transformation of primary cultures of shrimp (*Penaeus strylirostris*) lymphoid (Oka) organ with simian virus-40 (T) antigen. *Proc Soc Exp Biol Medic* 209:73–78.

Templeton NS, Laufer H. 1983. The effects of a juvenile hormone analog (Altosid ZR-515) on the reproduction and development of *Daphnia magna* (Crustacea: Cladocera). *Int J Invert Reprod* 6:99–110.

Thomas J. 1997. What is an international standard? *ASTM Standardization News*, December 1997. p 11.

Thompson AR, Gore FL. 1972. Toxicity of twenty-nine insecticides to *Folsomia candida*: Laboratory studies. *J Econ Entomol* 65:1255–1260.

Tighe-Ford DJ. 1977. Effects of juvenile hormone analogs on larval metamorphosis in the barnacle *Elminius modestus* Darwin (Crustacea: Cirripedia). *J Exp Mar Biol Ecol* 26:163–176.

Tinsley D, Johnson I, Boumphrey R, Forrow D, Wharfe JR. 1996. The use of direct toxicity assessment to control discharges to the aquatic environment in the United Kingdom. In: Tapp JF, Hunt SM, Wharfe JR, editors. Toxic impacts of wastes on the aquatic environment. Cambridge UK: Royal Society of Chemistry. p 36–43.

Tomlin CDS, editor. 1997. The pesticide manual: A world compendium. 11th ed. Farnham, Surrey, UK: British Crop Protection Council. 1606 p.

Toullec JY, Crozat Y, Patrois J, Porcheron P. 1996. Development of primary cell cultures from the penaeid shrimps *Penaeus vannamei* and *P. indicus*. *J Crustacean Biol* 16:643–649.

Toullec JY, Dauphin-Villemant C. 1994. Dissociated cell suspensions of *Carcinus maenas* Y-organs as a tool to study ecdysteroid production and its regulation. *Experimentia* 50:153–157.

Tripp MR, Bisignami LA, Kenny NT. 1966. Oyster amoebocytes in vitro. *J Invert Pathol* 8:137–140.

Tuberty, SR. 1998. Vitellogenesis in the red swamp crayfish, *Procambarus clarkii* (Dissertation). New Orleans LA: Tulane Univ; Department of Ecology, Evolution, and Organismal Biology. p 151.

Tyler-Schroeder DB. 1979. Use of the grass shrimp (*Palaemonetes pugio*) in a life-cycle toxicity test. In: Marking LL, Kimberly RA, editors. Aquatic toxicology. Philadelphia PA: ASTM. STP 667. p 159–170.

Uhler AD, Durell GS, Steinhauer WG, Spellacy AM. 1993. Tributyltin levels in bivalve mollusks from the east and west Coasts of the United States: Results from the 1988–1990 National Status and Trends Mussel Watch Project. *Environ Toxicol Chem* 12:139–153.

[USEPA] U.S. Environmental Protection Agency. 1993. Methods for measuring the acute toxicity of effluents and receiving waters to freshwater and marine organisms. Cincinnati OH: USEPA. EPA 600/4-90/027F.

[USEPA] U.S. Environmental Protection Agency. 1994a. Methods for measuring the toxicity of sediment-associated contaminants with estuarine and marine amphipods. Duluth MN: USEPA. EPA 600/R-94/025.

[USEPA] U.S. Environmental Protection Agency. 1994b. Short-term methods for estimating the chronic toxicity of effluents and receiving waters to freshwater organisms. 3rd ed. Cincinnati OH: USEPA. EPA 600/4-91-002.

[USEPA] U.S. Environmental Protection Agency. 1995. Short-term methods for estimating the chronic toxicity of effluents and surface waters to marine and estuarine organisms. 3rd ed. Cincinnati OH: USEPA. EPA-600/4-91/002.

[USEPA] U.S. Environmental Protection Agency. 1997. Office of pollution prevention and toxic substances (OPPTS) test guidelines. Series 850 on Ecological effects. Washington DC: USEPA.

[USEPA] U.S. Environmental Protection Agency. 1999a. Methods for measuring the toxicity and bioaccumulation of sediment-associated contaminants with freshwater invertebrates, 2nd ed. Duluth MN and Washington DC: USEPA. EPA-823-B-99-007.

[USEPA] U.S. Environmental Protection Agency. 1999b. Methods for measuring the toxicity of sediment-associated contaminants in long-term exposures with the marine amphipod *Leptocheirus plumulosus*. Washington DC: USEPA. (EPA series number pending.)

Van Gestel CAM, van Dis WA, van Breemen EM, Sparenburg PM. 1989. Development of a standardised reproduction test with the earthworm species *Eisenia fetida andrei* using copper, pentachlorophenol, and 2,4-dichloroaniline. *Ecotoxicol Environ Safety* 18:305-312.

Versteeg DJ, Stalmans M, Dyer SD, Jansen C. 1997. *Ceriodaphnia* and *Daphnia*: A comparison of their sensitivity to xenobiotics and utility as test species. *Chemosphere* 34:869–892.

Walsh G, McLaughlin L, Louie M, Deans C, Lores E. 1986. Inhibition of arm regeneration of *Ophioderma brevispina* by tributyltin oxide and triphenyltin oxide. *Ecotoxicol Environ Safety* 12:95–100.

Warren JT, Gilbert LI. 1988. Radioimmunoassay: Ecdysteroids. In: Gilbert LI, Miller TA, editors. Immunological techniques in insect biology. New York: Springer-Verlag. p 181–214.

Watts M, Pascoe D. 1996. Use of the freshwater macroinvertebrate *Chironomus riparius* (Diptera: Chironomidae) in the assessment of sediment toxicity. *Water Science Tech* 34:101–107.

Watts M, Pascoe D. 1998. Selection of an appropriate life-cycle stage of *Chironomus riparius* Meigen for use in chronic sediment toxicity testing. *Chemosphere* 36:1405–1413.

Weber DE, McKenney CL Jr, MacGregor MA, Celestial DM. 1996. Use of artificial sediments in a comparative toxicity study with larvae and postlarvae of the grass shrimp, *Palaemonetes pugio*. *Environ Pollut* 93:129–133.

Wells MJ, Young JZ. 1965. Split-brain preparations and touch learning in the octopus. *J Exp Biol* 43:565–579.

Wen CW, Kou CW, Chen SN. 1993. Cultivation of cells from the heart of the hard clam *Meretrix lusoria J Tiss Cult Meth* 15:123–130.

West CW, Ankley GT, Nichols JW, Elonen GE, Nessa DE. 1997. Toxicity and bioaccumulation of 2,3,7,8-tetrachlorodibenzo-*p*-dioxin in long-term tests with the freshwater benthic invertebrates *Chironomus tentans* and *Lumbriculus variegatus*. *Environ Toxicol Chem* 16:1287–1294.

Wharfe JR. 1996. Toxicity based criteria for the regulatory control of waste discharges and for environmental monitoring and assessment in the United Kingdom. In: Tapp JF, Hunt SM, Wharfe JR, editors. Toxic impacts of wastes on the aquatic environment. Cambridge UK: Royal Society of Chemistry. p 26–35.

Whitehurst IT. 1991.The *Gammarus-Asellus* ratio as an index of organic pollution. *Wat Res* 25:333–339.

Widbom B, Oviatt CA. 1994. The world prodigy oil spill in Narragansett Bay, Rhode Island: Acute effects on macrobenthic crustacean populations. *Hydrobiol* 291:115–124.

Widdows J, Donkin P. 1991. Role of physiological energetics in ecotoxicology. *Comp Biochem Physiol* 100C(1/2):69–75.

Wilby OK, Tesh JM. 1990. The *Hydra* assay as an early screen for teratogenic potential. *Toxicology In Vitro* 4:582–583.

Wiles JA, Krogh PH. 1998. Tests with collembolans *Isotoma viridis, Folsomia candida* and *Folsomia fimetaria*. In: Løkke H, van Gestel CAM, editors. Handbook of soil invertebrate toxicity tests. Chichester UK: J Wiley. p 157–179.

Williams AB. 1984. Shrimps, lobsters, and crabs of the Atlantic coast of the eastern United States, Maine to Florida., Washington DC: Smithsonian Institution Pr.

Williams K, Green D, Pascoe D. 1984. Toxicity testing with freshwater macroinvertebrates: Methods and application in environmental management. In: Pascoe D, Edwards RW, editors. Freshwater biological monitoring. Oxford UK: Pergamon. p 81–91.

Williams KA, Green DW, Pascoe D, Gower D. 1986. The acute toxicity of cadmium to different life stages of *Chironomus riparius* and its ecological significance for pollution control. *Oecologia* 70:362– 66.

Williams KA, Green DW, Pascoe D, Gower D. 1987. The effects of cadmium on oviposition and egg viability in *Chironomus riparius* (Diptera - Chironomidae). *Bull Environ Contam Toxicol* 38:86–90.

Williams PL, Dusenberg DB. 1990. Aquatic toxicity testing using the nematode *Caenorhabditis elegans*. *Environ Toxicol Chem* 9:1285–1290.

Wilson JEH, Costlow JD. 1986. Comparative toxicity of two Dimilin formulations to the grass shrimp, *Palaeomonectes pugio*. *Bull Environ Contam Toxicol* 36:858–865.

Wilson JEH, Cunningham PA, Evans DW. Costlow JD Jr. 1995. Using grass shrimp embryos to determine the effects of sediment on the toxicity and persistence of diflubenzuron in laboratory microcosms. In: Hughes JS, Biddinger GR, Mones E, editors. Environmental toxicology and risk assessment. Philadelphia PA: American Soc for Testing and Materials (ASTM). STP 1218. p 267–287.

Woin P, Larsson P. 1987. Phthalate esters reduce predation efficiency of dragonfly larvae, Odonata; *Aeshna*. *Bull Environ Contam Toxicol* 38:220–225.

Wood CE. 1967. Physioecology of the grass shrimp, *Palaemonetes pugio*, in the Galveston Bay estuarine system. *Contrib Mar Sci* 12:54–79.

Wood DE. 1995. Neuromodulation of rhythmic motor patterns in the blue crab *Callinectes sapidus* by amines and the peptide Proctolin. *J Comp Physiol A-Sensory Neural Behavioral Physiology* 177:335–349.

Wood DE, Derby CD. 1995. Coordination and neuromuscular control of rhythmic behaviors in the blue crab, *Callinectes sapidus. J Comp Physiol A-Sensory Neural & Behavioral Physiology* 177:307–319.

Wood DE, Gleeson RA, Derby CD. 1995. Modulation of behavior by biogenic amines and peptides in the blue crab, *Callinectes sapidus. J Comp Physiol A-Sensory Neural & Behavioral Physiology* 177:321–333.

Wood WB, editor. 1988. The nematode *Caenorhabditis elegans*. Plainview NY: Cold Spring Harbor Laboratory Pr.

Yamashita M. 1985. The embryonic development of the brittlestar, *Amphiopholis kochii*, in laboratory culture. *Biol Bull* 169:131–142.

Zar JH. 1984. Biostatistical analysis. 2nd ed. Englewood Cliffs NJ: Prentice-Hall.

Zaroogain G, Anderson S. 1995. Comparison of cadmium; nickel and benzo[a]pyrene uptake into cultured brown cells of the hard shell clam *Mercenaria mercenaria. Comp Biochem Physiol* 11C:109–116.

Zhou B, Jindra M, Hiruma K, Shinoda T, Seagraves WA, Malone F, Riddiford LM. 1998. Regulation of the transcription factor E75 by 20-hydroxyecdysone and juvenile hormone in the epidermis of the tobacco hornworm, *Manduca sexta*, during larval molting and metamorphosis. *Dev Biol* 193:127–138.

Zimmer M. 1999. The fate and effects of ingested hydrolyzable tannins in *Porcellio scaber. J Chem Ecol* 25:611–628.

CHAPTER 4

Field Assessment for Endocrine Disruption in Invertebrates

Peter Matthiessen (chair), Trefor Reynoldson (reporter), Zoe Billinghurst, David W. Brassard, Patricia Cameron, G.Thomas Chandler, Ian M. Davies, Toshihiro Horiguchi, David R. Mount, Jörg Oehlmann, Tom G. Pottinger, Paul K. Sibley, Helen M. Thompson, A. Dick Vethaak

Introduction

Invertebrates represent 95% of all faunal species and underpin a wide range of communities, including marine, freshwater, and terrestrial ecosystems. Invertebrates are a major food source for both carnivorous invertebrates and vertebrate species, in addition to being involved in plant fertilization, decomposition, and additional nondirect processes within ecosystems. Many invertebrate species exhibit complex life cycles that utilize disparate habitats. This results in their being exposed to more than a single source of contamination; for example, barnacles have lecithotrophic (yolk-feeding) larvae, a planktotrophic stage, and a filter-feeding adult. A number of other species have benthic larvae and free swimming or terrestrial adults.

Scope

The scope of this chapter covers approaches for evaluating potential effects of endocrine-disrupting compounds (EDCs) in both aquatic (marine and freshwater) and terrestrial invertebrates. It addresses what we currently know or suspect about effects in the natural world (particularly those at the individual and population levels), and the techniques required for their study.

The objectives of the Field Assessment Work Group were to

1) evaluate data on effects of EDCs in invertebrates under laboratory and field conditions;
2) assess the usefulness of current approaches to environmental monitoring, which can be used to determine exposure to and effects of EDCs;

Endocrine Disruption in Invertebrates: Endocrinology, Testing, and Assessment. Peter L. deFur et al., editors.

3) consider the background variability in invertebrate populations, life histories, and environmental conditions, so that it is possible to discriminate effects of EDCs above this baseline;
4) develop an approach to the detection of effects, assignment of causality, sources and sinks of potential EDCs, and the biological and chemical tools that will need to be developed or exploited for this purpose;
5) recommend suitable biomarkers and endpoints of exposure and effects for EDCs in invertebrates;
6) determine possible triggers or thresholds for field monitoring strategies and identification of causality of effects seen in the field;
7) develop a framework for monitoring in terms of what background, exposure, and effects information is required and how it should be collected; and
8) build on the recommendations of the Endocrine Modulators and Wildlife: Assessment and Testing (EMWAT) Workshop (Vethaak et al. 1997) concerning appropriate environmental monitoring strategies for EDCs (Tattersfield et al. 1997; Ankley et al. 1998).

Background

The invertebrate endocrine system

As was discussed in more detail in Chapter 2, the use of hormones to control and coordinate biochemical, physiological, and behavioral processes is common to all invertebrate taxa. Neuropeptide signaling mechanisms that use the peptide products of specialized neurosecretory cells are the predominant effectors among the endocrine systems so far characterized in invertebrates, but nonpeptide endocrine messengers are of importance in many groups. Both of these systems are potentially susceptible to interference by EDCs. Before considering the evidence for endocrine disruptive effects in invertebrates "in the field," it is appropriate to briefly summarize the major similarities and differences between the endocrine systems of the major invertebrate groups.

The endocrine systems of the 2 major arthropod classes, insects and crustaceans, are best documented. Insects, in addition to peptide hormones, use homosesquiterpenoid epoxides (the juvenile hormones [JHs]) and ecdysteroids (ecdysone, 20-hydroxyecdysone [20E]), neither of which is found in vertebrates (Cymborowski 1992). A range of vertebrate-type steroids (androgens, estrogens, progestogens, corticosteroids) have been detected in insects, but it remains to be demonstrated whether these have a functional role (Lafont 1991; Swevers et al. 1991).

In common with insects, crustaceans possess a wide range of peptide hormones and also use ecdysteroids (Fingerman 1997). In addition to ecdysteroids, the nonpeptide methyl farnesoate (MF) appears to act as a JH in crustacea (Laufer et al. 1993). As with insects, vertebrate-type steroids are detectable in crustacean tissues, but again their functional significance is yet to be demonstrated (Lafont 1991; Fingerman et al. 1993).

Mollusks utilize a wide range of peptide hormones to control and coordinate major processes (Joosse and Geraerts 1983; Joosse 1988). In contrast to the arthropods, ecdysteroids do not appear to play an important role in mollusks, but the presence of vertebrate-type steroids has been reported for a number of molluskan species, and in some cases, especially in the prosobranch gastropods, the evidence that these steroids play a functional role is strong (Reis-Henriques et al. 1990; Hines et al. 1996).

The echinoderms represent the invertebrate group in which the evidence for a significant role of vertebrate-type steroids is strongest. There is considerable published evidence suggesting that such steroids are synthesized by echinoderms and may have a role in the control of growth and reproduction (e.g., Harrington and Ozaki 1986; Voogt et al. 1992). Peptide signal systems are also important in echinoderms, and purine 1-methyladenine plays a role in the later stages of oocyte development (Cobb 1988; Smiley 1990).

Of the coelenterates, poriferans, acoelomates, aschelminths, and annelids, there are varying degrees of understanding about endocrine-type processes and isolated reports of steroid synthesizing activity (Thorndyke and Georges 1986; Takeuchi 1987; Lesh-Laurie 1988; Fairweather and Skuce 1995).

Susceptibility of invertebrate species to endocrine-disrupting compounds

It has been estimated that there are probably in the order of 60,000 organic pollutants present in fresh water and at least this number in the marine environment (Maugh 1987). These originate from domestic and industrial effluents, leachates from solid waste disposal sites, deliberate application of agrochemicals, agricultural and urban run-off, contaminated land, incineration, atmospheric fall-out, marine traffic, and accidental discharge.

Environmental contaminants can have an impact on the invertebrate endocrine system through 1 of 3 mechanisms:

1) by acting as a hormone mimic (agonist),
2) by acting as an anti-hormone (antagonist), and
3) by interfering with biochemical processes associated with the production, availability, or metabolism of hormones, or the modulation of receptors.

There is no evidence in any animal group, at present, of endocrine-disruptive effects being exerted through a peptide signaling system. However, if all nonpeptide signaling systems are potential targets for EDCs, then every invertebrate group must be considered to be potentially susceptible. Given the complex and multifunctional roles performed by those nonpeptide hormones whose significance is best understood in invertebrates, the likely impact of interference at any particular site within the endocrine system is difficult to predict. Nonetheless, it can safely be assumed that the effects of exposure to EDCs will include alterations in one or more of reproduction, molting, feeding, and behavior.

Invertebrate species may be particularly sensitive to EDCs during critical developmental periods such as molting or metamorphosis and during the vigorously growing larval stages (see Chapter 3). In particular, the vulnerability of aquatic invertebrates to general pollutants suggests that they will also be susceptible to exposure to EDCs. In addition, invertebrates include a number of ecologically and commercially important species, the collapse of which may have significant biological or economic impacts (e.g., oyster fisheries).

Use of invertebrates in monitoring for the effects of endocrine-disrupting compounds in the field

Attributes specific to the use of invertebrates

The investigation and development of biomarkers of exposure and effects, including higher-level impacts of endocrine disruption (ED) in invertebrates, is a necessary supplement to existing monitoring programs which tend to emphasize effects on vertebrates. The effect of EDCs on invertebrates can be observed and monitored through the full life cycle of the organisms: early life stages, juveniles, adults, and egg stages (see Chapter 3). Initially, genetically similar individuals can be cultured in order to generate and validate biomarkers of exposure and can be used as sentinel species for in situ testing in field trials. This may enable the effects of potential EDCs to be determined for utility in field studies. In addition, the use of invertebrate species facilitates transgenerational impacts to be investigated over a relatively short time period.

Sessile invertebrate species or life stages (e.g., mussels, bryozoans, barnacles, tunicates, and sponges) provide ideal organisms to employ in field studies (Rosenberg and Resh 1993). These may address the presence or absence of ED at specific sites, as they provide information on the site in which they are found, with minimal risk of immigration as larval forms. Care must be taken with mobile species, as exposure may have occurred at early life stages and not at the site where the adult is found and the signs of exposure observed.

Concerns associated with the use of invertebrates in monitoring

Significant difficulties arise when the study of invertebrates is introduced into the field of ED, a number of which may also be relevant to vertebrate studies. The impacts of EDCs may differ according to the specific life stage at which exposure occurs. This may be further complicated by periods during which levels of endogenous hormones vary (e.g., during metamorphosis and molting). Changes in the ecology of the life stages may also introduce complications (McKenney and Matthews 1990), and the generally small mass of invertebrate individuals presents limitations on the amount of tissues for biomarker and residue measurements, etc.

In invertebrates, as with certain vertebrate species, there are a significant number of species that naturally exhibit sequential or simultaneous hermaphrodism (Villalejo Fuerte et al. 1996; McCartney 1997). When population and individual effects of reproductive EDCs are sought, hermaphrodism may introduce complications into determining the endpoints and ultimate effects of EDCs on the species. In addition, it is possible that impacts of reproductive EDCs are more likely to be reversible in hermaphrodites than are those observed in nonhermaphroditic species.

Furthermore, given the substantial differences that exist between the vertebrate and invertebrate endocrine systems, chemicals that have been identified as not having an effect in vertebrates may nonetheless impact on hormonal systems that are unique to particular groups of invertebrates, and vice versa (e.g., Charniaux-Cotton 1962; King 1964). In addition, EDCs may not produce a classic concentration response but may generate a biphasic or sigmoidal response through a range of concentrations and at different life stages of different species. Endocrine disruption may occur following exposure to low concentrations of certain substances, with nonendocrine toxicity effects being apparent at higher levels.

The physiological and endocrine processes that may be affected in invertebrates are less well understood than are those for many vertebrate species (Chapter 2). Consequently, caution should be exercised in interpreting effects of exposure to compounds that are manifested as changes in invertebrate growth patterns, reproductive performance or the occurrence of abnormalities. These need not necessarily be a result of ED (e.g., Billinghurst et al. 1998).

Differentiating between ED and other toxic effects is problematic, as chemicals can interfere with an animal's endocrine system before toxic effects become apparent, thus complicating the interpretation of "effect." Furthermore, as biological systems degenerate following toxic exposure, the integrity of the endocrine system is likely to collapse, although this would not generally be described as ED.

Evidence for Endocrine Disruption in Field and Laboratory Studies

The case histories that follow document the main examples of ED. There are remarkably few of these from the field situation, and they are dominated by the case of the antifouling-paint ingredient tributyltin (TBT), whose range of ED effects in mollusks has had repercussions for mollusk populations worldwide. The following discussion may appear unbalanced due to the relative lack of field data on other ED impacts in invertebrates, but this reflects the state of knowledge at the present time. Furthermore, the understanding of TBT and its effects is unusually comprehensive, so dwelling disproportionately on this interesting case is felt to be justified by the insights that are gained into possible modes of action in other invertebrate species and with other chemicals.

Organotins and mollusks

Gastropod mollusks

Morphological expression of imposex and intersex

The effects of TBT on gastropods are the most complete example of the effects of an EDC on marine invertebrates. Although the effects were originally observed as morphological changes in individuals (Smith 1981; Bryan et al. 1986), there is now an understanding of the effects at all levels of organization, from the molecular to the population (Bryan et al. 1988; Spence et al. 1990) and possibly the community, all of which present opportunities for observation of effects in field samples.

At the molecular level, TBT interferes with hormone metabolism (Matthiessen and Gibbs 1998), increasing the levels of androgens in the snails. It is possible to measure these levels and also the activity of the aromatase enzyme, but these sophisticated approaches are not normally considered for field monitoring programs.

Within dioecious prosobranch gastropods, TBT compounds are known to induce 2 different masculinization phenomena: imposex and intersex. Their common feature is the gradual and concentration-dependent increase of virilization intensities in females, with the final result of female sterility. While in imposex-affected species the entire female genital system is conserved but superimposed (thus the term "imposex") by male organs like a penis and/or vas deferens, in the periwinkle *Littorina littorea*, the only intersex-affected species described to date, the female pallial organs are modified toward a male morphological structure in lower intersex stages (1 and 2) (see section "Indices of imposex and intersex" below for details of imposex and intersex indices) and then supplanted by the corresponding male formation, a prostate gland, in stages 3 and 4. These organ-level phenomena can be described by using evolutive schemes (Figures 4-1 and 4-2), and are the basis for the quantification of imposex and intersex.

The final result of intersex and imposex development is the sterilization of females. *Littorina littorea* is definitively sterilized in stages 2, 3, and 4, either because the oocytes and capsular material leak into the mantle cavity (stage 2) or the glands responsible for the formation of the planktonic egg capsule are at least partially missing. Imposex-affected prosobranchs are sterilized in stages 5 and 6. In stage 5, the oviduct is blocked by the formation of a prostate (Stage 5a) or by proliferating vas deferens tissue (stage 5b), resulting in the accumulation of abortive egg capsules in the pallial oviduct (stages 6a, b). Another possibility is the inhibition of the ontogenetic closure of the pallial oviduct section, which normally infolds from the mantle epithelium. Therefore, this oviduct section is split ventrally in adult stage 5c females, with comparable reproductive consequences to the intersex stage 2 of the periwinkle.

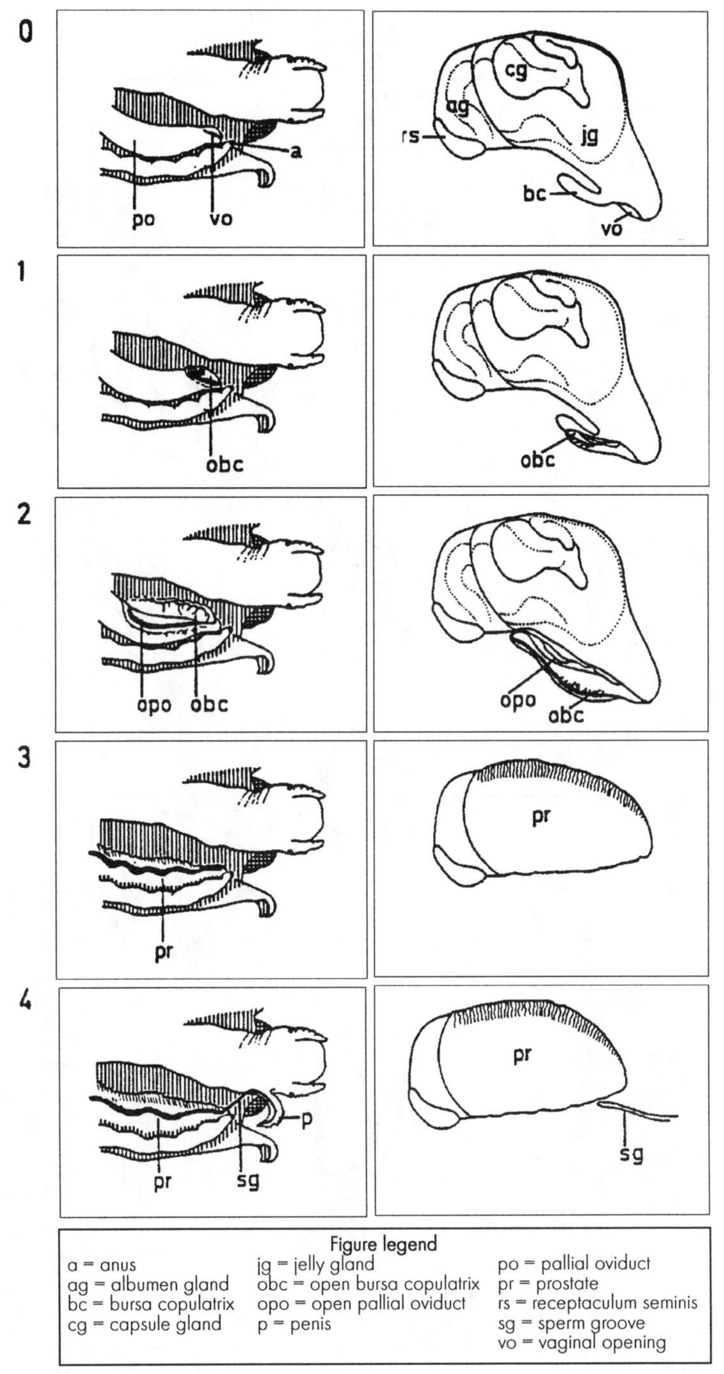

Figure 4-1 *Littorina littorea* (female) intersex development scheme with 4 different stages. Dorsal views with opened mantle cavity (left) and lateral views of the pallial section of the genital tract (right) (from Oehlmann 1998, by permission).

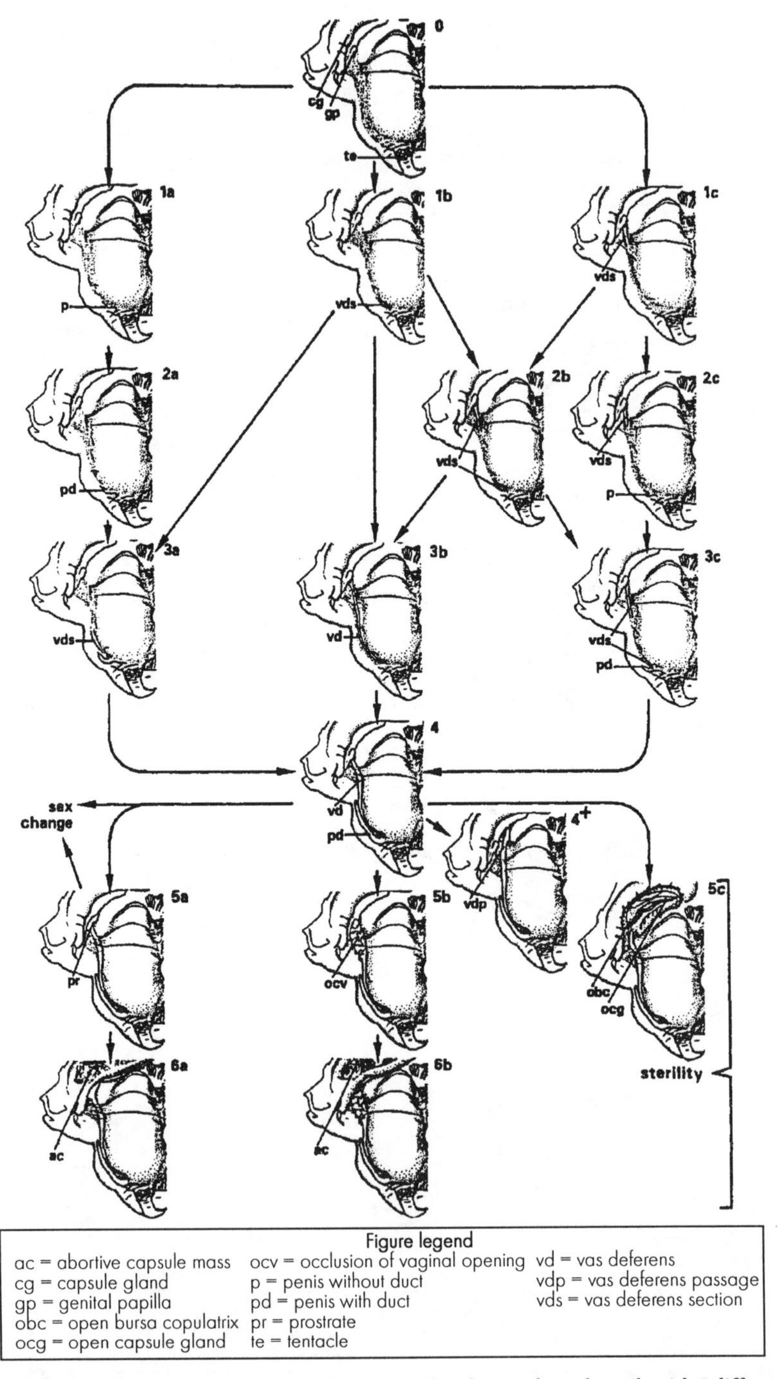

Figure 4-2 General scheme of imposex development in female prosobranch snails with 6 different stages (dorsal views with opened mantle cavity). In *Nucella lapillus,* all stages occur, with the exception of 4+ and 5c (from Oehlmann 1998, by permission).

The degree of development of the vas deferens in prosobranch snails such as the dogwhelk *Nucella lapillus* is a direct expression of the impact of imposex on the reproductive capability of the population, the consequences of which have been observed as distortions of the sex ratio, reductions in the numbers of juveniles and their recruitment to the adult population, and reduction and elimination of populations in severely affected locations.

The ability to demonstrate clear links between effects at the organ or individual level and the population level is strongly dependent on the rather low fecundity of dogwhelks, on their limited mobility on beaches, and on the benthic mode of life of hatchlings, all of which limit dispersion (Crothers 1985). Populations of some other gastropods, which have shown similar effects on reproductive capacity (Bauer et al. 1995, 1997) but which have planktonic larval stages, have not generally shown such marked effects at the population level, as it has been possible for recruitment to take place in heavily contaminated populations by larvae produced some distance away.

Indices of imposex and intersex

As a necessary precursor to the use of imposex (and later, intersex) observations in field monitoring programs, it was necessary to develop numerical indices of effect. For *Littorina littorea*, the intersex index (ISI, average intersex stage in a given sample), the incidence of sterile females, and the mean length of the prostate gland in females can be used; if ISI > 1 in *Littorina*, some females will have been sterilized (Bauer et al. 1995, 1997). For imposex-exhibiting species, the vas deferens sequence index (VDSI, average imposex stage in a sample of *Nucella*; VDSI > 4 indicates that some females are sterile), the relative penis size index (RPSI, measured as a percentage of the male penis size in *Nucella*), and relative penis length index (RPLI, used in *Buccinum*), imposex incidence, or the average penis length of females can be used (Gibbs et al. 1987; Oehlmann 1994). The ISI and VDSI provide very good estimates of the reproductive capability of females in analyzed populations (Stroben et al. 1996).

Chemical substances responsible for imposex in gastropods

The possible mechanisms of imposex and intersex induction in prosobranch snails have been recently reviewed by Matthiessen and Gibbs (1998). Féral and Le Gall (1982, 1983) reported that TBT inhibited the release of a neuroendocrine factor from the pleural ganglia, which was responsible for the suppression of penis formation in females, resulting in the development of imposex in TBT-exposed female *Ocenebra erinacea*, although the factor had not been identified. Because TBT is reported to accumulate specifically in cerebral ganglia in the dogwhelk (Bryan et al. 1993), it is possible that disturbance of the neuroendocrine system by TBT toxicity is involved in the development of imposex, perhaps because the neuroendocrine factor directly or indirectly mediates steroid metabolism or production.

It seems clear that TBT exposure results in an increase of endogenous testosterone titers (Spooner et al. 1991), but there is a debate about the mechanisms that are responsible for this. While Oehlmann and Bettin (1996) and Bettin et al. (1996) have found evidence for a competitive inhibition of the cytochrome P450-dependent aromatase, Ronis and Mason (1996) favor an interference with the phase II-metabolism resulting in a reduced excretion of testosterone (although a reduction of aromatase activity by 30 to 40% has also been demonstrated in their own experiments). Furthermore, Oberdörster, Rittschof, and McClellan-Green (1998) have found decreased production of testosterone metabolites in field-collected snails with imposex. Matthiessen and Gibbs (1998) conclude that, at the moment, the aromatase inhibition hypothesis of Bettin et al. (1996) is the better established.

It has been accepted that imposex is induced almost typically by TBT used in antifouling paints, which is based on results of laboratory experiments using the dogwhelk *Nucella lapillus* (Bryan et al. 1987). In the rock shell *Thais clavigera*, however, triphenyltin (TPhT) as well as TBT promoted the development of imposex during injection experiments (Horiguchi et al. 1997a). The potency of TPhT for promoting the development of imposex in the rock shell is estimated to be approximately the same as that of TBT (Horiguchi et al. 1997a). Results of injection experiments using 18 kinds of organotin compounds showed that 4 organotins (tripropyltin [TPrT], TBT, tricyclohexyltin [TCHT] and TPhT) promoted the development of imposex in the rock shell (Horiguchi et al. unpublished data). The estimated order of potency for promoting imposex in the rock shell is as follows: TPhT = TBT > TCHT > TPrT (Horiguchi et al. unpublished data). In the case of the dogwhelk, TPhT did not promote the development of imposex, although TPrT and tetrabutyltin slightly promoted it (Bryan et al. 1988). It is also reported that monophenyltin (MPhT)-induced imposex in *Ocenebra erinacea* (Hawkins and Hutchinson 1990). It is possible that sensitivity to organotins inducing or promoting the development of imposex differs among gastropod species, although little is known about such interspecific differences.

Imposex-affected species exhibit in general a higher TBT sensitivity than the intersex-developing snail *Littorina littorea*, where the aquatic threshold concentration for development of intersex characteristics is 5.0 ng TBT-Sn/L compared with < 0.5 ng TBT-Sn/L for imposex development in *Nucella lapillus* (Oehlmann 1998). The most sensitive species known currently seems to be *Ocinebrina aciculata*, with a threshold concentration of 0.1 ng TBT-Sn/L.

In contrast to intersex, with *Littorina littorea* being the only known affected species currently, imposex is widespread; Table 4-1 summarizes the approximately 150 different species that exhibit this virilization phenomenon. On the other hand, some prosobranch snails do not develop imposex even when exposed to high aqueous TBT concentrations or sampled in highly contaminated areas. Within these species are the limnic archaeogastropod *Theodoxus fluviatilis*, the mesogastropods *Bithynia tentaculata*, *Hydrobia ventrosa*, and *Potamopyrgus antipodarum*, and the buccinid species *Columbella rustica*. Other marine species such as *Capulus ungaricus* and

Table 4-1 List of known imposex-affected prosobranch species

Mesogastropoda

Ampullariidae
- *Ampullaria (Pachylabra) cinerea*
- *Ampullaria gigas*
- *Ampullaria polita*
- *Marisa cornuarietis*
- *Pila globosa*
- *Pomacea canaliculata*

Viviparidae
- *Campeloma rufum*

Hydrobiidae
- *Hydrobia ulvae*

Rissiodae
- *Rissoa auriscalpium*
- *Rissoa ventricosa*
- *Rissoa violacea*
- *Rissoina allanae*
- *Rissoina boucheti*
- *Rissoina bruguierei*
- *Rissoina nivea*

Cypreaidae
- *Trivia aperta*
- *Trivia arctica*
- *Trivia costata*
- *Trivia millardi*
- *Trivia monacha*
- *Trivia verhoefi*

Lamellariidae
- *Lamellaria perspicua*

Pterotracheidae
- *Pterotrachea coronata*

Aporrhaiidae
- *Aporrhais pespelecani*

Cassididae
- *Galeodea tyrrhena*

Cymatiidae
- *Fusitriton oregonensis*
- *Monoplex echo*
- *Charonia sauliae sauliae*

Strombidae
- *Strombus luhuanus*

Naticidae
- *Neverita didyma*

Tonnidae
- *Tonna luteostoma*

Neogastropoda

Galeodidae
- *Buscycon carica*
- *Buscycon contrarium*
- *Melongena corona*

Muricidae
- *Calotrophon ostrearum*
- *Ceratostoma burnetti*
- *Ceratostoma foliatum*
- *Cronia margariticola*
- *Cronia pothuauii*
- *Dicathais orbita*
- *Drupella fragum*
- *Drupella ochrostoma*
- *Drupella rugosa*
- *Ergalatax contractus*
- *Eupleura caudata*
- *Forreria belcheri*
- *Haustrum haustorium*
- *Lepsiella albomarginata*
- *Lepsiella scobina*
- *Lepsiella vinosa*
- *Morula granulata*
- *Morula marginalba*
- *Morula marginatra*
- *Morula musiva*
- *Murex (Truncolariopsis) trunculus*
- *Murex brandaris*
- *Murex florifer dilectus*
- *Murex pomum*
- *Naquetia capucina*
- *Nucella canaliculata*
- *Nucella emarginata*
- *Nucella freycineti*
- *Nucella heyseana*
- *Nucella lamellosa*
- *Nucella lima*
- *Nucella lapillus*
- *Ocenebra erinacea*
- *Ocenebra lumaria*
- *Ocenebra lurida*
- *Ocinebrina aciculata*
- *Plicopurpura patula*
- *Rapana venosa venosa*
- *Thais bronni*
- *Thais clavigera*
- *Thais haemastoma*
- *Thais kieneri*
- *Thais luteostoma*
- *Thais mancinella*
- *Thais orbita*
- *Thais savignyi*
- *Urosalpinx cinerea*
- *Urosalpinx perrugata*
- *Urosalpinx tampaensis*
- *Vasum turbinellus*
- *Xymene ambiguus*

Coralliophilidae
- *Coralliophila abbreviata*
- *Coralliophila lamellosa*

Buccinidae
- *Babylonia japonica*
- *Buccinum middendorffi*
- *Buccinum opisthoplectum*
- *Buccinum undatum*
- *Colus gracilis*

Buccinidae (continued)
- *Colus halli*
- *Cominella virgata*
- *Japeuthria ferra*
- *Kelletia lischkei*
- *Neptunea arthritica arthritica*
- *Neptunea phoenecia*
- *Pisania tinctus*
- *Pusiostoma mendicaria*
- *Searlesia dira*
- *Searlesia fuscolabiata*
- *Volutharpa ampullacea perryi*

Melongenidae
- *Hemifusus tuba*

Nassidae
- *Bullia rhodostoma*
- *Cyclope neritea*
- *Hinia incrassata*
- *Hinia reticulata*
- *Hinia trivittata*
- *Hinia vibex*
- *Ilyanassa obsoleta*
- *Reticunassa festiva*

Columbellidae
- *Anachis avara*
- *Mitrella lunata*

Fasciolariidae
- *Fasciolaria hunteria*
- *Fasciolaria lilium*
- *Fusinus perplexus perplexus*
- *Fusinus syracusanus*
- *Pleuropoca gigantea*
- *Taron dubius*

Cancellariidae
- *Sydaphera spengleriana*

Marginellidae
- *Marginella apiciana*

Olividae
- *Amalda (Baryspira) australis*
- *Olivella biplicata*

Conidae
- *Conus anemone*
- *Conus coronatus*
- *Conus dorreensis*
- *Conus klemae*
- *Conus lischkeanus*
- *Conus marmoreus bandanus*
- *Conus mediterraneus*
- *Conus sponsalis*
- *Kurtziella cerina*
- *Lora (Propebela) turricula*
- *Raphitoma reticulata*
- *Viriconus lividus*
- *Virroconus ebraeus*
- *Virroconus fulgetrum*

Terebridae
- *Terebra dislocata*
- *Terebra protexta*

Sources: Updated from Fioroni et al. 1991, The pseudohermaphroditism of prosobranchs morphological aspects, *Zool Anz* 226:1–26, with permission from Bustav Fischer Verlag Jenna GmbH; and Horiguchi et al. 1997b, Imposex in sea snails, caused by organotin (tributyltin and triphenyltin) pollution in Japan: a survey. *Appl Organometal Chem* 211:451–455, © John Wiley & Sons Limited, reproduced with permission.

Turritella communis are not able to develop imposex because males are aphallic and do not possess a vas deferens (Oehlmann 1998).

Field survey data

A large number of reports of field surveys and assessments are available in the literature. As a representative example, we present data from Sullom Voe, Shetland Islands, Scotland (Harding et al. 1997). Sullom Voe is a narrow inlet approximately 10 km long, leading southwest from Yell Sound into the Shetland mainland. There is a large oil terminal on the promontory of Calback Ness, with 3 loading jetties for oil tankers on the sheltered southwest side of the peninsula. Surveys of imposex in *Nucella* have been carried out in Sullom Voe and adjacent parts of Yell Sound since 1987, at approximately 2-year intervals. In 1995, low background VDSI values were found at the northerly limits of Yell Sound. These values increased toward the oil terminal. The population closest to the jetties had a VDSI of 5.17. In the earliest surveys, populations of dogwhelks had been found immediately adjacent to the jetties but consisted entirely of old adult males and sterile females. In subsequent surveys, dogwhelks were absent at these sites. VDSI values exceeded 4 at all sampling sites in Sullom Voe, showing that many females were sterile.

RPSI values showed a very similar pattern of distribution, with values of < 0.1% at the most northerly stations in Yell Sound and generally around 20 to 30% inside Sullom Voe. The data clearly indicate that the main sources of TBT in the area are the tanker vessels visiting the oil terminal, particularly around the loading jetties (Figures 4-3 and 4-4).

Repeated surveys can be used to investigate temporal trends in the degree of imposex and levels of contamination. Between 1987 and 1996, there has been a general decline in RPSI levels in Sullom Voe. There has also been a decline in VDSI values, particularly at stations 14 through 20 to the east of Sullom Voe. It is suggested that these improvements are probably the result of the decline in the use of free-association TBT paints and in the number of vessels using the port. In designing monitoring programs and interpreting temporal data, it is necessary to take into account that the typical life span of *Nucella* is around 5 to 6 years and that the morphological expression of imposex is not reversible. Imposex in adults may therefore reflect the effects of exposure as juveniles and obscure any reductions in the environmental contamination by TBT that may have taken place subsequently. The collection of supporting information, particularly chemical analysis of tissues for tin and TBT, can assist in interpreting the data.

In Japanese waters, imposex was found both in areas with high and with low shipping traffic intensity. The latter are likely to receive TPhT compounds not only from antifouling paints but additionally from neighboring agricultural areas, where they are used as fungicides and for the control of algae and snails in rice paddies (Horiguchi et al. 1994). Experimental evidence regarding the ability to develop imposex due to TPhT exposure is at present contradictory, since both negative and

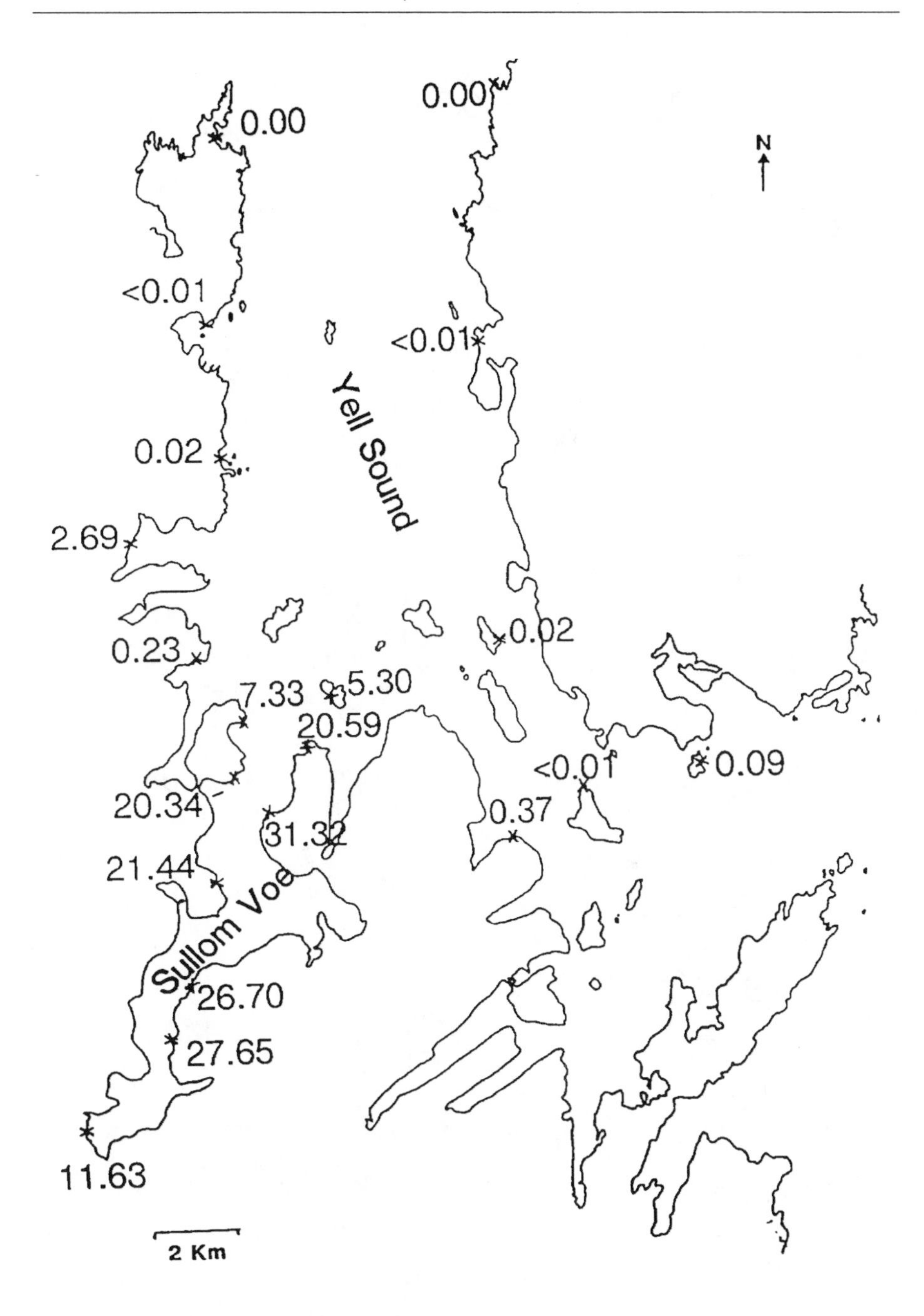

Figure 4-3 RPSIs in toothed adult *N. lapillus* from populations in Sullom Voe and Yell Sound surveyed in 1995. (RPSI = relative penis size index, "the mean bulk of the female penis expressed as percentage of the mean bulk of the male penis. The bulk of a penis is calculated as the cube of i length" [Gibbs et al. 1987; reprinted with permission from Cambridge University Press].)

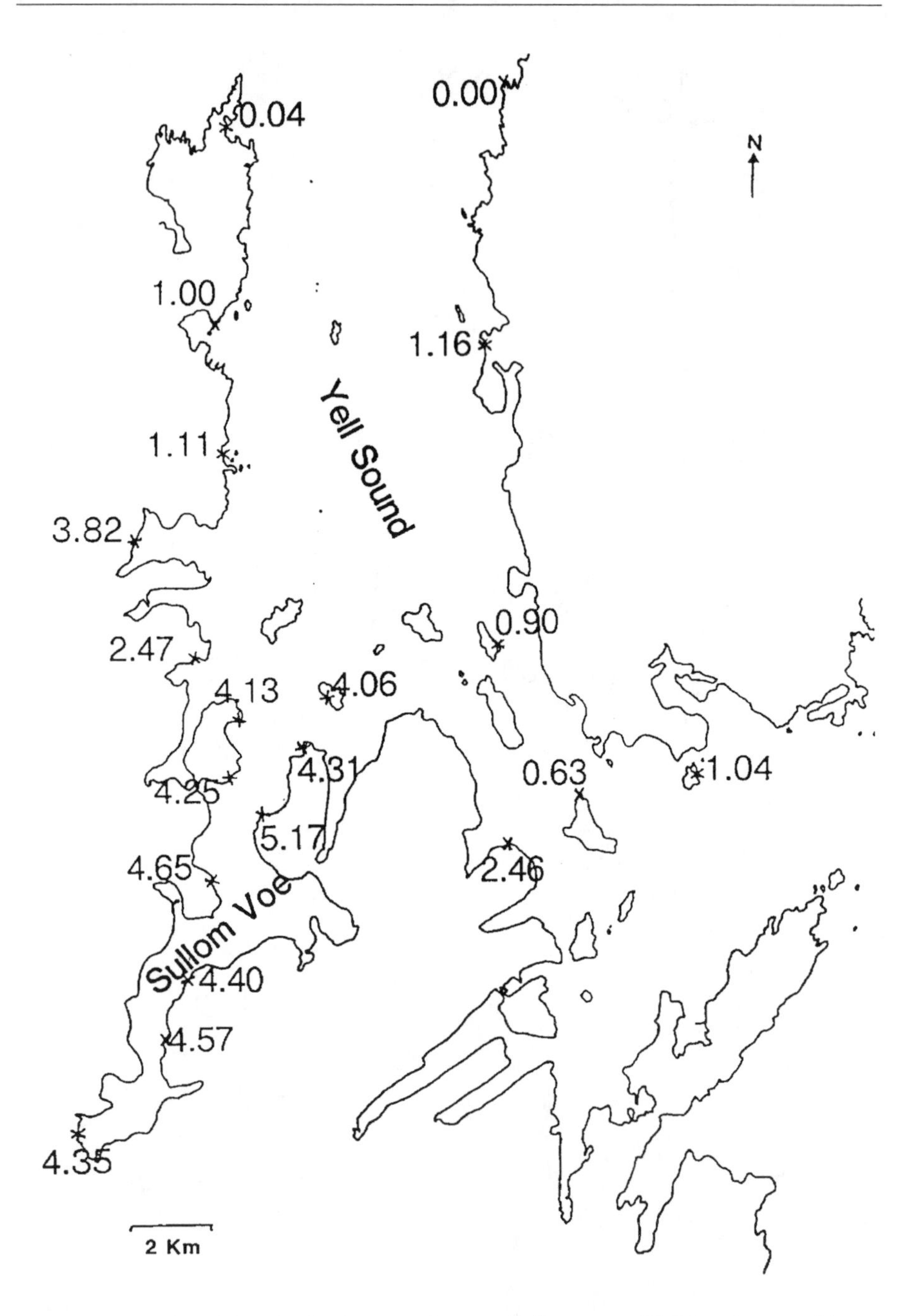

Figure 4-4 VDSIs in adult *N. lapillus* from populations in Sullom Voe and Yell Sound surveyed in 1995. (VDSI = vas deferens sequence index, the stage of vas deferens development in the female [Gibbs et al. 1987; reprinted with permission from Cambridge University Press])

positive experimental study results have been reported, possibly depending on the species tested (Bryan et al. 1988; Horiguchi et al. 1997a). In the latter case, it can now be stated that TPhT at least enhances imposex development in the rock shell (*Thais clavigera*). For this species, TPhT exposure resulted in an enhancement of imposex intensities, which was comparable to the effects of TBT within the same concentration range. Therefore, a possible role for TPhT in the development of imposex seems quite likely in this species. Recent and yet unpublished results have shown that in *Thais clavigera*, TCHT and tripropyltin compounds are also capable of inducing imposex (Horiguchi et al. 1994, 1997b, 1998). Triphenyltin compounds are used as biocides in antifouling paints and also as fungicides in agriculture. They are not used in antifouling paints or agriculture in Japan, although their use continues in some countries at relatively low levels. TCHT compounds have been used as fungicides but are now prohibited in various areas, including Europe and Japan.

Imposex surveys can therefore be undertaken at a wide range of geographical scales. The Sullom Voe surveys have been carried out over a total distance of up to 30 km. Surveys have been reported around single harbors in which sampling sites may be separated by much smaller distances, perhaps only 100 to 300 m (Minchin et al. 1995, 1996), and also on regional scales, e.g., around the coasts of the North Sea (Harding et al. 1999) and Japan, where the distance between sites may be several tens of km or more. It has been demonstrated that *Nucella* from populations separated by more than 10 degrees of latitude (Gibbs et al. 1991) show a consistency of response to TBT exposure. Imposex measurements can therefore be used to provide interpretable information on the effects of TBT in both spatial and temporal surveys.

Bivalve mollusks

In addition to gastropods, several species of bivalve mollusks exhibit biological effects in response to TBT exposure. In the early 1980s, considerable concern arose in some areas where the Pacific oyster *Crassostrea gigas* was cultivated. The oysters showed unusual shell morphology, in which the valves became greatly thickened and developed internal chambers. In extreme cases, the oysters became ball-shaped, and the meat yield was greatly reduced to the extent that the oysters were no longer commercially acceptable. The effect was quantified through a shell thickness index. Using this index, field surveys were carried out in Arcachon, France and in the UK, which demonstrated that the deformity became less severe with distance from marinas. Subsequent research showed that the effect was caused by exposure to TBT, and the deployment of juvenile Pacific oysters in net bags for a period of weeks was used widely as a method for assessing the TBT contamination status in estuarine and in-shore waters (Waldock and Thain 1983; Alzieu et al. 1986). It was also shown that at high concentrations of TBT, the female reproductive systems were altered and spermatogenesis could occur. Various effects of TBT on reproduction have been reported in a range of bivalves including *Ostrea edulis, Mytilus edulis, Scrobicularia plana,* and *Mercenaria mercenaria* (Thain et al. 1986; Laughlin et al. 1988, 1989;

Lapota et al. 1993; Ruiz, Bryan, and Gibbs 1995; Ruiz, Bryan, and Wigham 1995) and have been shown to be linked to alterations of the endocrine system. It is not clear whether shell thickening in Pacific oysters is a consequence of some form of ED or arises from interaction between TBT and other aspects of the biochemistry of the oysters. It may be noted that not all strains of Pacific oyster are equally susceptible to shell thickening caused by TBT. The strain prevalent in northern France and the UK seems particularly sensitive. Strains cultivated on the Mediterranean coasts of France and Spain, and in Japan (Okoshi et al. 1987), seem much less susceptible.

Community-level effects of tributyltin

There is relatively little information on this issue, but it seems likely that the endocrine-disrupting effects of TBT in mollusks have consequent effects for some marine communities. Experiments have been conducted on the Isle of Man (Spence et al. 1990) in which all *Nucella lapillus* individuals were removed in an attempt to simulate the consequences of severe imposex. In essence, the results showed a subsequent increase in the population density of limpets and to a lesser extent those of mussels and barnacles (on which dogwhelks prey). The increase of limpets in turn led to a decline in macrophytes because of heavy grazing pressure.

Detailed temporal trend studies of benthic and epibenthic community structure (Matthiessen et al. 1999; Rees et al. 1999; Waldock et al. 1999) in the Crouch estuary, UK, have been conducted since 1987, when TBT use was banned on the many small vessels moored in the area. Other polluting inputs to the estuary have been negligible historically (including TBT inputs from large vessels), so this was a useful natural experiment. The outcome is that over a 5 to 10 year period, benthic and epibenthic species diversity increased considerably in those upper- and mid-estuarine areas where TBT concentrations were high in 1987, but which subsequently declined to relatively low levels. There were no significant changes in the originally less-contaminated areas of the outer estuary, giving some confidence that the effects were at least partly due to the decline in TBT inputs. Recovery was seen not only in several mollusk species (including the flat oyster *Ostrea edulis*) but also in crustaceans and ascidians, although it is not known whether these latter groups were originally affected directly or indirectly.

In summary, field data of this (or any) type cannot be taken as proof that the endocrine-disrupting effects of TBT have led to changes in marine communities, but such a conclusion does not seem unreasonable.

Effects of controls on the use of tributyltin

Once it became apparent in the mid 1980s that TBT from antifouling paints was a potent toxicant to a wide variety of marine organisms, including some of commercial importance, regulations began to be introduced at national and international levels to restrict the input of TBT to the aquatic, and particularly the marine, environment. In preparing controls, distinctions were generally made between terrestrial uses of TBT (e.g., in timber treatment), in aquaculture (for preventing the settlement of fouling organisms on netting cages), and as an antifouling agent on small boats and larger vessels. In response to varying national and international perspectives, controls differed between states and regions, and there is still widespread use of TBT on large vessels.

In some areas, the controls were effective and there were measurable improvements in water quality and biological responses, but in other regions, environmental concentrations remain high and effects on mollusks are detectable. Some surveys clearly showed a recovery from imposex in formerly heavily affected populations (e.g., Evans et al. 1994; Smith 1996; Harding et al. 1997). Dogwhelk populations in a Scottish sea loch, which had previously been strongly affected by TBT, now show very little indication of imposex. Imposex has declined in areas where the main input of TBT was previously from small boats. Areas around harbors used by larger vessels that continue to use TBT paints generally show higher levels of imposex (Harding et al. 1997). This reflects continuing releases from vessels as well as dissolution of TBT from contaminated seabed sediment. In other European regions, the situation seems to be even more serious. At distances of several kilometers from harbors in France and Ireland, dogwhelk populations are unable to reproduce because of the existing level of TBT exposure (e.g., Oehlmann et al. 1993; Minchin et al. 1995, 1996, 1997; Huet et al. 1996).

Imposex has been reported in the common whelk *Buccinum undatum* from the open North Sea, where 20 years ago no signs of this lesion were observed, and this is apparently related to the intensity of commercial shipping traffic in the area (Hallers-Tjabbes et al. 1994). In the southern part of the North Sea and in the western part of the Baltic, TBT concentrations are so high that up to 100% of female periwinkle (*Littorina littorea*) populations are sterile (e.g., Oehlmann et al. 1997). Even after stopping further inputs to smaller harbors, concentrations in the sediment are still high, by a factor of up to 47,000 above the no-effect concentration for *Littorina littorea* (Bauer et al. 1997). Binding of TBT to the sediments and remobilization from them is now the main source of contamination. In an estuary in Delaware, the frequency of imposex in *Ilyanassa obsoleta* is still high, although its severity has been reduced considerably over the last 12 years (Curtis and Kinley 1998).

Shell thickening of oysters in Arcachon declined in parallel with reductions in the concentration of TBT in the water, although even today between 38 and 46% of oyster shells are deformed in this area (Sarradin et al. 1994) and in other European regions. However, an improvement can be observed on hard-bottom coasts with strong water exchange. In soft-bottom areas, the situation either has not improved or is getting worse in areas of low water exchange like the Baltic Sea (Oehlmann et al. 1997).

In Japan, where all marine uses of organotins as antifoulants were stopped in the early 1990s, TPhT has not been detected in seawater for 2 years, but TPhT concentrations in bottom sediments and in biota (fish and shellfish) are still detectable. While TBT concentrations in seawater seem to be decreasing, TBT concentrations in bottom sediments and in biota (fish and shellfish) remain and seem to be high in areas of high shipping activity (Environmental Agency of Japan 1998). Foreign vessels larger than 25 m in length, which can still use TBT antifouling paints, may contribute to this contamination as well as to the persistence of organotin compounds in bottom sediments.

Imposex in the rock shell, *Thais clavigera*, has been extensively observed in a countrywide survey in Japan. Incidence (percentage occurrence) of imposex and observed values of RPLI have been high in this species. Individuals that are sterile because of oviduct blockage caused by vas deferens formation have also been observed frequently in some areas. Imposex symptoms have been severe in the rock shell, although mass extinction has not occurred because of its relatively longer planktonic (veliger) larval stage (Horiguchi et al. 1994, 1997b, 1998).

In other Asian countries, there has been less regulation of organotin compounds such as TBT and TPhT. Information about organotin contamination and imposex in gastropods is limited from Asian countries, although some reports from Malaysia, Singapore, and Korea reveal imposex and contamination by TBT and TPhT (Ellis and Pattisina 1990; Tan 1997). Further extensive surveys of imposex and organotin contamination are necessary in Asian countries.

Endocrine disruption in other invertebrates

Details of laboratory and field studies of ED in marine, freshwater, and terrestrial invertebrates are provided in Table 4-2. In these studies, heavy metals have been shown to have significant impacts on crustaceans; processes such as molting, growth, and reproduction have been impacted. Limb regeneration and abnormal pigmentation are also significant effects of heavy metal exposure. In contrast to the mollusks, however, there is minimal evidence of impacts of TBT in crustaceans. This may be because the crustaceans appear to have a greater ability than mollusks to metabolize TBT (Lee 1985, 1991).

Details of the impacts of vertebrate sex steroids, alkylphenols, polychlorinated biphenyls (PCBs), pentachlorophenol (PCP), heavy metals, and several pesticides on marine, freshwater, and terrestrial invertebrates are given in Table 4-2. The impacts of insecticides having developmental and growth effects at low exposure levels are apparent, and these are correlated with the timing of exposure.

Endocrine disruption in marine invertebrates

Limited field data are available on ED in marine invertebrate species in the natural environment, with the exception of the well-documented cases of TBT exposure in mollusks (See “Organotins and mollusks,” p 204). High prevalence of intersex in

Table 4-2 Representative laboratory and field studies in which endocrine disruption may be occurring. In the less well-established cases where no mechanism is suggested, non-endocrine mechanisms are also possible, but the observations are at least consistent with ED.

Phylum	Species or life stage	Contaminant	Concentration range	Major effects observed	Lab or field	Possible mechanism	Measured variables	Reference
Annelid	*Platynereis dumerilii, Nereis sucinea*	Volatile organic substances from fractionated distillation of crude oil	0.3 mg L^{-1} per compound	Induction of characteristic spawning behavior	Lab	Interference with natural chemical signalling processes	Behavioral bioassays	Beckmann et al. 1995
Crustacean	*Daphnia magna*, all stages, multigeneration	Diethylstilbestrol (DES)	0.031–0.5 mg L^{-1}	Reduced molting frequency, reduced fecundity, disrupted testosterone metabolism	Lab	Inhibition of steroid hydroxylase? Antagonistic interaction with 20-OHecdysone?	Survival, molting, fecundity, brood size, length, testosterone metabolism	Baldwin et al. 1995
Crustacean	*Daphnia magna*, adults	4-nonylphenol	25, 50, 100 µg L^{-1}	Reduced fecundity, disrupted testosterone metabolism	Lab	Decreased metabolic elimination of testosterone?	Survival, fecundity, androgen accumulation, elimination of conjugates, effects on F2 generation	Baldwin et al. 1997
Crustacean	*Balanus amphitrite*, larval stage	4-nonylphenol	0.01–1.0 µg L^{-1}	Increased production of larval storage protein	Lab	May represent promotion of an estrogen-dependent process	Production of cyprid major protein	Billinghurst et al. 1999
Crustacean	*Daphnia magna*, juveniles	Cadmium	5, 10, 20 µg L^{-1}	Elevated ecdysteroid levels	Lab	None suggested	Whole animal ecdysteroid concentrations	Bodar et al. 1990
Crustacean	*Daphnia magna*	Cadmium	0.5–50 µg L^{-1}	Delayed reproduction, increased brood size, reduced size of neonates	Lab	None suggested	Mortality, brood size, start of reproduction, size of neonates	Bodar et al. 1988
Crustacean	*Daphnia magna*	Phthalate esters	1 mg L^{-1}	No effects on survival, growth, fecundity	Lab	None suggested	Survival, growth, production of offspring	Brown et al. 1998
Crustacean	*Rhithropanopeus harrisii*, larval stages to adult	(S)-methoprene (JH analog)	100 µg L^{-1}	Mortality, increased intermolt duration	Lab	Methoprene may function as a JH analog in crustaceans	Mortality, developmental duration	Celestial and McKenney 1994
Crustacean	*Daphnia magna*, prenatal–adult	Nonylphenol	0.018–0.32 mg L^{-1}	Mortality of adults and offspring, reduced brood size, reduction in length	Lab	No mechanisms suggested. Nonylphenol is an estrogen mimic in vertebrates.	Survival, total and live offspring, length	Comber et al. 1993

Table 4-2 *continued*

Phylum	Species or life stage	Contaminant	Concentration range	Major effects observed	Lab or field	Possible mechanism	Measured variables	Reference
Crustacean	*Daphnia pulex, Eurytemora affinis* neonate–adult	Dieldrin	1–10 µg L^{-1}	Mortality, time to first reproduction, reduced fecundity, increased time to maturity, reduced number of broods	Lab	None suggested	Survival, fecundity, brood size, number of broods, day of first reproduction	Daniels and Allan 1981
Crustacean	*Daphnia magna,* juvenile–adult	Endosulfan (organochlorine pesticide)	0.12–0.31 mg L^{-1}	Reduced growth, decreased survival, reduction in brood size and number of broods, delayed maturation	Lab	Unknown but site of action of the pesticide is the CNS	Survival, growth, total young per female, maximum number of broods, mean brood size, number of broods, time to first reproduction	Fernandez-Casalderrey et al. 1993
Crustacean	*Daphnia magna*	Endosulfan (organochlorine), Diazinon (organophosphate)	0.15–0.62 mg L^{-1} (endosulfan) 0.23–0.9 µg L^{-1} (diazinon)	Reduced rates of filtration and ingestion	Lab	Unknown but site of action of the pesticides is the CNS	Filtration and ingestion rates	Fernandez-Casalderrey et al. 1994
Crustacean	*Uca pugilator*	Cadmium, PCBs, naphthalene	—	Abnormal coloration	Lab	Inhibition of black pigment-dispersing hormone (BPDH) synthesis	Color	Fingerman and Fingerman, 1978; Hanumante et al. 1981; Staub and Fingerman 1984a, 1984b (Cited by Fingerman et al. 1996)
Crustacean	*Daphnia pulex*	Simazine (herbicide)	4, 20 mg L^{-1}	Impairment of reproduction, reduction in molt frequency	Lab	Unknown	—	Fitzmayer et al. 1982
Crustacean	*Daphnia magna*	Selenium	Sublethal	Suppression of growth and reduced egg production	Lab	Unknown	Changes in length, egg production	Johnston 1987
Crustacean	*Daphnia magna,* adults	Diuron (herbicide/algicide)	0.2 mg L^{-1}	Reduced fecundity, arrested development of juveniles	Lab	No mechanism suggested	Enumeration of adults and neonates	Kersting 1975

Table 4-2 *continued*

Phylum	Species or life stage	Contaminant	Concentration range	Major effects observed	Lab or field	Possible mechanism	Measured variables	Reference
Crustacean	*Mysidopsis bahia* all life stages	Methoprene (JH analog)	2–125 μg L^{-1}	Mortality, reduced growth, delayed brood release, reduced brood size	Lab	Likely to be a JH-related endocrine effect	Mortality, growth, maturation, reproduction	McKenney and Celestial 1996
Crustacean	*Palaemonetes pugio*, larval stages	Methoprene (JH analog)	0.1–1000 mg L^{-1}	Mortality, inhibition of larval development	Lab	Likely to be a JH-related endocrine effect	Mortality, molting frequency, development	McKenney and Matthews 1990
Crustacean	*Macrobrachium kistnensis* (freshwater prawn) *Barytelphusa cunicularis* (crab)	Cadmium	—	Hyperglycemia	Lab	Reduced synthesis of crustacean hyper-glycemic hormone (CHH)	Blood glucose, hormone titers	Nagabhushanam and Kulkarni, 1981b; Machale et al. 1989 (cited by Fingerman et al. 1996)
Crustacean	*Daphnia magna*, all life stages	Pentachlorophenol (broad spectrum biocide)	0.062–0.5 mg L^{-1}	Mortality, reduction in fecundity, reduced elimination of testosterone metabolites	Lab	Possibly alteration of the metabolic clearance of endogenous steroid hormones	Survival, molt frequency, maturation, fecundity, in vivo testosterone metabolism	Parks and LeBlanc 1996
Crustacean	*Caradina rajadhari* (freshwater prawn), mature	Tributyltin (TBT)	0.015–0.04 mg L^{-1}	Retardation of regenerative limb growth and molting, increase in weight and calcium content of cast exuviae	Lab	Inhibition of chitinolytic enzymes / Ca resorption	Weight and Ca content of cast exuvia	Reddy et al. 1991, 1992; Nagabhushanam et al. 1990
Crustacean	*Procambrus clarkii*, crayfish, adult	Naphthalene	10 mg L^{-1}	Suppression of ovarian growth	Lab	Possible inhibition of gonad stimulating hormone release	Ovarian index (GSI)	Sarojini et al. 1994
Crustacean	*Daphnia pulex*	Atrazine (herbicide)	1–5, 10, 20 mg L^{-1}	Reduced fecundity and growth	Lab	Unknown	Production of offspring	Schober and Lampert 1977

Table 4-2 *continued*

Phylum	Species or life stage	Contaminant	Concentration range	Major effects observed	Lab or field	Possible mechanism	Measured variables	Reference
Crustacean	*Daphnia galeata mendotae*, adult	Nonylphenol	10–100 μg L^{-1}	Reduction in fecundity, developmental abnormalities	Lab	No specific mechanism suggested	Production of males, females and ephippia, developmental abnormalities	Shurin and Dodson 1997
Crustacean	*Uca pugilator* (fiddler crab), adult	Tributyltin (TBT)	0.5–500 μg L^{-1}	Retardation of limb regeneration and molting, abnormalities in the regenerates	Lab	No specific mechanism suggested; interference with developmental processes	Growth of limb buds following autonomy	Weis et al. 1987
Crustacean	*Uca pugilator* (fiddler crab), adult	Cadmium, mercury, lead	0.1, 1.0 mg L^{-1}	Retarded limb regeneration (Cd and Hg only)	Lab	Depression of metabolic rate suggested as cause	Growth of limb buds following autonomy	Weis 1976
Crustacean	*Uca pugilator* (fiddler crab), adult	Zinc	1, 3, 5 mg L^{-1}	Retarded limb regeneration	Lab	Depression of metabolic rate suggested as cause	Growth of limb buds following autonomy	Weis 1980
Crustacean	*Daphnia magna*, adult	Diethylstilbestrol (DES), endosulfan	0.05–0.20 mg L^{-1}	Increased duration of intermolt period	Lab	Antagonistic competition for the endogenous molt hormone receptor	Mortality, sex ratio, molting periodicity	Zou and Fingerman 1997a
Crustacean	*Daphnia magna*, neonates	Lindane, 4-octylphenol	0.01–0.04 mg L^{-1} (4-octylphenol); 0.05–0.2 mg L^{-1} (lindane)	No effect on molting	Lab	—	Mortality, intermolt period	Zou and Fingerman 1997a
Crustacean	*Daphnia magna*, neonates	PCB29, Aroclor 1242, or diethyl phthalate (DEP)	0.05–0.2 (PCB29); 0.05,0.10 (Aroclor 1242); 5.6–22.4 (DEP) mg L^{-1}	Increased time to complete four molts	Lab	Presumed to act as antagonists of endogenous ecdysteroids	Mortality, intermolt period	Zou and Fingerman 1997b
Echinoderm	*Sclerasterias mollis* (starfish), adult	Estradiol-17β, estrone	Daily injections	Increased ovarian estrone and progesterone levels, increased pyloric caeca estrone levels, incorporation of protein into oocytes	Lab	Ovaries and pyloric caeca may be estrogen target organs	Ovary and pyloric caeca weight, oocyte size, steroid analysis of tissues	Barker and Xu 1993

Table 4-2 *continued*

Phylum	Species or life stage	Contaminant	Concentration range	Major effects observed	Lab or field	Possible mechanism	Measured variables	Reference
Echinoderm	*Asterias rubens*, adult	Cadmium	25 μg L^{-1} (long-term) 200 μg L^{-1} (short-term)	Delayed germinal vesicle breakdown during induced spawning, abnormal development of fertilized oocytes, abnormal embryo development, reduced ovarian growth, reduced steroid levels and P450 levels in pyloric caeca	Semi-field (long-term)/ lab (short-term)	Adverse effects on steroid levels impacting on reproductive processes	Germinal vesicle breakdown during induced spawning, gamete fertilizability, embryonic development, ovarian structure, steroid titers in pyloric caeca and gonads, MFO activity, in-vitro steroid metabolism	den Besten et al. 1989; den Besten, van Donselaar et al. 1991; den Besten, Elenbaas et al. 1991
Echinoderm	*Asterias rubens*	PCBs (Clophen A50)	Administered through PCB-exposed mussels	Abnormal development of fertilized oocytes, abnormal embryo development, reduced steroid levels in pyloric caeca, elevated steroid levels in gonads, reduced P450 levels in pyloric caeca	Semi-field	Adverse effects on steroid levels impacting on reproductive processes	Steroid titers, embryo development	den Besten et al. 1989; den Besten, Elenbaas et al. 1991
Echinoderm	*Strongylocentrotus intermedius*	Cadmium	0.1–5.0 mg L^{-1}	Low fertilization success, formation of anomalous sex cells, arrested embryogenesis	Lab	unknown	Artificial fertilization, development of fertilized eggs, gonad histology	Khristoforova et al. 1984
Echinoderm	*Paracentrotus lividus* (sea urchin), developing eggs	Pentachlorophenol (broad spectrum biocide)	0.1–0.8 mg L^{-1}	Alteration in embryonic development and differentiation, reduced echinochrome synthesis	Lab	No mechanism identified	Fertilization success, development, enzyme activities in egg extracts	Ozretic and Krajnovic-Ozretic 1985
Echinoderm	*Asterias rubens*, adults	Estradiol-17β	Administered by injection	Increased estrone levels in ovaries, increased oocyte diameters	Lab	Suggested that there has been enhanced incorporation of material into the oocytes	Numerous parameters associated with ovarian function	Schoenmakers et al. 1981
Echinoderm	*Asterias rubens*, adult	Cadmium, zinc	100 μg L^{-1} (cadmium) 240 μg L^{-1} (zinc)	Increased steroid metabolism, elevated testosterone and progesterone levels	Lab	Effects of metals directly on enzyme activity or as a cofactor?	Steroid levels in gonad and pyloric caeca, steroid metabolism in tissue in vitro	Voogt et al. 1987

Table 4-2 *continued*

Phylum	Species or life stage	Contaminant	Concentration range	Major effects observed	Lab or field	Possible mechanism	Measured variables	Reference
Insect	*Chironomus tentans*, all life stages	4-nonylphenol	12.5–200 μg L^{-1}	Reduced survival at high concentrations. No other effects	Lab	No apparent estrogen-related effects	Mortality, growth, emergence success and emergence pattern, sex ratio, fecundity, egg viability	Kahl et al. 1997
Insect	*Macromia cingulata* larvae	Tannery and paper pulp mill effluent	5, 10, 15, 20, 25%	Shortened time to first molt (tannery), arrested larval molting (paper pulp)	Lab	Possible JH mimics within the effluents	Time of molting	Subramanian and Varadaraj 1993
Insect	*Chironomus riparius*, larvae–adult	Phthalate esters (various)	100, 1000, 10000 mg kg^{-1} dry weight of sediment	No effects on survival, development or emergence	Lab		Collection of emergent adults, determination of sex	Brown et al. 1996
Insect	*Chironomus* spp.	Industrial effluent containing metals	Ambient at site	Higher level of mentum deformities in exposed larvae in both field and lab conditions	Field and lab	No mechanisms suggested	Examination of menta for deformities	Dickman and Rygiel 1996
Insect	*Chironomus* gr. *thummi*, larvae	Organic and inorganic pollutants	Ambient at site	Prevalence of morphological deformities related to pollutants	Field	No mechanisms suggested	Assessment of deformities in mandible, premandible, menta and antennae	De Bisthoven et al. 1995
Insect	*Chironomus* spp.	Possible exposure to pollutants	Ambient at site	Deformed mouth parts, heavily pigmented head capsules, unusually thick head capsule and body wall	Field	No mechanisms suggested	Morphological examination	Hamilton and Saether 1971
Mollusk	*Lymnaea stagnalis*, adult	DDT (insecticide) MCPA (herbicide)	50, 500 μg L^{-1} (DDT); 10, 100 mg L^{-1} (MCPA)	Reduced fecundity	Lab	No mechanism suggested	Mortality, numbers of eggs in egg masses	Woin and Bronmark 1992
Mollusk	*Mytilus edulis*, adults	Cadmium	100 μg L^{-1}	Inhibits gonadal follicle development, stimulates spawning frequency	Lab	No mechanism suggested	Mortality, number of gametes released, gonadal histology	Kluytmans et al. 1988
Mollusk	*Paramphiascella hyperborea*, *Stenhelia gibba*, *Halectinosoma* spp.	Sewage outfall	Ambient at site	Intersexuality	Field	Not strongly correlated with proximity to outfall. Cause unknown.	Morphology	Moore and Stevenson 1991, 1994

harpacticoid copepods has been reported from the vicinity of a sewage outfall in the Firth of Forth, Scotland (Moore and Stevenson 1991, 1994). However, no conclusive causal relationship between the sewage outfall and the levels of intersex was determined. Similarly, the lobster *Homarus americanus* has been observed with ovotestes in Nova Scotia, but it was not concluded whether this was a natural or site-related phenomenon (Sangalang and Jones 1997).

In laboratory studies, heavy metals have been observed to have significant impacts on certain hormonally controlled processes in echinoderms, such as reduced oogenesis and disrupted gonadal development. Results from field studies are, however, inconclusive. In a num0ber of surveys in the North Sea, sea stars (*Asterias rubens*) were collected at different sites in order to study the quality of their offspring and to perform biomarker measurements. No pollution gradient-related effects were observed on oocyte maturation or on the early development of sea star embryos. However, slight effects were observed on the microsomal cytochrome P450 monooxygenase system in the pyloric caecae of female sea stars collected from polluted sites within the influence of rivers like the Elbe (Germany), the Rhine, the Western Schelde (The Netherlands), and the Humber (UK) (den Besten et al. 1996; Postma and Valk 1996, 1997). Because in semi-field studies with *Asterias rubens* it was shown that rates of P450-dependent steroid metabolism can be decreased as a result of exposure to contaminants (see Chapter 2), these decreased P450 activities found in the North Sea field study may signal potential adverse effects.

Endocrine disruption in freshwater invertebrates

As far as we are aware, there are no documented examples that demonstrate EDC effects on freshwater invertebrates in the natural environment. This probably reflects the lack of effort so far expended in seeking evidence of such effects. There are, however, a substantial number of laboratory-based studies that identify the effects of putative EDCs on freshwater invertebrate species. These are presented in Table 4-2 and discussed below.

Insects

Some herbicides have been shown to adversely affect endocrine-mediated processes in aquatic insects. For example, the bipyridilium herbicides diquat dibromide and paraquat dichloride, which inhibit cytochrome P-450 enzymes, have been demonstrated to exert effects on growth and development in *Neobellieria bullata* (Diptera) larvae, including prolongation of the first instar stadium, disturbance of the first molt, and delay of pupation in larvae (Darvas et al. 1990).

More than 100 phytoecdysteroids, plant-derived substances that display hormone-like activity, have been identified and in many cases have been shown to demonstrate hormonal activity in insects (Adler and Grebenok 1995). Effluents from pulp and paper mills, which contain phytoecdysteroids, were shown to influence molting in aquatic insects. Molting was arrested in dragonfly larvae (*Macromia cingulata*) exposed to paper and pulp mill effluent (Subramanian and Varadaraj 1993), and exposure of the same species to tannery effluent resulted in a shortening of the time to first molt, suggesting that components of the effluents had interfered with ecdysteroid metabolism.

Vertebrate-type steroids occur widely in insects (Bradbrook et al. 1990) but fulfill no established endocrine function. It is therefore unsurprising that some of the compounds known to exhibit estrogenic activity in vertebrates apparently have no physiological effects, even at high doses, in insects. For example, exposure to phthalate ester plasticizers (not all of which are known EDCs) had no detectable effect on the development of the midge *Chironomus riparius* (Brown et al. 1996), and no effects on survival, growth, or reproduction were observed in *Chironomus tentans* exposed to tetrachlorodibenzo-*p*-dioxin (TCCD; West et al. 1997). Nor did TCDD exert any effect on pupation in the mosquito *Aedes aegyptii* (Miller et al. 1973). Similarly, nonylphenol did not affect the growth, survival, emergence, sex ratio, fecundity, or egg viability of *Chironomus tentans* (Kahl et al. 1997).

Deformities of the mouthparts and other head capsule features of chironomid larvae have been suggested to arise as a consequence of exposure of the organisms to heavy metals and pesticides (Hamilton and Saether 1971). Particular types of deformities in *Chironomus thummi* have been associated with the presence of heavy metals, phthalates, and DDT (De Bisthoven et al. 1995). Morphological deformities in key structures in Chironomidae (e.g., mouthpart deformities) and Oligochaeta (chaetal deformities) have been used as indicators of environmental stress from contaminants for many years. While the underlying mechanism for these deformities remains elusive, they appear to originate in relation to developmental processes, and this has led to their consideration as indicators of stress from EDCs. Indeed, increased incidences of chironomid deformities have been observed in sediments contaminated by known EDCs (e.g., dichlorodiphenyldichloroethylene [DDE], heavy metals, sediments below pulp mills), and ecological indices to assess the ecological relevance of these deformities have been developed (Milbrink 1983; Kosalwat and Knight 1987; Dickman et al. 1990; Warwick 1989, 1991). Despite the extensive research on deformities, however, large variation in the frequency of occurrence, poor control in many studies, and a lack of understanding of causative agents suggest that the use of deformities to assess contaminants, including EDCs, currently must be considered qualitative and subjective (Johnson et al. 1993).

Crustaceans

The effects of heavy metals on crustaceans include disruption of molting and limb regeneration, alteration of blood glucose level, effects on color change, and impairment of reproduction (Weis 1978; Weis et al. 1992; Fingerman et al. 1996, 1998). Cadmium (Bodar et al. 1990) and selenium (Schultz et al. 1980) adversely affect the process of molting in crustaceans. In *Daphnia magna*, selenium exposure delays the onset of proecdysis, whereas cadmium lengthens the intermolt cycle, reduces the number of successful ecdyses, and causes a reduction in body weight and size of the neonates. Cadmium elicits an increase in ecdysteroid level in *Daphnia* (Bodar et al.

1990) suggesting that the observed effects of cadmium exposure on molting could be the result of changes in ecdysteroid titers. These authors pointed out that the metallo-enzymes involved in the molt process might also be directly affected by the essential-metal substitution characteristics of cadmium.

A hyperglycemic response is observed in the freshwater prawn *Macrobrachium kistnensis* following cadmium exposure (Nagabhushanam and Kulkarni 1981b). This effect may arise as a consequence of disruption of CHH activity; cadmium stimulates the release of CHH, while at the same time inhibiting its synthesis (Fingerman et al. 1996).

TBT is reported to exert some effects in crustaceans. Although TBT is acutely toxic to daphnids, sublethal concentrations elicited no adverse effects on molting or reproduction. Testosterone metabolism was, however, enhanced in animals exposed to concentrations approaching lethal levels (Oberdörster, Rittschof, and LeBlanc 1998; see Chapter 2 for a discussion of the relevance of testosterone to crustacean endocrinology). TBT-exposed freshwater prawns, *Caradina rajadhari,* exhibited a significant retardation of regenerative limb growth (Reddy et al. 1991), of molting (Reddy et al. 1992), and of inhibition of calcium resorption from the exoskeleton (Nagabushanam et al. 1990). TBT also retards limb regeneration in fiddler crabs, *Uca pugilator* (Weis et al. 1987), as well as affecting activity and behavior (Weis and Perlmutter 1987).

No definitive role has been established for vertebrate-type steroids in crustaceans, but various studies have demonstrated the existence of physiological and biochemical functions that may be perturbed by environmental estrogens or other putative endocrine disrupters. For example, exogenous testosterone has a variety of androgenic effects in prawns (Nagabhushanam and Kulkarni 1981a). Also, the synthetic estrogen diethylstilbestrol (DES) reportedly inhibits molting in juvenile, but not adult, *Daphnia magna* (Baldwin et al. 1995) possibly by competing antagonistically for the 20E receptor. Alternatively, adults may be insensitive to DES because they are better able to metabolize the compound. However, effects in juveniles were observed only at very high dose levels, suggesting the effects may be pharmacological rather than physiological.

The alkylphenol ethoxylates are ubiquitous non-ionic surfactants that degrade to alkylphenols such as nonylphenol. The alkylphenols are well established as estrogen mimics in vertebrate systems. In *Daphnia magna*, exposure to nonylphenol increased the accumulation of exogenous testosterone (Baldwin et al. 1997). However, as noted above, it remains to be resolved whether testosterone has a role within the crustacean endocrine system, so the functional significance of this observation is unclear. It has been suggested that nonylphenol would also suppress the metabolic elimination of ecdysteroid conjugates (Baldwin et al. 1997).

Shurin and Dodson (1997) investigated the effects of nonylphenol in sex determination and development in *Daphnia*. As is the case for many cladocerans, female

Daphnia partition their reproductive effort between females, males, and resting eggs according to environmental conditions. Exposure of females to nonylphenol resulted in a change in fecundity in terms of the 3 types of offspring. The production of females and of resting eggs was affected most markedly, while the production of males was less sensitive to nonylphenol. Prenatal exposure to nonylphenol resulted in the occurrence of morphological abnormalities in a proportion of the juvenile *Daphnia.*

Neither the synthetic estrogen DES nor the estrogenic pesticide endosulfan affected sex differentiation in *D. magna*, although they both inhibited molting (Zou and Fingerman 1997a, 1997b). Given the absence of evidence for the involvement of vertebrate-type steroids in the crustacean endocrine system, it is not clear whether this effect is mediated by the endocrine system or represents a toxic effect exerted by a more conventional route. Additional reported effects of endosulfan in *Daphnia* include reduced growth, decreased survival and mean total young per female, an increase in the time to first reproduction, and a reduction in the mean number of broods (Fernandez-Casalderrey et al. 1993).

Schober and Lampert (1977) showed that the herbicide atrazine has a marked effect on fecundity and growth in *D. pulex* at concentrations considerably below those that affect survival. Atrazine also impaired reproduction and development in the malacostracan *Gammarus fasciatus* (Macek et al. 1976). Another triazine herbicide, simazine, gives rise to reduced frequency of molting in *D. pulex* (Macek et al. 1976), and the urea herbicide diuron affects reproduction in *Daphnia* (Kersting 1975). The mechanisms underlying these effects are not known but the evidence is suggestive of endocrine-related disturbances.

Mollusks

In spite of the attention that has been given to the effects of TBT and related compounds on marine mollusks, there appears to have been only little work done in freshwater systems. The occurrence of TBT in limnic waters and sediments and its effects on freshwater mollusks have been reviewed by Schulte-Oehlmann et al. (1996) and Schulte-Oehlmann (1997). According to this information, TBT concentrations in European lakes and rivers have reached maximum values of up to 930 ng/L in water and up to 340 μg/g (wet weight) in sediments (River Elbe). In sewage treatment plant (STP) effluents containing domestic waste water, concentrations of up to 20.3 μg TBT/L have been observed, and up to 61.8 μg TBT/L in STP effluents containing industrial waste water. In general, the degree of organotin contamination of freshwaters is more poorly characterized than for coastal waters, though it is reported that levels of TBT in fresh water have generally declined (Fent 1996).

The decline of TBT-sensitive prosobranch gastropod species in some European areas (e.g., *Theodoxus fluviatilis*, *Bithynia tentaculata*, and *Potamopyrgus antipodarum)* has been attributed by Schulte-Oehlmann et al. (1996) to the existing degree of organotin contamination in these areas although other EDCs may have contributed to the general decline of these species.

An additional report of possible ED-related effects in freshwater mollusks concerns the snail *Lymnaea stagnalis*, in which sublethal concentrations of the organochlorine insecticide DDT caused a substantial reduction in fecundity (Woin and Bronmark 1992).

Endocrine disruption in terrestrial invertebrates

The development of insect growth regulators (IGRs) designed as endocrine disrupters for use in pest control has provided information about the effects of these compounds on a number of terrestrial invertebrate species (Griffith et al. 1996). Ecdysteroid receptor agonists such as the nonsteroidal tebufenozide, for example, induce symptoms of hyperecdysonism in terrestrial insects, delaying postembryonic development and inducing malformations and nymphal–adult intermediates in such species as the lepidopteran *Plutella xylostella* and in *Oncopeltus fasciatus* (Wing 1988; Darvas et al. 1992). There are major terrestrial invertebrate classes, however, for which there are no apparent data on the effects of exposure to these groups of chemicals, e.g., earthworms and mollusks. There are also no reported incidents of effects of nonpesticidal endocrine disrupters on terrestrial invertebrates. However, the effects of a number of other compounds with unknown mechanisms on invertebrate growth and development have been reported, and these are shown in Table 4-2. Table 4-3 outlines the reported effects of known exposure of predominantly insect species to known invertebrate endocrine disrupters.

Chitin synthesis inhibitors

Diflubenzuron and related chitin synthesis inhibitor-based IGRs have been widely misreported as EDCs. However, although at high levels they may ultimately affect circulating hormone levels within invertebrates, due to morbid effects, they are not directly EDCs (see Chapter 2).

Other effects on terrestrial invertebrates

Reproductive effects have been observed in earthworms exposed to a range of metals and organic compounds, but no linkage to their endocrine system has been made (Table 4-2). It also has been shown (Drobne and Strus 1996) that zinc contamination of leaf litter prolongs the molt cycle and reduces molt frequency in the isopod *Porcellio scaber*. It was postulated that zinc interfered with the rates of 20E synthesis, degradation, or molecular interaction, but hormone levels were not measured. Polycyclic aromatic hydrocarbons (PAHs) have been shown to stimulate the incidence of gravidity in female isopods (*Oniscus asellus*), an effect that may be attributable to interference with hormonal regulation (Van Brummelen et al. 1996). However, effects such as these have not been reported under field conditions, and none have been clearly identified as ED.

Table 4-3 Confirmed cases where endocrine disruption has been identified as the mechanism causing observed effects

Compound	Concentration	Species	Effect	Lab or field	References
Triorganotin compounds (primarily tributyltin [TBT])	0.2 ng/L (as Sn)	*Nucella lapillus*	Imposex	Lab and field	Gibbs et al. 1987; Oehlmann 1994
	0.1 ng/L (as Sn) and higher	150 Prosobranch gastropods	Imposex	Lab and field	Oehlmann et al. 1996
	5 ng/L (as Sn) and higher	*Littorina littorea*	Intersex	Lab and field	Bauer et al. 1995, 1997
	>1 ng/L (as Sn)	*Crassostrea gigas*	Shell thickening	Lab and field	Alzieu et al. 1991
	>10 ng/L (as Sn)	Various bivalves including *Crassostrea gigas, Mytilus edulis, Ostrea edulis*	Induction of spermatogenesis in ovaries	Lab and field	Alzieu et al. 1991
	Normally >2ng/L (as Sn)	Various Prosobranch gastropods	Induction of spermatogenesis in ovaries	Lab and field	Gibbs et al. 1988; Oehlmann et al. 1996
Ecdysone agonists					
Methoxyfenozide	0.063 mg/kg	European corn borer *Ostrinia nubilalis*	Ovicidal, inhibit larval growth, delayed pupation, decreased adult emergence, interrupt feeding, premature and lethal molting cycle induced	Lab	Trisyono and Chippendale 1997
RH 5849	10 μg/g insect	Common cutworm *Spodoptera litura* larvae	Died without ecdysis, rectal prolapse	Lab	Tateishi et al. 1993
RH 5849	50 μg/g insect	Larval lepitopterans	Lethal premature molt	Lab	Salgado 1992
RH 5849	50–100 mg/kg	Variety of insect orders	Delayed postembryonic development, molting disturbances, nymphal adult intermediates, malformed structures (wings, mandibles) sterility, decreased bodyweight	Lab	Darvas et al. 1992

Table 4-3 *continued*

Compound	Concentration	Species	Effect	Lab or field	References
RH 5992	?	Lepidopteran *Helicoverpa zea* larvae	Reduced fertility males, no effect females	Lab	Carpenter and Chandler 1994
Tebufenozide	?	Lacewings *Micromus tasmaniae* larvae treated	Decreased total no. eggs, peak egg production, oviposition period F1 generation	Lab	Rumpf et al. 1998
Tebufenozide	?	Two spotted mite *Tetranychus urticae,* European red mite	Decreased populations	Field	Valentine et al. 1996
Tebufenozide	0.016 mg/kg	European corn borer *Ostrinia nubilalis*	Ovicidal, inhibit larval growth, delayed pupation, decreased adult emergence, interrupt feeding, premature and lethal molting cycle induced	Lab	Trisyono and Chippendale 1997
Ecdysteroid antagonists					
1-substituted imidazoles	?	*Bombyx mori* larvae	Precocious metamorphosis	Lab	Nakamura et al. 1998
1-substituted imidazoles	1 µg/g	Wild silkmoth *Antherae yamamai* first instar larvae	Terminated diapause	Lab	Nakamura et al. 1998
Azadirachtin	?	Earwig *Labidura riparia,* adults	Decreased ovarian development, degenerated follicles, decreased vitellogenin, vitellin not deposited in ovary	Lab	Sayah et al. 1996
Azadirachtin	203 µg/g body wt	German cockroach *Blattella germanica* adults	Decreased body weight	Lab	Prabhakaran and Kamble 1996
Azadirachtin	?	Earwig *Labidura riparia* adults	Inhibition of vitellogenesis, decreased ovarian ecdysteroid levels, degeneration corpus allatum	Lab	Sayah et al. 1998

Table 4-3 *continued*

Compound	Concentration	Species	Effect	Lab or field	References
KK-42	?	Gypsy moth *Lymantria dispar* eggs	Prevents normal diapause	Lab	Bell 1996
KK42 (imidazole erivative)	10–5M	Mealworms *Tenebrio molitor*	Decreased growth and development of ovaries, decreased no. oocytes and basal follicles	Lab	Soltani et al. 1997
Juvenile hormone analogs					
Fenoxycarb	?	Honeybee larvae	Crescent shaped mark on eye, larval and pupal mortality	Lab	Marletto et al. 1997
Fenoxycarb	?	Winterform pear psylla *Cacopsylla pyricola*, adults	Prompts ovarian development in diapausing adults, decreased egg hatch	Field	Horton and Lewis 1996
Fenoxycarb	60 pg/L	Silkworm *Bombyx mori*, 5th instar larvae	Decreased food intake, decreased growth rate	Lab	Leonardi et al. 1996
Fenoxycarb	10 mg/kg	Termites *Coptotermes formosanus* colonies	Decreased colony size, death undifferentiated individuals, reduced differentiation into workers	Lab	Jones and Lenz 1996
Fenoxycarb	?	Pear psylla *Cacopsylla pyri* adult	Accelerates ovarian maturation, rapid breakdown of winter diapause, inc mating rate males, eliminate winter behavior arrest in males	Lab	Lyoussoufi et al. 1994
Fenoxycarb	field rate	Coleopteran *Stethorus punctum* larvae	Disrupted larval–pupal molt	Field	Biddinger and Hull 1995
Fenoxycarb	0.25–1% bait	Pharoah ant *Monomorium pharoanis* colonies	Decreased egg production, delayed production winged reproductives, intermediate castes produced	Lab	Williams and Vail 1993

Table 4-3 *continued*

Compound	Concentration	Species	Effect	Lab or field	References
Fenoxycarb	0.25–5 µg/g larvae	Silkworm *Bombyx mori* larvae	Reached last larval instar but failed to spin or pupate, increased weight, decreased growth silk glands, high levels hemolymphatic juvenile hormone	Lab	Plantevin et al. 1991
Fenoxycarb	?	Eastern spruce budworm *Choristoneura fumiferana,* eggs, larvae, adults	Failure of eggs to hatch, lethal morphometric changes in treated 5th instar larvae, treated females laid eggs failed to hatch, treated males sterile	Lab	Hicks and Gordon 1992
Fenoxycarb	?	Codling moth *Cydia pomonella* adults	Decreased fertility, male sterility	Lab	Charmillot and Pasquier 1992
Fenoxycarb	field rate	Parasitoid fly *Cryptochaetum iceryae* eggs, larvae	Decreased parasitism, decreased development	Lab	Mendel et al. 1994
Fenoxycarb	field rate	Coccinellid *Rodolia cardinalis,* eggs, larvae	No larval development	Lab	Mendel et al. 1994
Fenoxycarb	field rate	Coleopteran *Chilocorus bipustulatus,* eggs, larvae	Prevented egg hatch	Lab	Mendel et al. 1994
Fenoxycarb	?	Lacewings *Micromus tasmaniae* larvae	Decreased longevity, decreased total no. eggs laid, decreased oviposition period	Lab	Rumpf et al. 1998
Fenoxycarb	?	*Corcyra cephalonica*	Reduced egg hatch	Lab	Bhargava and Urs 1993
Fenoxycarb	?	German cockroach *Blattella germanica* juvenile	Twisted wing, affected adults	Lab	Kaakeh et al. 1997
Fenoxycarb	0.5%	Pharoah ant *Monomorium pharaonis* laying queens	Decreased egg production, decreased brood levels	Lab	Kao and Su 1995

Table 4-3 *continued*

Compound	Concentration	Species	Effect	Lab or field	References
Fenoxycarb	?	Pear psylla *Cacopsylla pyricola* adults	Decreased egg hatch	Lab	Higbee et al. 1995
Fenoxycarb	1–100 µg/g insect	German cockroaches *Blattella germanica* nymphs	Ecdysal failure, developmental delays, morphogenetic wing twisting, reproductive inhibition	Lab	Reid et al. 1994
Fenoxycarb	?	Two spotted mite *Tetranychus urticae,* European red mite	Decreased populations	Field	Valentine et al. 1996
Fenoxycarb	?	*Coccinellids Adalia bipunctata, Coccinella septempunctata,* eggs, larvae, adults	Drastic reduction in fecundity	Lab	Olszak et al. 1994
Hydroprene	?	*Corcyra cephalonica*	Reduced egg hatch	Lab	Bhargava and Urs 1993
Hydroprene	?	German cockroach *Blattella germanica*	Twisted wing, affected adults	Lab	Kaakeh et al. 1997
Methoprene	?	*Corcyra cephalonica*	Reduced egg hatch	Lab	Bhargava and Urs 1993
Methoprene	?	Psyllids Psyllidae, pirate bug last larval instar	Wing and external genitalia malformations	Lab	Baldassari et al. 1997
Methoprene	2 mg/kg wheat	*Rhyzopertha dominica* adults	Reduced fecundity	Lab	Daglish and Pulvirenti 1998
Methoprene	?	Mosquito *Aedes aegypti* larvae	Decreased longevity in adult females	Lab	Sawby et al. 1992
Pyriproxifen	Field rate	Coccinellids *Chilocorus nigrita, Cryptolaemus montrouzieri,* adults	Failure of egg hatch	Lab	Hattingh and Tate 1995
Pyriproxyfen	Field rate	Parasitoid fly *Cryptochaetum iceryae,* eggs and larvae	Decreased parasitism, decreased development	Lab	Mendel et al. 1994

Table 4-3 *continued*

Compound	Concentration	Species	Effect	Lab or field	References
Pyriproxyfen	?	German cockroach *Blattella germanica* juvenile	Twisted wing, affected adults	Lab	Kaakeh et al. 1997
Pyriproxyfen	1%	Pharoah ant *Monomorium pharaonis* laying queens	Decreased egg production, decreased brood levels	Lab	Kao and Su 1995
Pyriproxyfen	?	Pear psylla *Cacopsylla pyricola* adults	Decreased egg hatch	Lab	Higbee et al. 1995
Pyriproxyfen	400 l/ha	*Chrysoperla externa*, eggs, larvae	No metamorphosis in third instar larvae	Lab	Velloso et al. 1997
Pyriproxyfen	?	Predatory bug *Podisus maculiventris*, 5th instar nymphs	Severe deformities at ecdysis	Lab	De Clercq et al. 1995
Pyriproxyfen	0.1 mg/kg	*Aedes albopictus*, pupae	Inhibition adult emergence	Lab	Takagi et al. 1995
Pyriproxyfen	0.96 µg/ insect	Horn fly *Haemotobia irritans* adults	Inhibits development of young	Lab	Bull and Meola 1993
Pyriproxyfen	Field rate	Coccinellid *Rodolia cardinalis*	No larval development	Lab	Mendel et al. 1994
Pyriproxyfen	Field rate	Coleopteran *Chilocorus bipustulatus*, eggs, larvae	Prevented egg hatch	Lab	Mendel et al. 1994
Pyriproxyfen	1–100 µg/g insect	German cockroaches *Blattella germanica* nymphs	Ecdysal failure, developmental delays, morphogenetic wing twisting, reproductive inhibition	Lab	Reid et al. 1994
R-20458	0.0001–0.1%	Diamond back moth *Plutella xylostella* prepupae and pupae	Pupal–adult intermediates, malformed adults	Lab	Khan and Urs 1992

Table 4-3 *continued*

Compound	Concentration	Species	Effect	Lab or field	References
R394	0.195 μg/g	Silkworm *Bombyx mori* larvae	Decreased ecdysteroid titres, delayed initiation of spinning, non-spinning syndrome	Lab	Trivedy et al. 1996
ZR-777	0.0001–0.1%	Diamond back moth *Plutella xylostella* prepupae and pupae	Pupal–adult intermediates, malformed adults	Lab	Khan and Urs 1992

Temporal extent of exposure to, and effects of, EDCs in invertebrates

Exposure

The EDC exposures that invertebrates will experience are driven by the physical and chemical characteristics of the substances in question, the nature of their sources, and the mode of life of the organisms. For example, aquatic filter feeders are exposed to persistent and hydrophobic EDCs adsorbed to suspended solids and in the dissolved state, although the availability of the EDCs from the 2 phases may differ. Herbivorous species may receive much less exposure to compounds on suspended particles. Invertebrate EDCs appear to be drawn from extremely diverse chemical groups and sources, so it is not possible to generalize about the temporal or spatial extent of exposure. However, it is apparent that some of these materials (e.g., TBT) have experienced worldwide distribution as a result of their persistence or wide use, so there are cases in which the exposure experienced by certain invertebrates must be very widespread and long term. In general, persistent chemicals may present a high risk for causing effects because of the long-term exposure that they produce. In the case of endocrine disrupters, even quickly degradable substances may be considered as potentially hazardous because short-term exposure in very sensitive life stages may result in severe long-term effects.

Effects

Developmental and morphological effects caused by EDCs in invertebrates (e.g., changes such as imposex and exoskeleton abnormalities) are generally permanent, in many cases even when the polluting stimulus has been removed. On the other hand, available data suggest that biochemical effects such as altered hormone titers may return to normal in due course. In general, EDC-impacted invertebrate populations have faster recovery times than many vertebrates (due to their shorter life cycles), although field data (Waldock et al. 1999; Rees et al. 1999) suggest that marine invertebrate community recovery from the effects of TBT may take at least a decade. However, it must be remembered that some EDCs are very persistent, so even when their inputs are restricted, residues in sedimentary and other sinks may, if remobilized, continue to exert their effects for many years.

Conclusions About the Scale and Implications of Effects

Is there a problem in invertebrates?

Most field assessments of aquatic and terrestrial ecosystems with invertebrates have examined either population- or community-level effects. The focus of most invertebrate studies is at this biological scale. Population- and community-level studies with invertebrates provide useful information for monitoring and assessment, at appropriate temporal and spatial scales, by documenting the source and extent of environmental degradation. However, studies at this scale do not usually provide definitive information on the cause of the problem, nor do they provide information on the mechanism. Identifying both causation and mechanism requires studies at other biological scales (Chapters 2 and 3). To identify ED as the mechanism for any observed effect in either populations or communities will require studies that include components from different biological scales and demonstrate interaction at those scales.

Few studies have clearly demonstrated effects operating through a known causative agent, resulting in ED that modifies individuals and results in population- and community-level changes. The best example is that of TBT in mollusks, as described above. Its widespread distribution has been associated with locally severe effects on marine mollusk populations. While there are many examples of ED in terrestrial invertebrates, these studies are associated with pesticide field trials. There are few studies of wild terrestrial populations where ED has been identified as a serious problem.

In summary, there are only a few field examples of invertebrate ED. Most field studies do not focus on mechanisms but rather on high-level response. It is suspected, however, that there are many real-world examples in which ED is affecting invertebrate populations and communities. The conclusion that effects of EDCs on invertebrates in the field are likely widespread, though undetected, stems from several lines of evidence and reasoning. First, the conservatism in basic mechanisms of endocrine function suggests that if various vertebrate endocrine systems are affected (see Kendall et al. 1998), then other animals exposed to the same or similar compounds may be affected similarly. Second, invertebrates are exposed to a great variety of compounds known to alter hormone function in both vertebrates and invertebrates, although the evidence on invertebrates is mostly from laboratory research (Chapters 2 and 3). Scientists know of invertebrate ED through more than 30 years of research and testing to develop EDCs intentionally for insect control programs. Finally, invertebrates garner far less attention than the larger and more charismatic vertebrates, despite the overwhelming ecological and economic (not to mention zoological) importance of invertebrates. Not only could effects on invertebrate populations remain undetected, but such a pattern has been documented for ED in TBT-exposed snails throughout the world. Furthermore, many invertebrate populations have changed dramatically, with enormous consequences and little notice until effects on society are severe. Examples include introductions of non-native species, infections of honeybees with mites, and pesticide resistance in crop pests, to name a few.

Do we require field monitoring focused on EDC effects in invertebrates?

While the full scope of EDC effects on natural invertebrate populations is uncertain, some issues are clear:

1) EDC effects on invertebrates can occur (e.g., TBT).
2) Invertebrates possess endocrine pathways that are theoretically susceptible to disruption.
3) Focused laboratory studies have illustrated effects that are consistent with ED (see Chapter 3).
4) Current toxicity tests will have to be modified or expanded, to a greater or lesser extent, before they can be used as sensitive and diagnostic screens for ED effects (see Chapter 3).

From these facts, one must conclude that it is very possible at present for chemicals with ED activity against invertebrates to be released into the environment undetected. For this reason, it follows that monitoring of biological responses in the field must be a component of EDC assessment, or important effects will likely be missed. This is underscored by the reality that those cases of documented EDC effects, in invertebrates or vertebrates, have come to light as the direct result of field observations.

Natural field communities incorporate several important attributes that are difficult to simulate in laboratory studies of physiology or toxicology, including these:

1) Integration of exposure over time, chemicals, media, and pathways. Little knowledge of the causative chemicals or their environmental fate is required, and the concentrations and pathways of exposure are comprehensive and realistic.
2) Incorporation of important biogeochemical factors. Environmental variables such as microbial transformation are implicitly incorporated.
3) Field assemblages include all relevant species. There is no concern over the representativeness of species selected for study, as might be the case for laboratory testing.
4) Field populations reflect all relevant life stages and processes. There is no concern over missing critical elements of life history, as there might be for species that cannot be tested over their entire life cycle in the laboratory.
5) Incorporation of ecologically significant behavioral effects. Because the initiation and/or expression of many behaviors are under endocrine control, behavioral changes may be a likely mode of EDC effects; while difficult to quantify and interpret in many laboratory assays, field populations incorporate the influence of behavioral effects.
6) Representation of higher-order biological interactions. Population-level or interspecific interactions are represented.

7) Field responses define ecological significance. Expression of biological effects in field populations is a typical measure of the ecological significance of toxicological responses.

Conversely, it should also be pointed out that field studies inevitably suffer from a number of confounding effects that can hinder accurate interpretation. There will be a variety of these nontreatment effects (e.g., temperature variations, habitat differences, seasonal differences, variations in food supply, etc.), which should be minimized by good survey design and which must be borne in mind when the data are being assessed.

Accepting that field studies are a key component of EDC assessment, the next obvious question is how to design field studies that will be sensitive to EDC effects. Design of field monitoring is critically dependent on the objectives. In the proceedings from a previous workshop on EDCs, Vethaak et al. (1997) outlined a multilevel approach to field monitoring for EDC effects in wildlife (Table 4-4). This approach differentiated among study types based on the level of existing knowledge of the problem or suspected problem. When the potential effects of EDCs on invertebrates are being considered, this approach serves as a useful basis for the design of field studies (Figure 4-5).

Table 4-4 EMWAT Workshop proposed scheme for monitoring effects of endocrine-disrupting compounds (from Tattersfield et al. 1997)

Objective	Knowledge required	Techniques
Determination of presence of effects in invertebrate populations in a geographic location	None	Community attributes, e.g., number of taxa Population effects, e.g., sex ratios Individual effects, e.g., gonad size
Identification of causality	Population-level effects have been established	Combination of analytical measures and verification of effects, using • knowledge of sources and causality, • gradient studies, and • TIE approach.
Validation of field observations	Confirmation/ verification	In situ bioassays or laboratory tests, using • known concentrations of suspect chemicals, • chemical fractionation of environmental samples, and • environmentally relevant mixtures.

Field studies to detect EDC effects can be broken down into 2 primary types. In the most generic case, monitoring studies may be undertaken to identify biological effects when there is no a priori knowledge of the causative chemicals, endocrine mechanisms, sensitive organisms, or likely effects. Alternatively, field studies may be

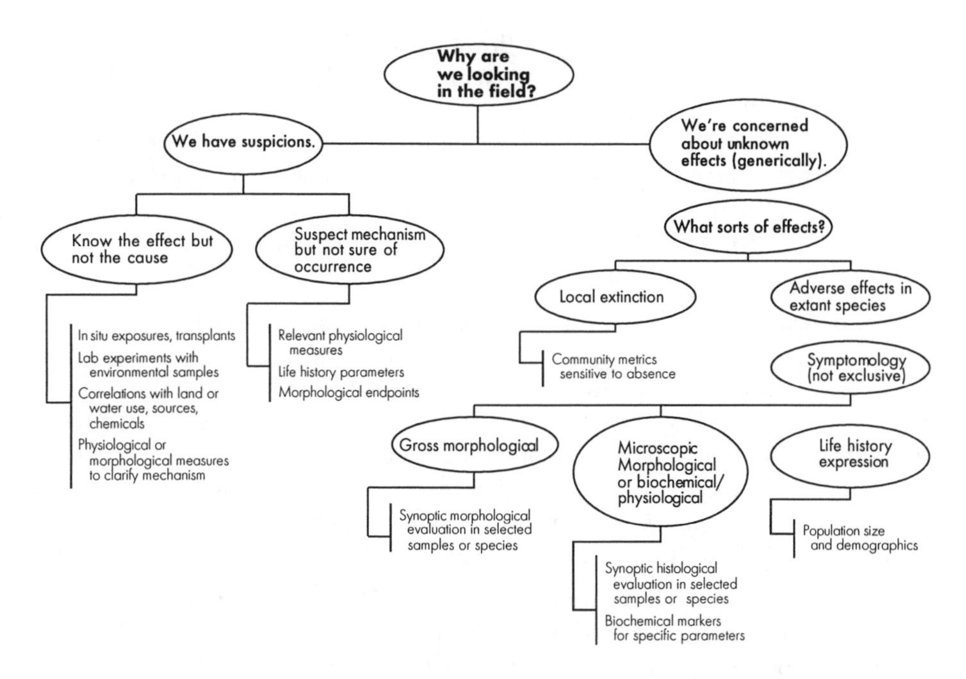

Figure 4-5 Conceptual diagram of field monitoring goals and strategies for endocrine-disrupting chemicals (modified from Vethaak et al. 1997)

of interest when there are indications of an EDC-related problem, but the specifics of that problem are not understood (Figure 4-5, left side). As an example, a particular biological response may have been observed (e.g., a morphological change), but the causes and/or consequences of that response are not known. Field studies of several types may be helpful in this instance. In situ exposures (e.g., caged organisms) may be employed for firmly associating the field exposure with the observed effect, or in establishing the spatial distribution of the causative factors. Similarly, measurements of occurrence in field populations may help establish associations with particular land or water use patterns, known sources of contaminants, or specific chemical gradients. Physiological or morphological measures (e.g., hormone titers) on organisms showing symptoms in the field may help elucidate the toxic mechanism.

Perhaps most important, field sites may provide samples (water, sediment, soil, tissue) that can be assayed for EDC activity using appropriate laboratory bioassays; if the EDC activity can be expressed in a laboratory assay, then it becomes possible to apply biologically directed fractionation (such as Toxicity Identification Evaluation [TIE]) techniques to identify the specific chemicals responsible for the biological effects. TIE has proven to be a powerful tool for identifying the causes of biological effects in water and sediment and recently has been applied to physiological responses, including those that are EDC-related (Desbrow et al. 1998).

In other cases, knowledge of the properties of a particular chemical may give reason to suspect potential EDC activity, but the actual expression of effects in the field may be uncertain. In this case, gradients of chemical concentration found in association with a particular source may provide a valuable opportunity to study the responses in a wide range of organisms. Appropriate monitoring variables will be determined by the nature of the suspected mechanism but may include physiological or histological changes, gross morphology, or particular life-history endpoints such as population size or demographics.

A more difficult situation may arise in the generic case, when the design of a monitoring program requires sensitivity to EDC effects, but when the nature of those effects and the chemicals that may cause them are unknown (Figure 4-5, right side). Without knowing the biological mechanisms underlying EDC effects, it is very difficult to ensure that a particular monitoring design will have the desired sensitivity. Despite these uncertainties, it is possible to discuss some general design considerations based on the range of biological effects that might be expected from EDCs, along with the insights from those cases where EDC effects have already been found.

As a first pass, it may be useful to consider potential EDC effects in 2 categories: 1) those that will have caused local extinction of a species and 2) those that will be observed in species still extant at the collection site. Detection of the first case obviously requires data collection and analysis approaches that are sensitive to the absence of particular taxa, even if those taxa are only minor components of the overall invertebrate community. In this sense, integrative indices of invertebrate communities, such as the diversity indices often used for monitoring of benthic invertebrate communities, may have relatively little power unless effects are observed on dominant taxa (see "Currently available population- and community-level endpoints for EDCs," p 249).

When the EDC effects are not lethal, or have not become severe enough to eliminate the species, detecting EDC exposure may rely on measuring specific changes in individuals or the demographics of a population (e.g., dominance of older, reproductively senescent individuals). The outward signs of EDC stress may occur in several forms. Effects such as imposex in gastropods and intersex in copepods are observable as gross morphological changes of the body form. In addition to gross morphological effects, EDC effects may be expressed as histological changes in specific tissues, such as the conversion of ovarian tissue to testicular tissue or vice versa. Without knowledge of the specific malformations that may be involved, it is impossible to determine comprehensively all relevant structures or tissues for morphological analysis, although the sex organs, sexually dimorphic structures, and other features known to be under endocrine control are likely candidates.

Changes in physiological or biochemical variables are likely to be associated with most EDC effects, though they will probably be even more difficult to detect in monitoring programs that do not have specific target chemicals or mechanisms. However, as more is discovered regarding the mechanisms behind known EDC effects, it may be possible to define variables that are commonly associated with EDCs.

Finally, EDC effects may be expressed as alterations in population size or demographics, such as unusual age-class structures (e.g., as with gastropod imposex) or altered sex ratios and sex intergrades (Moore and Stevenson 1991, 1994). Typical designs for many invertebrate monitoring programs, such as those used for benthic macroinvertebrates in freshwater systems, do not document changes of these types. Variables other than taxonomic counts, including demographic effects (e.g., life stage), would likely increase the power of invertebrate community sampling to detect the influence of EDCs (see "Currently available population- and community-level endpoints for EDCs," p 249).

While it is realistic to expect that EDCs may affect variables such as age-class structure and presence or absence of particular taxa, it is clear that such responses are not unique to endocrine-mediated toxic mechanisms, or indeed to pollution alone. As such, additional studies will be required in virtually all cases to define the specific causes of observed effects (see "Is there a problem in invertebrates," p 235) and determine whether they are in fact EDC-induced. Not until our understanding of EDC effects improves substantially is it likely that EDC-specific monitoring measures will become available universally.

A further complication arises when the spatial extent of contamination of the EDC stressor is unknown, as it will typically be initially. Where EDC sources are localized and concentration gradients are therefore present (which may be true, but unknown), spatial patterns in presence or absence of particular taxa may provide the needed context to detect the absence of individual taxa. However, if the gradients are not large within the study area, the detection of missing species would rely more heavily on prior knowledge of the species assemblage that should exist at a particular site. This issue is equally troublesome when other variables such as morphological characteristics are assessed; frequencies of deformities in unstressed populations are poorly understood for almost all invertebrates, although background rates of chironomid mouthpart deformities and asymmetries have recently been determined at Great Lakes reference sites.

Current Invertebrate Monitoring Programs

Marine environments

In recognition of the prominence given in the North Sea Quality Status Report (1993) to concern over the biological effects of contaminants, the Oslo and Paris Commissions (OSPAR) have developed a strategy to incorporate effects measurements into their Joint Assessment and Monitoring Program (JAMP) for the European convention area. This strategy includes assessment of the contaminant-specific effects of TBT and includes imposex in *Nucella* and *Buccinum* and intersex in

Littorina as monitoring targets. In preparation for this program, OSPAR recognized that it was necessary to prepare guidelines for participating laboratories (JAMP 1998). The guidelines indicate that in coastal areas, imposex should be monitored in *Nucella*, and where *Nucella* is absent, *Littorina* should be examined for intersex. Where both species occur sympatrically, each should be monitored for sexual development. In areas south of the range of *Nucella*, shell thickening in Pacific oysters should be measured. Imposex in *Buccinum* should be monitored in offshore areas, particularly around shipping lanes. The guidelines include the following:

1) detailed and internationally agreed analytical protocols;
2) sampling strategies to meet defined, quantified objectives;
3) the requirements for supporting information, such as chemical analyses;
4) bases for the interpretation of data; and
5) quality assurance procedures, including training workshops, internal quality control, and external quality assurance.

The OSPAR Guidelines provide items 1 to 4 for both *Nucella* and *Littorina*, and a training workshop has been held covering all 3 species. External quality assurance for *Nucella* and *Littorina* is available through the Quality Assurance in Marine Environmental Monitoring (QUASIMEME) program. It is anticipated that the basis of interpretation and external quality assurance for *Buccinum* will be available soon.

The Environmental Agency of Japan has carried out biological monitoring programs since 1978 to survey the present status of chemical pollution in Japan. Organochlorine compounds, such as dioxins and furans, PCBs and DDTs, organotin compounds such as TBT and TPhT, and other chemical substances have been detected in biological specimens collected by the program. Biological specimens consist of fish, shellfish, and birds that have been stored frozen as a specimen bank program of the Japanese National Institute for Environmental Studies. Specimen banking is useful for retrospective descriptions of the contamination caused by newly occurring substances in the environment. Such specimens, including shellfish (mussels, oysters, and short-necked clams) are usually available for chemical analysis but not for histopathological examination. Only gross morphological observation can be used. Improvement of the methods for sample collection, fixation, and archiving is therefore necessary for the examination of the adverse effects of endocrine disrupters in aquatic organisms.

There have been few Japanese monitoring programs to evaluate possible adverse effects of endocrine disrupters on benthic or planktonic communities. The only one in progress involves continuous field studies of imposex in gastropods conducted by the National Institute for Environmental Studies. Both periodic sampling at a reference site and at contaminated sites (e.g., marinas and dockyards), as well as countrywide sampling for gastropod specimens, have been carried out in this program since 1990. Gastropod species collected are mainly the rock shell *Thais clavigera*, which has a broad distribution over most of Japan. The rock shell is easily collected in the rocky inshore zone and is very sensitive to TBT and TPhT. Consequently, the rock shell is considered a sentinel or indicator species for organotin pollution from antifouling paints in Japan. As not only biological or ecological observations but also chemical analysis are included in the program, relatively comprehensive information is being obtained (e.g., population abundance, size-age structure, imposex symptoms by size-age, spawning behavior and tissue concentrations of TBT and TPhT, etc.).

The U.S. National Oceanic and Atmospheric Administration's (NOAA) National Status and Trends Program (NSandT) monitors, on a national scale, chemical contamination and biological responses of invertebrates to contamination. The Mussel Watch analyzes mussels and oysters collected from 200 sites throughout the coastal and estuarine U.S. for contaminants such as heavy metals (zinc, cadmium) and organochlorines (DDT, PCBs, etc.) (National Research Council [NRC] 1990; USEPA 1998; NOAA 1999). Similar national marine programs are run in many other parts of the world, including European countries. Such programs study both benthic and planktonic communities and measure a variety of individual-, population-, and community-level changes, as well as chemical contamination. However, none are focused on EDCs as such, although they might well be responsive to such substances.

Freshwater and estuarine environments

The U.S. Environmental Protection Agency's (USEPA) Environmental Monitoring and Assessment Program (EMAP) in the Office of Research and Development (ORD) integrates its own research with environmental monitoring conducted by NOAA and state agencies to assess the condition of watersheds and estuaries (USEPA 1998). Monitoring of invertebrate populations was conducted in EMAP's Condition of the mid-Atlantic Estuaries Project. ORD is currently conducting an ecological assessment on the Little Miami River Watershed in Southwestern Ohio (USEPA 1999a). ORD also monitors macroinvertebrates in freshwater streams for use as indicators of stream conditions (USEPA 1999b). USEPA's Great Lakes National Program Office conducts biannual invertebrate monitoring surveys of the Great Lakes in from 8 to 20 different locations. Plankton, mysids and *Diporeia* amphipods, and other invertebrates are periodically sampled from the Great Lakes (USEPA 1999c).

The U.S. Geological Survey's (USGS) Biomonitoring of Environmental Status and Trends (BEST) program monitors the effects of environmental contaminants on biological resources. For example, this program intensively monitors invertebrates in the upper Mississippi River and its tributaries (USGS 1999a, 1999b, 1999c)

The Illinois Department of Natural Resources (IL DNR) Critical Trends Assessment Program (CTAP) monitors aquatic and terrestrial invertebrate populations with its RiverWatch and ForestWatch projects. The RiverWatch project samples benthic macroinvertebrates (including aquatic insects, mussels, clams, snails, and worms) in 200 streams. Many other states have similar monitoring programs (IL DNR 1999).

In Canada, there are no national-scale programs using invertebrates for monitoring ecosystems, and most environmental monitoring comes under the jurisdiction of the provincial governments. Two provinces, Alberta and Ontario, have used benthic invertebrate community structure for assessing aquatic ecosystems, but the focus of their programs has been on general monitoring rather than on addressing specific mechanisms such as ED. More recently, Environment Canada has initiated the Environmental Effects Monitoring (EEM) program which is assessing the adequacy of regulations governing various industrial sectors. The first to be examined under this program was the pulp and paper industry, and one of the components of the assessment was benthic invertebrate community structure, focusing on changes in communities in response to effluent discharge (Environment Canada 1997). The next sector to be examined under the EEM program will the metal mining industry.

Most European countries conduct routine freshwater invertebrate monitoring programs that include inventories of invertebrate benthic fauna. Many of these activities are coordinated through international organizations such as the International Rhine Commission (IRC), the International Commission for the Protection of the Elbe (ICPE), and the International Commission for the Protection of the Meuse (ICPM). In the United Kingdom, the River Invertebrate Prediction and Classification System (RIVPACS) is used nationally to compare invertebrate community structure at a given location with that to be expected in a pristine habitat of the same type (Wright et al. 1989). However, as with most marine monitoring programs, this approach can reveal pollution-related impacts without providing evidence about causative mechanisms.

Terrestrial environments

There is considerable monitoring of terrestrial invertebrates in the U.S. and Europe, often in association with pesticide applications to agricultural lands. This research focuses on the effects of pesticides on beneficial and nontarget arthropods. The effects of pesticides on populations of predators, parasitoids, pollinators, and commercially important decapods (crayfish) are frequently studied in the U.S. (e.g., Woolwine et al. 1997; French et al. 1998). The U.S. Forest Service (USFS) and many land grant universities also monitor terrestrial invertebrates adjacent to or under the forest canopy of pesticide treated areas.

Strengths of current invertebrate monitoring

The strengths of current invertebrate monitoring programs are discussed in "Endocrine disruption in other invertebrates" (p 216). In general, field monitoring offers advantages over laboratory testing in that it involves organisms in their natural environment, exposed to realistic (actual) concentrations of contaminants and their metabolites over time and space. Field monitoring offers insights into actual biological or chemical interactions that cannot be duplicated by laboratory tests and predictive models.

Numerous accounts are available of the assessment of effluent impacts by studying localized in situ invertebrate assemblages, by introducing enclosures containing test organisms, and by laboratory exposures of test organisms to water samples collected from study sites (Pinder and Pottinger 1998).

Weaknesses of current monitoring

It will be apparent from the foregoing discussion that most invertebrate monitoring programs are not designed to look specifically for the effects of endocrine disrupters, and indeed have little if any diagnostic capability. In other words, they are able to detect gross biological changes at the community level but cannot generally identify the causes. Furthermore, they rarely investigate any type of biochemical or cellular change and frequently do not even analyze age structure or sex ratios. With a few exceptions, invertebrate monitoring efforts are concentrated on the near-field effects of anthropogenic contaminants (e.g., effects of pesticide applications on nontarget insects near fields; TBT effects in coastal areas close to shipping activity) and rarely consider far-field changes that may be important for persistent and bioaccumulative substances. As in vertebrate monitoring, such programs also may be limited by ethical concerns surrounding the destructive sampling of rare species.

Unassessed historical data

Historical data and archived biological material could potentially provide invaluable information on the status of invertebrate populations in the past, with which current status could be compared. A number of countries have conducted routine surveys of aquatic and terrestrial invertebrates for several decades, and they hold data on at least species abundance and distribution.

This is useful in a general sense, but of far more value is archived biological material that could be used to give information on structural and biochemical abnormalities in historical populations. Many countries (e.g., Great Lakes reference material held in the U.S. and Canada) store extensive reference collections of invertebrates preserved in formalin or alcohol, which have already been used, for example, to investigate background rates of chironomid deformities. Such reference collections would be suitable for investigating most types of morphological abnormality caused by EDCs, including histological lesions. It is also possible to use formalin- or alcohol-preserved material to investigate the degree of DNA diversity in reference collections, which could indicate whether historical populations contained greater genetic variability than more recently when EDCs may have forced species through population "bottlenecks" (i.e., whether genetic variability has been reduced because present populations have descended from only a few individuals).

Even more valuable would be archived cryo-preserved biological samples, which could be used for studying various biochemical markers such as vitellin induction or enzyme activity. Several countries (e.g., Germany, Japan, and the U.S.) have stored such invertebrate material for about 20 years, and it may thus become possible to establish historical trends in a variety of endocrine-related endpoints (once these are developed for general use).

There are a number of established databases that record population levels of invertebrate species in the terrestrial environment. These are both national (e.g., the UK Natural Environment Research Council's Biological Database) and more local (e.g., long-term individual site surveys such as for Sites of Special Scientific Interest designated by English Nature in the UK). These are primarily directed at species of high public concern, e.g., butterflies and bumblebees in the UK. Germany undertakes specimen banking for chemical analysis, biomarkers, and populations, and monitors earthworm (oligochaetes), lugworm (*Arenicola marina*), zebra mussel (*Dreissena polymorpha*), and mussel (*Mytilus edulis*) populations. Many population-level surveys are related to the conservation status of individual species. For example, in the U.S. there are concerns about population levels of some species of spiders listed in the Endangered Species Act or the International Union for the Conservation of Nature (IUCN) Red List, and investigations include surveys of statewide incidence.

In the terrestrial environment, much can probably be gained by considering data from experiments on pesticide effects, e.g., the identification of key species such as spiders, predatory mites, parasitoids, and earthworms for pesticide registration purposes by European Standard Characteristics of Beneficials Regulatory Testing (ESCORT; Barrett et al. 1995). These may include assessments of the implications of the effects resulting from IGR exposure on population levels. Critical population reductions of 30% have been set for ecologically significant effects on nontarget arthropods in off-crop situations.

Priority Groups for Monitoring

Selection of invertebrate taxa for field monitoring of EDC effects should be prioritized toward those groups having

1) well-characterized endocrine systems wherever possible,
2) ecological importance in ecosystem maintenance,
3) rapid reproductive cycles,
4) sessile or semi-sessile behaviors,
5) widespread distribution, and/or
6) economic importance.

Currently, only 2 invertebrate groups, the insects and crustaceans, have endocrine systems sufficiently well characterized to be of widespread value in EDC field monitoring programs (see Chapter 2). Furthermore, as an extensive database exists for toxicological effects in a relatively small subset of all invertebrate taxa (Chapter 3), EDC monitoring programs targeted at these taxa may prove most useful for environmental management.

Crustaceans of widespread use in current testing regimes are amphipods, cladocerans, and mysid shrimp; major terrestrial arthropod models include parasitic wasps, mites, carabid beetles, honeybees, and lacewings. Currently, widespread long-term chemical status and trends monitoring is ongoing only for mussels and oysters (e.g., NOAA's MusselWatch program).

A few insect species, e.g., butterflies, are the subject of long term monitoring programs. These programs could be modified to include biomarkers or bioindicators of EDC effects such as vitellogenin/vitellin, ecdysone, and JH disruptions.

Marine gastropods have proven an ideal monitor and sentinel species group for EDC effects of TBT. Their utility for other EDCs is not yet known. Possible sentinel species for the marine environment could be selected from common and easily collected mollusks, crustaceans (planktonic and benthic), coelenterates, echinoderms, annelids, and nematodes. For freshwater environments, all the latter taxa except echinoderms are available, plus insects. In terrestrial environments, possible sentinels include insects, mollusks, isopods, annelids, arachnids, and nematodes. However, it must be noted that we do not yet have enough knowledge of the endocrinology and sensitivity of particular species to make confident recommendations for their routine use.

Currently Available Biomarkers and Organism-level Endpoints for Endocrine-disrupting Compounds

Endpoints in aquatic organisms

There are a limited number of biomarkers currently available to measure exposure to EDCs and their effects in invertebrate species. Morphological variables are the most commonly used biomarkers in crustaceans and mollusks. Examples of these abnormalities in crustaceans include exoskeleton deformities, precocious juveniles, limb aberrations, genital aberrations (e.g., "chest" hair in female Daphnids), and intersex in harpacticoid copepods (e.g., Moore and Stevenson 1991, 1994). TBT-induced intersex and imposex in marine snails and shell thickening in oysters provide clear examples of biomarkers of exposure and effect to a known EDC, i.e., TBT. Gonadal histology and morphometry are useful tools for the detection of reproductive abnormalities, including stages of intersexuality. Additional histological investigation of other target organs or the entire organism may provide a good evaluation of the various EDC endpoints alluded to in previous chapters.

Abnormal development of secondary sexual characteristics will also provide a useful endpoint for the effects of EDCs. Gross abnormalities are not usually found in the field, as they correlate with reduced fitness and increased predation; therefore, these abnormalities may be most pertinent in controlled field experiments.

In addition to morphological biomarkers, a limited number of biochemical measurements are in current use. Vitellin titers and steroid hormone levels can be measured in crustacean and molluskan species (where applicable) and will potentially provide biochemical markers for exposure to EDCs. For example, in benthic amphipods and copepods, there may be some scope for ecdysone titers to be utilized, and levels of cyprid major protein have been shown to increase in cirripede larvae following laboratory exposure to certain EDCs. The use of the alkali-labile phosphate (ALP) technique (Gagne and Blaise 1998) provides a useful nonspecies-specific method for the determination of hepatocyte production of vitellogenin. This has been adapted for study of vitellogenin-type proteins in bivalves (Blaise et al. 1998). Modification of enzyme production, e.g., aromatase CYP 19 inhibition in snails and CYP1A activity in sea stars (den Besten, Elenbaas et al. 1991), could be a useful tool in monitoring the impacts of EDCs in field studies. There are also CYP 3A-like proteins that metabolize steroids in blue crab, CYP 45 and 2L in lobster, and CYP 1A, 3A, 4A and 11A-like proteins in mollusks, most of which are modulated by xenobiotics (James and Shiverick 1984; Lee 1993; Wootton et al. 1995; Brown et al. 1997; Snyder 1998). Finally, Phase II enzymes are important in the elimination of steroids, and glutathione-S-transferase has been extensively investigated in crustaceans and mollusks (Lee et al. 1988; Yadwad 1989).

The use of sensitive wildlife individuals in caging or explant studies should be encouraged. These will provide a more appropriate linkage between laboratory experiments, the use of laboratory-raised species in field tests, and real-world situations. Examples of caging experiments have included a range of crustacean and mollusk species, as well as explants of tunicates, sponges, and bryozoans allowed to settle on plates for ease of manipulation. In addition, spat of sessile species can be settled in the laboratory and immediately placed in open water on tiles.

Terrestrial organism endpoints

Markers of the effects of ED on terrestrial invertebrates have been limited essentially to arthropods. These are most readily assessed by investigating the effects of known endocrine disrupters such as the IGRs.

Morphological abnormalities caused by ecdysone agonists include decreased adult emergence, death without ecdysis, lethal premature molt and nymphal–adult intermediates with malformations of wings and mandibles (Table 4-3). Appropriate markers for the effects of anti-ecdysteroids are more difficult to determine, with effects such as precocious metamorphosis and interference with diapause in adults (Table 4-3). Juvenile hormone analogs (JHAs) have developmental effects such as deformities, e.g., crescent-shaped mark on the eyes of developing honeybee pupae and the disruption of larval–pupal molt. The cocoon-spinning ability of silkworms has been shown to be affected by JHAs. The German cockroach nymph has been observed with developmental malformations such as twisted wings and ecdysal failure and the psyllids have shown external genitalia malformations. Diamondback moths have been observed to produce pupal–adult intermediates following exposure (Khan and Urs 1992).

Ecdysteroid titers in terrestrial invertebrates have been used in a number of species (e.g., the silkworm) to assess the effects of chemicals in the laboratory. Azadirachtin, an ecdysteroid antagonist, has been reported to decrease vitellin levels in earwigs and to affect ecdysteroid hormone levels. It has been shown to result in effects on ovarian development in the earwig, with degenerated follicles and reduced deposition of vitellin in the ovary. In addition, it has been shown to decrease the growth and development of ovaries with decreased oocyte and basal follicle numbers in mealworms (Sayah et al. 1996, 1998).

Ecdysteroid and other receptor binding assays are available (see Chapter 3) and could be applied as screening for specific EDCs, or in a TIE approach with either environmental media or tissue extracts.

For specific effects (e.g., JH agonists), standard methods have been developed; for example, for IGR use on flowering crops, a standard test exists for the honeybee (Oomen et al. 1992). In addition many of the test methods developed for pesticide risk assessment, e.g., by the International Organization for Biological and Integrated Control of Noxious Animals and Plants (IOBC), can be adapted readily for use with individual- or population-level studies. Examples include tests established for a wide range of arthropod fauna. In addition, the use of barriered plots would also allow assessment of effects on other invertebrate populations such as earthworms and snails.

Currently Available Population- and Community-level Endpoints for Endocrine-disrupting Compounds

Surveys of population- and community-level endpoints identify problems, but diagnostic capability is often limited and often too late (i.e., the effect has already occurred). In addition, these measures are not sensitive, so there is a need for more sensitive biomarkers.

In recent years, numerous community-based metrics (biotic indices of diversity, evenness, etc.) have been developed to assess impacts caused by anthropogenic stressors. These are most commonly applied in benthic community studies and in some cases constitute the foundation of biological monitoring programs. Programs can be designed to investigate changes in community composition and structure at various spatial and temporal scales, provided there is some underlying understanding of the scale of the effect to allow appropriate selection of control or reference sites.

While useful as tools to reduce complex multivariate data and to assess general environmental impacts, the community metrics are likely to be rather insensitive to the absence of minor taxa. Specifically, many indices either lose (through the generation of single summary indices) or ignore (e.g., through being used at relatively high taxonomic levels) information that may be critical in the detection of trends in the abundance of nondominant taxa. Nevertheless, the use of multivariate techniques such as principal components analysis (PCA) and multidimensional scaling (MDS) can be useful to assess community-level change.

Classical community analysis is unlikely to provide a diagnostic discrimination of impacts due solely to ED, although with hindsight, it has been possible to recognize the impact of TBT on benthic communities in the Crouch estuary (see "Effects of controls on the use of TBT," p 214), as the fauna recovered after TBT use was prohibited on small boats. Moreover, the detection of impacts at community level normally suggests that environmental conditions are already significantly adversely affected, to a degree that might be considered undesirable. They do not provide an "early warning" of adverse changes.

One possibility for community-level detection of EDCs may be the use of community-size spectra, i.e., the spectrum of body sizes present in the range of species at a given location. In freshwater environments (as opposed to marine environments, where distinct size-spectra are less obvious), it has long been recognized that body size and biomass of organisms can be used as surrogates for trophic position and responsiveness to environmental variation, including stressors (Peters 1983). This property reflects the fact that most biomass in the limnetic zones of lakes falls within 3 or 4 distinct size classes (Sprules et al. 1983), with little biomass residing between these classes. The distinctiveness of each size class is defined by similarities in morphology, physiology, life-history characteristics, and sensitivity to environmental variation, relative to adjacent classes (Neill 1994). Changes in community structure dynamics within each class could potentially be used to assess response to environmental stressors, including EDCs.

Current invertebrate population-level endpoints for EDCs are focused most strongly on frequency of imposex, frequency of intersex (Féral 1980; Saavedra Alvarez and Ellis 1990), population indices of imposex intensity (e.g., RPSI and VDSI) deviations of sex ratio from the norm, and population age-structure analysis (Bryan et al. 1987; Gibbs et al. 1987; Smith and McVeagh 1991; Stewart et al. 1992; Wilson et al. 1993; Oehlmann et al. 1993; Moore and Stephenson 1994; Oehlmann 1994; Bauer et al. 1995, 1997; Minchin et al. 1995, 1996, 1997; Oehlmann and Bettin 1996; Harding et al. 1997; Poloczanska and Ansell 1999).

EDCs acting at the level or time of sexual differentiation in an invertebrate may skew population sex ratios in favor of males or females depending on the mechanisms involved, which are often species-specific. Normal sex ratios for many invertebrates are 1:1, but there are frequent exceptions. For example, many harpacticoid copepod populations exhibit normal sex ratios of 4 to 5 females per male in estuarine/marine systems, and insects may have many males per reproductive female. Thus for EDC-mediated sex ratio effects to be resolved, the baseline unperturbed sex ratios of populations must be known. Once known, then field monitoring of population sex ratios may provide a simple but robust indication of EDC activity acting at an ecologically important level.

For most invertebrates, population age-structure characterization is more difficult to assess than sex ratios. However, age-structure analysis coupled with sex-ratio determinations provides a robust indicator of EDC effects acting upon organism growth, molting, age-specific survival, and recruitment success. For example, if an invertebrate population is dominated by an early larval metamorphosis stage in the presence of an EDC, it may indicate EDC-mediated suppression of growth and/or metamorphosis to subsequent life stages. Similarly, if EDC-exposed populations are dominated by older individuals, then it may indicate local reproductive depression or failure and/or a failure in larval attraction, settlement, or recruitment (e.g., dogwhelk population age-structure under TBT exposure). Data analysis methods similar to those used widely in marine and freshwater science to investigate the passage of cohorts of individuals through a series of developmental stages may reveal abnormalities in the normal development (rate and survival) of cohorts such as might arise from exposure to EDCs.

For population-level EDC effects, a more robust analytical tool is life-table analysis and modeling, e.g., Lotka-Volterra models (Lotka 1922, 1943; Volterra 1926) and Leslie matrices (Leslie 1945), where effects can be characterized by changes in life-expectancy, fertility, and intrinsic population growth rates (r) (see Chapter 3). These variables allow multigenerational modeling of population growth under EDC exposure in the field or more often in the laboratory. Unfortunately, life-table characterizations are logistically difficult and are known for only a few aquatic and marine invertebrates, e.g., freshwater daphnids (Frank et al. 1957; Day and Kaushik 1987; Bechmann 1994) and gastrotrichs (Hummon 1974), estuarine copepods (Allan and Daniels 1982; Green et al. 1995), and several terrestrial insects (e.g., *Oncopeltus* milkweed bugs, Dingle 1968; aphids, Root and Olson 1969; and *Tribolium* flour beetles, Mertz 1971). Sessile field populations (e.g., mussels, oysters, and barnacles) may be most amenable to life-table characterization and subsequent analysis under EDC field exposures. Culturable motile fauna have proven useful in laboratory modeling of toxicant life-table effects (*sensu* Green and Chandler 1996), but they are problematic in field studies because of migration or emigration and subsequent difficulties in tracking individuals or individual cohorts through time.

A range of other approaches to population or community studies might have application to endocrine disrupters, and these are illustrated in Table 4-5.

Table 4-5 Other approaches to population or community studies potentially applicable to endocrine disrupters

Response	Type of response	Background	Field approach
Reproductive success	Reproductive effect that could arise from exposure to endocrine-disrupting compounds (EDCs)	Animals affected by some forms of endocrine disruption would be expected to show reduced reproductive capacity or success. This may be observed in the F1 generation.	Affected adults from field populations are transferred to the laboratory, and reproductive success and of the quality of the juveniles are assessed.
Population genetics	Genetic response to a range of effects leading to restriction of reproduction, which may be caused by EDCs	Populations that have previously been restricted in their reproduction can show reduced genetic diversity. Published examples include copepods close to a U.S. Superfund–contaminated sediment site and around an oilrig in the Gulf of Mexico.	Organisms of single species would be collected along contaminant gradients and measurements of genetic diversity would be made.
Insecticidal resistance	Indicator of prior exposure to groups of chemicals that include potential EDCs	Repeated use of pesticidal compounds can lead to resistance (e.g., increased LC50 values) in the remaining populations. There are many examples in insects and some in crustacean parasites.	Field population samples would be tested in the laboratory for sensitivity to selected potential EDCs. The results would indicate whether these populations had previously been exposed to EDCs.
Diversity of DNA	Indicator of changes in diversity which may be result of exposure to EDCs	EDC effects in populations may result in reduced diversity. A relatively new method of assessing diversity involves the application of degradation gradient gel electrophoresis (DGGE) to wild populations.	This method is most commonly applied to microbial communities. It may be possible to apply it to other relevant groups of organisms.

Major Research Needs

A large number of areas for further research need to be addressed before real-world monitoring programs for ED effects in invertebrates can be designed and initiated. These have been divided into 4 areas according to their application for

1) general monitoring for impacts on populations or communities;
2) monitoring of specific problems in wild populations, i.e., effects of a known chemical or investigations of identified population impacts, e.g., TBT;
3) active (or in situ) monitoring using transplanted or caged organisms; and
4) rapid screening tools for chemical fractionation studies and similar approaches.

General monitoring

To assess whether impacts are occurring in populations and communities, a number of techniques require further development:

1) Development of criteria for the selection of appropriate sentinel species for use in the marine, freshwater, and terrestrial environments. Although some attributes will be similar to those determined for use in laboratory studies (Chapter 3), there are likely to be further selection criteria specific to field monitoring. These attributes will be apparent from the results of research needs outlined below, e.g., taxonomic specificity of responses. In general, sentinel species should not be used singly but should be grouped so as to minimize the risk of missing effects due to EDCs.
2) An understanding of the specificity of responses between taxonomic groups. A fundamental part of this is the need for more information on invertebrate endocrinology in individual taxa, particularly of non-arthropod species, and an understanding of the significance of vertebrate steroids in crustaceans. This will assist in elucidating biochemical, morphological, or behavioral responses to chemicals across taxa, e.g., the impact of ecdysteroid agonists in nematodes. Further information would allow the identification of appropriate species and endpoints and improve interpretation of data.
3) Development of appropriate biochemical and morphological markers for use in field studies. At present we have a very limited suite of markers (equivalent, for example, to vitellogenin induction or secondary sexual characteristics in fish) for use in invertebrate monitoring studies. This is due to our limited understanding of biochemical, morphological, or behavioral responses to endocrine disrupters. These need to be developed for sentinel species and may include markers for assessment of the genetic diversity of populations. This research should include validation of the markers in studies at known pristine and contaminated sites.
4) Improved understanding of the population dynamics of invertebrates as they apply to endocrine disrupters. This may be progressed through the development of predictive models of the effects on the individual, population, and community of exposure to an endocrine disrupter and their validation using field data from reference and contaminated sites.

Specific monitoring

A number of research needs are concerned with assessing further the effects of endocrine disrupters on invertebrate populations where there is a specific pollution problem:

1) Improved ability to link effects of classes of endocrine disrupters from the molecular to the population or community level. This is probably best approached through the use of microcosm or mesocosm studies of known classes of endocrine disrupters and will include the identification of appropriate reference chemicals and endpoints (Chapter 3).

2) Establishment of mechanisms of effects and generic models. At present, there are many examples of changes that may be related to ED, but there is insufficient understanding of the mechanism to allow confirmation of this, e.g., heavy metals. Greater understanding of mechanisms, e.g., aromatase inhibition, may allow the development of generic models capable of predicting changes at higher levels of organization.

Active monitoring

Active or in situ monitoring using caged or transplanted individuals (see Chapter 3) may be a powerful tool. To develop in situ monitoring requires the following research:

1) Determination of appropriate sentinel species. Depending on the environment, and on which species are amenable to caging, organisms can be reared to the same stage in large numbers under laboratory conditions and give a rapid response, i.e., short reproductive cycle. Some examples of possible species include the crustacean *Gammarus*, the annelids *Lumbriculus* and *Tubifex*, and the mollusk *Potamopyrgus* in the aquatic environment, and the insects *Chrysoperla*, *Aphidius*, *Typholodromus* and the annelid *Eisenia* in the terrestrial environment.
2) Identification of suitable biomarkers for classes of ED, e.g., ecdysone analogues, in these sentinel species. This depends on developing an understanding of the endocrinology of such species.
3) Validation of the testing strategies using reference chemicals in microcosms or mesocosms. The sensitivity and robustness of proposed in situ testing methods can be evaluated under semi-controlled conditions in micro- and mesocosms using reference EDCs with well-understood characteristics (see Chapter 3).

Rapid bioassays

These are used for screening large numbers of environmental samples. For example, in fractionation studies, bioassay tools with high throughput and sensitivity are required.

1) Need for genetically engineered cell lines containing hormone receptors. Similar to the current yeast-based estrogen receptor assay, these are required to allow rapid screening for receptor-mediated effects.
2) Need for a range of molecular markers for effects, e.g., a nonspecies-dependent vitellin/vitellogenin assay. These may include genetic markers of increased protein transcription (see Chapter 3).
3) A greater understanding of the link between molecular changes and impacts at the individual or population level. This would allow prediction of potential effects, e.g., aromatase inhibition, changes in porphyrin profiles, and metallothionein induction.

Conclusions

This chapter has reviewed the available information from the field on ED in terrestrial and aquatic invertebrates. We conclude that very few well-established examples of ED are currently known (mollusks affected by TBT, insects affected by growth-regulating insecticides), but this is probably because little effort has so far been allocated to the subject. It certainly should not be assumed that the incidence of effects in invertebrates as a whole is genuinely rare or insignificant. The case of TBT's masculinizing effects in about 150 species of mollusks is an example that shows how apparently trivial biochemical changes can have drastic effects up to the population and community levels, and there is no reason to suppose that such far-reaching changes are in any sense unique. For this reason and others described in this chapter (e.g., "Is there a problem in invertebrates?," p 235), it is likely that other cases of ED in feral populations or individuals will be identified.

There is somewhat more extensive information on ED in invertebrates exposed to chemicals in laboratory studies, although even in these, one is rarely presented with conclusive information about the mechanisms of action. However, such studies show that, in addition to mollusks and insects, ED effects can be produced in a variety of crustaceans and echinoderms exposed inter alia to alkylphenols, heavy metals, and IGRs and other pesticides at environmentally relevant concentrations. There are also more-or-less well-founded suspicions that several other phyla may be capable of responding in this way. It therefore seems that ED effects are to be expected across a wide range of invertebrate phyla with very diverse endocrine systems.

The general strategy, described in "Do we require field monitoring focused on EDC effects in invertebrates?" (p 237), that should be used when designing monitoring programs for ED effects in invertebrates has been developed from the outline scheme proposed by the EMWAT workshop (Vethaak et al. 1997). It can be divided into techniques for use in

1) general monitoring when impacts are unknown,
2) targeted monitoring when following up on existing knowledge about possible ED effects,
3) active monitoring with, for example, caged sentinel species, and
4) bioassay-led fractionation procedures and other rapid screening techniques designed to throw light on causality.

Very little is currently known about the biochemical changes that most ED-affected invertebrates may experience, so the scope for development of biochemical and cellular biomarkers for use in monitoring programs will be limited until invertebrate endocrinology is better understood (Chapter 2). However, current knowledge suggests that there is considerable opportunity for the development and use of morphological and reproductive indicators of ED. These indicators are rarely unequivocal about the causative mechanism, implying the need for follow-up diagnostic studies when such effects are observed. Once suitable biochemical and morphological biomarkers begin to emerge as robust tools, there may be substantial scope to establish historical trends in effects through the use of archived invertebrate tissues, which exist in several countries.

This chapter has developed a series of research requirements that, if addressed, will enable field monitoring programs to identify the effects of EDCs with greater reliability and precision. Carefully targeted monitoring programs of this type are urgently needed because of our perception that ED effects in invertebrates are probably widespread but generally undetected at present.

Recommendations

The following recommendations for research to underpin invertebrate ED monitoring have been made in this chapter:

General monitoring (see "General monitoring" under "Major research needs," p 253, for details)

- Development of criteria for the selection of appropriate sentinel species for use in the marine, freshwater, and terrestrial environments
- Development of an understanding of the specificity of ED responses between taxonomic groups
- Development of appropriate biochemical and morphological markers for use in field studies
- Improved understanding of the population dynamics of invertebrates as they apply to endocrine disrupters

Specific monitoring (see "Specific monitoring" under "Major research needs," p 253, for details)

- Improved ability to link effects of classes of endocrine disrupters from the molecular to the population or community level
- Establishment of mechanisms of effects and generic models

Active monitoring (see "Active monitoring" under "Major research needs," p 254, for details)

- Determination of appropriate sentinel species
- Identification of suitable biomarkers for classes of endocrine disruption
- Validation of the testing strategies using reference chemicals in microcosms or mesocosms

Rapid bioassays (see "Rapid bioassays" under "Major research needs," p 254, for details)

- Need for genetically engineered cell lines containing hormone receptors
- Need for a range of molecular markers for effects
- A greater understanding of the link between molecular changes and impacts at the individual or population level

References

Adler JH, Grebenok RJ. 1995. Biosynthesis and distribution of insect-molting hormones in plants: a review. *Lipids* 30:257–262.

Alzieu C, Michel P, Tolosa I, Bacci E, Mee LD, Readman JW. 1991. Organotin compounds in the Mediterranean: a continuing cause of concern. *Mar Environ Res* 32:261–270.

Alzieu C, Sanjuan J, Deltreil JP, Borel M. 1986. Tin contamination in Arcachon Bay: effects on oyster shell anomalies. *Mar Pollut Bull* 17:494–498.

Allan JD, Daniels JM. 1982. Life table evaluation of chronic exposure of *Eurytemora affinis* (Copepoda) to kepone. *Mar Biol* 66:179-184.

Ankley G, Mihaich E, Stahl R, Tillitt D, Colborn T, McMaster S, Miller R, Bantle J, Campbell P, Denslow N, Dickerson R, Folmar L, Fry M, Giesy J, Gray LE, Guiney P, Hutchinson T, Kennedy S, Kramer V, LeBlanc G, Mayes M, Nimrod A, Patino R, Peterson R, Purdy R, Ringer R, Thomas P, Touart L, Van Der Kraak G, Zacharewski T. 1998. Overview of a workshop on screening methods for detecting potential (anti-) estrogenic/androgenic chemicals in wildlife. *Environ Toxicol Chem* 17:68–87.

Baldwin WS, Graham SE, Shea D, LeBlanc GA. 1997. Metabolic androgenization of female *Daphnia magna* by the xenoestrogen 4-nonylphenol. *Environ Toxicol Chem* 16:1905–1911.

Baldwin WS, Milam DL, LeBlanc GA. 1995. Physiological and biochemical perturbations in *Daphnia magna* following exposure to the model environmental estrogen diethylstilbestrol. *Environ Toxicol Chem* 14:945–952.

Barker MF, Xu RA. 1993. Effects of estrogens on gametogenesis and steroid levels in the ovaries and pyloric caeca of *Sclerasterias mollis* (Echinodermata: Asteroidea). *Invert Reprod Develop* 24:53–58.

Barrett KL, Grandy N, Harrison EG, Hassan S, Oomen P. 1995. Guidance document on regulatory testing procedures for pesticides with non-target arthropods. Brussels, Belgium: Society of Environmental Toxicology and Chemistry (SETAC)-Europe. 51 p.

Bauer B, Fioroni P, Ide I, Liebe S, Oehlmann J, Stroben E, Watermann B. 1995. TBT effects on the female genital system of *Littorina littorea*: a possible indicator of tributyltin pollution. *Hydrobiologia* 309:15–27.

Bauer B, Fioroni P, Schulte-Oehlmann U, Oehlmann J, Kalbfus W. 1997. The use of *Littorina littorea* for tributyltin (TBT) effect monitoring: results from the German TBT survey 1994/1995 and laboratory experiments. *Environ Pollut* 96:299–309.

Bechmann RK. 1994. Use of life tables and LC50 tests to evaluate chronic and acute toxicity effects of copper on the marine copepod *Tisbe furcata* (Baird). *Environ Toxicol Chem* 13:1509–1517.

Beckmann M, Hardege JD, Zeeck E. 1995. Effects of volatile fractions of crude oil on spawning behaviour of nereids (Annelida, Polychaeta). *Mar Environ Res* 40:267–276.

Bell RA. 1996. Manipulation of diapause in the gypsy moth *Lymantria dispar*. L. by application of KK42 and precocious chilling of eggs *J Insect Physiol* 42(6):557–563.

Bettin C, Oehlmann J, Stroben E. 1996. TBT induced imposex in marine neogastropods is mediated by an increasing androgen level. *Helgoländer Meeresunters* 50:299–317.

Bhargava MC, Urs KCD. 1993. Ovicidal effects of three insect growth regulators on *Corcyra cephalonica*. *Indian J Plant Prot* 21(2):195–197.

Biddinger DJ, Hull LA. 1995. Effects of several types of insecticides on the mite predator *Stethorus punctum* (Coleoptera: Coccinellidae) including insect growth regulators and abamectin. *J Econ Entomol* 88(2):358–366.

Billinghurst Z, Clare AS, Fileman T, McEvoy J, Readman J, Depledge MH. 1998. Inhibition of barnacle larval settlement by the environmental estrogen 4-nonylphenol and the natural estrogen 17β-estradiol. *Mar Pollut Bull* 36 (10):833–839.

Billinghurst Z, Clare AS, Matsumura K, Depledge MH. 1999. Cypris major protein: a potential biomarker of xeno-estrogen exposure in barnacles. *J Mar Pollut:* In press.

Blaise C, Gagne F, Pellerin J, Hansen, PD. 1998. Measurement of a vitellogenin-like protein in the haemolymph of *Mya arenaria* (Saquenay Fjord, Canada): a potential biomarker for endocrine disruptions. *Biomarkers*: In press.

Bodar CWM, Van Leeuwen CJ, Voogt PA, Zandee DI. 1988. Effect of cadmium on the reproduction strategy of *Daphnia magna*. *Aquat Toxicol* 12:301–310.

Bodar CWM, Voogt PA, Zandee DI. 1990. Ecdysteroids in *Daphnia magna*: their role in moulting and reproduction and their levels upon exposure to cadmium. *Aquat Toxicol* 17:339–350.

Bradbrook DA, Clement CY, Cook B, Dinan L. 1990. The occurrence of vertebrate-type steroids in insects and a comparison with ecdysteroid levels. *Comp Biochem Physiol* 95B:365–374.

Brown D, Croudace CP, Williams NJ, Shearing JM, Johnson PA. 1998. The effect of phthalate ester plasticizers tested as surfactant stabilized dispersions on the reproduction of the *Daphnia magna*. *Chemosphere* 36:1367–1379.

Brown D, Thompson RS, Stewart KM, Croudace CP, Gillings E. 1996. The effect of phthalate ester plasticisers on the emergence of the midge (*Chironomus riparius*) from treated sediments. *Chemosphere* 32:2177–2187.

Brown DJ, Clarke GC, VanBeneden RJ. 1997. Halogenated aromatic hydrocarbon-binding proteins identified in several invertebrate marine species. *Aquat Toxicol* 37: 71-78.

Bryan GW, Bright DA, Hummerstone LG, Burt GR. 1993. Uptake, tissue distribution and metabolism of ^{14}C-labelled tributyltin (TBT) in the dog-whelk, *Nucella lapillus*. *J Mar Biol Ass UK* 73:889–912.

Bryan GW, Gibbs PE, Burt GR. 1988. A comparison of the effectiveness of tri-n-butyltin chloride and five other organotin compounds in promoting the development of imposex in the dog-whelk, *Nucella lapillus*. *J Mar Biol Ass UK* 68:733–744.

Bryan GW, Gibbs PE, Burt GR, Hummerstone LG. 1987. The effects of tributyltin (TBT) accumulation on adult dog-whelks, *Nucella lapillus* : Long-term field and laboratory experiments. *J Mar Biol Assoc UK* 67(3):525–544.

Bryan GW, Gibbs PE, Hummerstone LG, Burt GR. 1986. The decline of the gastropod *Nucella lapillus* around south-west England: evidence for the effects of tributyltin from antifouling paints. *J Mar Biol Assoc UK* 66:611–640.

Bull DL, Meola RW. 1993. Effects and fate of the insect growth regulator pyriproxyfen after application to the horn fly (Diptera: Muscidae). *J Econ Entomol* 86 (6):1754–1760.

Carpenter JE, Chandler LD. 1994. Effects of sublethal doses of two insect growth regulators on *Helicoverpa zea* (Lepidoptera: Noctuidae) reproduction. *J Entomol Sci* 29 (3):428–435.

Celestial DM, McKenney CL. 1994. The influence of an insect growth regulator on the larval development of the mud crab *Rhithropanopeus harrisii*. *Environ Pollut* 85:169–173.

Charmillot P-J, Pasquier D. 1992. Modification of fertility in codling moth *Cydia pomonella* L. by adult contact with an insect growth regulator and an inhibitor IGR and IGI. *Entomol Exp Appl* 63(1):87–93.

Charniaux-Cotton H. 1962. Androgenic gland of crustacea. *Gen Comp Endocrinol Suppl* 1L:241–247.

Cobb JLS. 1988. Neurohumors and neurosecretion in echinoderms: a review. *Comp Biochem Physiol* 91C:151–158.

Comber MHI, Williams TD, Stewart KM. 1993. The effects of nonylphenol on *Daphnia magna*. *Wat Res* 27:273–276.

Crothers JH. 1985. Dogwhelks: an introduction to the biology of *Nucella lapillus* (L.). *Field Stud* 6:291–360.

Curtis LA, Kinley JL. 1998. Imposex *Ilyanassa obsoleta* still common in a Delaware estuary. *Mar Pollut Bull* 36:97–101.

Cymborowski B. 1992. Insect endocrinology. Amsterdam, The Netherlands: Elsevier.

Daniels RE, Allan JD. 1981. Life table evaluation of chronic exposure to a pesticide. *Can J Fish Aquat Sci* 38:485–494.

Darvas B, Polgár L, Tag El-Din MH, Eröss K, Wing KD 1992. Developmental disturbances in different insect orders caused by an ecdysteroid agonist RH 5849. *J Econ Entomol* 85(6):2107–2112.

Darvas B, Zdarek J, Timar T, El-Din MHT. 1990. Effects of the bipyridylium herbicides diquat dibromide and paraquat dichloride on growth and development of *Neobellieria bullata* (Diptera:Sarcophagidae) larvae. *J Econ Entomol* 83:2175–2180.

Day K, Kaushik K. 1987. An assessment of the chronic toxicity of the synthetic pyrethroid, fenvalerate, to *Daphnia galeata mendotae*, using life tables. *Environ Pollut* 44:13–26.

De Bisthoven LJ, Huysmans C, Ollevier F. 1995. The in situ relationships between sediment concentrations of micropollutants and morphological deformities in *Chironomus thummi* larvae (Diptera, Chironomidae) from lowland rivers (Belgium): a spatial comparison. In: Cranston PS, editor. Chironomids: from genes to ecosystems. Canberra, Australia: Commonwealth Scientific and Industrial Research Organisation (CSIRO).

De Clerq P, De Cock A, Tirry L, Vinuela E, Degheele D. 1995. Toxicity of diflubenzuron and pyriproxyfen to the predatory bug *Podisus maculiventris*. *Entomol Exper et Applicata* 74(1):17–22.

den Besten PJ, van Donselaar EG, Herwig HJ, Zandee DI, Voogt PA. 1991. Effects of cadmium on gametogenesis in the sea star *Asterias rubens* L. *Aquat Toxicol* 20:83–94.

den Besten PJ, Elenbaas JML, Maas JR, Dieleman SJ, Herwig HJ, Voogt PA. 1991. Effects of cadmium and polychlorinated biphenyls (Clophen A50) on steroid metabolism and cytochrome P-450 monooxygenase system in the sea star *Asterias rubens* L. *Aquat Toxicol* 20:95–110.

den Besten PJ, Herwig HJ, Zandee DI, Voogt PA. 1989. Effects of cadmium and PCBs on reproduction of the sea star *Asterias rubens*: aberrations in the early development. *Ecotoxicol Environ Saf* 18:173–180.

den Besten PJ, Valk S de, Dubbeldam M, Wanningen H. 1996. Biomar-I project: biomarkers for reproductive toxicity in marine invertebrates. Amsterdam, The Netherlands: AquaSense Laboratory. Technical Report no. 96.0409

Desbrow C, Routledge EJ, Brighty G, Sumpter JP, Waldock M. 1998. Identification of estrogenic chemicals in STW effluent. 1. Chemical fractionation and in vitro biological screening. *Environ Sci Technol* 32:1549–1558.

Dickman M, Lan Q, Matthews B. 1990. Teratogens in the Niagara River watershed as reflected by chironomid (Diptera:Chironomidae) labial plate deformities. *Water Pollut Res J Can* 24:47–79.

Dickman M, Rygiel G. 1996. Chironomid larval deformity frequencies, mortality, and diversity in heavy-metal contaminated sediments of a Canadian riverine wetland. *Environ Internat* 22:693–703.

Dingle H. 1968. Life history and population consequences of density, photo-period, and temperature in a migrant insect, the milkweed bug *Oncopeltus*. *American Naturalist* 102:149–163.

Drobne D, Strus J. 1996. Moult frequency of the isopod *Porcellio scaber*, as a measure of zinc-contaminated food. *Environ Toxicol Chem* 15:126–130.

Ellis DV, Pattisina LA. 1990. Widespread neogastropod imposex: a biological indicator of global TBT contamination? *Mar Pollut Bull* 21:248–253.

Environment Canada. 1997. Pulp and paper aquatic environmental effects monitoring requirements. Annex 1 to EEM/1997/1.

Environmental Agency of Japan. 1998. A summary of results on environmental monitoring for organotin compounds conducted in 1996. In: Environmental Health Department, editor. Chemicals in the environment. Tokyo, Japan: Environmental Agency of Japan. p 255–273 (In Japanese).

Evans SM, Hawkins ST, Porter J, Samosir AM. 1994. Recovery of dogwhelk populations on the Isle of Cumbrae, Scotland following legislation limiting the use of TBT as an antifoulant. *Mar Pollut Bull* 28:15–17.

Fairweather I, Skuce PJ. 1995. Flatworm neuropeptides: present status, future directions. *Hydrobiologia* 305:309–316.

Fent K. 1996. Ecotoxicology of organotin compounds. *Crit Rev Toxicol* 26:1–117.

Féral C. 1980. Variations dans l'évolution du tractus génital mâle externe des femelles de trois Gastéropodes Prosobranches gonochoriques de stations Atlantiques. *Cah Biol Mar* 21:479–491.

Féral C, Le Gall S. 1982. Induction expérimentale par un polluant marin (le tributylétain), de l'activité neuroendocrine contrôlant la morphogenèse du pénis chez les femelles d'*Ocenebra erinacea* (Mollusque Prosobranche gonochorique). *C R Acad Sc Paris* 295, Série III:627–630.

Féral C, Le Gall S. 1983. The influence of a pollutant factor (tributyltin) on the neuroendocrine mechanism responsible for the occurrence of a penis in the females of *Ocenebra erinacea*. In: Lever J, Boer HH, editors. Molluskan neuro-endocrinology. Oxford NY: North-Holland. p 173–175.

Fernandez-Casalderrey A, Ferrando MD, Andreu-Moliner E. 1993. Effects of endosulfan on survival, growth and reproduction of *Daphnia magna*. *Comp Biochem Physiol* 106C:437–441.

Fernandez-Casalderrey A, Ferrando MD, Andreu-Moliner E. 1994. Effect of sublethal concentrations of pesticides on the feeding behaviour of *Daphnia magna*. *Ecotoxicol Environ Saf* 27:82–89.

Fingerman M. 1997. Crustacean endocrinology: A retrospective, prospective, and introspective analysis. *Physiol Zool* 70:257–269.

Fingerman M, Devi M, Reddy PS, Katyayani R. 1996. Impact of heavy metal exposure on the nervous system and endocrine-mediated processes in crustaceans. *Zool Stud* 35:1–8.

Fingerman SW, Fingerman M. 1978. Influence of polychlorinated biphenyl preparation Aroclor 1242 on color changes of the fiddler crab *Uca pugilator*. *Mar Biol* 50:37–45.

Fingerman M, Jackson NC, Nagabhushanam R. 1998. Hormonally-regulated functions in crustaceans as biomarkers of environmental pollution. *Comp Biochem Physiol* 120C:343–350.

Fingerman M, Nagabhushanam R, Sarojini R. 1993. Vertebrate-type hormones in crustaceans: localization, identification and functional significance. *Zool Sci* 10:13–29.

Fioroni P, Oehlmann J, Stroben E. 1991. The pseudohermaphroditism of prosobranchs; morphological aspects. *Zool Anz* 226:1–26.

Fitzmayer KM, Geiger JG, van den Avyle MJ. 1982. Effects of chronic exposure to simazine on the cladoceran, *Daphnia pulex*. *Arch Environ Contam Toxicol* 11:603–609.

Frank PW, Boll CD, Kelly RW. 1957. Vital statistics of laboratory cultures of *Daphnia pulex* DeGeer as related to density. *Physiol Zool* 30:287–305.

French BW, Elliot N, Berberet R. 1998. Reverting conservation reserve program lands to wheat and livestock production: effects on ground beetle assemblages. *Environ Entomol* 27:1323–1335.

Gagne F, Blaise C. 1998. Estrogenic properties of municipal and industrial wastewaters evaluated with a rapid and sensitive chemoluminescent in situ hybridization assay (CISH) in rainbow trout hepatocytes. *Aquat Toxicol*: In press.

Gibbs PE, Bryan GW, Pascoe PL. 1991. TBT-induced imposex in the dogwhelk, *Nucella lapillus* : Geographical uniformity of the response and effects. *Mar Environ Res* 32(1–4):79–87.

Gibbs PE, Bryan GW, Pascoe PL, Burt GR. 1987. The use of the dog-whelk, *Nucella lapillus*, as an indicator of tributyltin (TBT) contamination. *J Mar Biol Ass UK* 67:507–523.

Gibbs PE, Pascoe PL, Burt GR. 1988. Sex change in the female dog-whelk, *Nucella lapillus*, induced by tributyltin from antifouling paints. *J Mar Biol Ass UK* 68: 715-731.

Green AS, Chandler GT. 1996. Life-table evaluation of sediment-associated chlorpyrifos chronic toxicity to the benthic copepod, *Amphiascus tenuiremis*. *Arch Environ Contam Toxicol* 31:77–83.

Green AS, Chandler GT, Coull BC. 1995. Age-specific survival analysis of an infaunal meiobenthic copepod, *Amphiascus tenuiremis*. *Mar Ecol Prog Ser* 129:107–112.

Griffith MB, Barrows E, Perry S. 1996. Effects of aerial application of diflubenzuron on emergence and flight of adult aquatic insects. *J Econ Entomol* 89:442–446.

Hallers-Tjabbes Ten CC, Kemp JF, Boon JP. 1994. Imposex in whelks (*Buccinum undatum*) from the open North Sea: relation to shipping traffic intensities. *Mar Pollut Bull* 28:311–313.

Hamilton AL, Saether OA. 1971. The occurrence of characteristic deformities in the chironomid larvae of several Canadian lakes. *Can Entomol* 103:363–368.

Hanumante MM, Fingerman SW, Fingerman M. 1981. Antagonism of the inhibitory effect of the polychlorinated biphenyl preparation, Aroclor 1242, on color changes of the fiddler crab, *Uca pugilator*, by norepinephrine and drugs affecting neurotransmission. *Bull Environ Contam Toxicol* 26:479–484.

Harding MJC, Davies IM, Bailey SK, Rodger GK. 1999. Survey of imposex in dogwhelks (*Nucella lapillus*) from North Sea coasts. *Appl Organometal Chem*: In press.

Harding MJC, Rodger GK, Davies IM, Moore JJ. 1997. Partial recovery of the dogwhelk (*Nucella lapillus*) in Sullom Voe, Shetland, from tributyltin contamination. *Mar Env Res* 44:285–304.

Harrington FE, Ozaki H. 1986. The effect of estrogen on protein synthesis in Echinoid coelomocytes. *Comp Biochem Physiol* 84B:417–421.

Hattingh V, Tate B. 1995. Effects of field weathered residues of insect growth regulators on some coccinellidae (Coleoptera) of economic importance as biocontrol agents. *Bull Entomol Res* 85 (4):489–493.

Hawkins LE, Hutchinson S. 1990. Physiological and morphogenetic effects of monophenyltin trichloride on *Ocenebra erinacea* (L.). *Func Ecol* 4:449–454.

Hicks BJ, Gordon R. 1992. Effects of the juvenile hormone analog fenoxycarb on various developmental stages of the eastern spruce budworm *Choristoneura fumiferana clemens* Lepidoptera Tortricidae. *Can Entomol* 124(1):117–123.

Higbee BS, Horton DR, Krysan JL. 1995. Reduction of egg hatch in pear psylla (Homoptera: Psyllidae) after contact by adults with insect growth regulators. *J Econ Entomol* 88(5):1420–1424.

Hines GA, Bryan PJ, Wasson KM, McClintock JB, Watts SA. 1996. Sex steroid metabolism in the antarctic pteropod *Clione antarctica* (Molluska: Gastropoda). *Invert Biol* 115:113–119.

Horiguchi T, Hyeonseo C, Shiraishi H, Shibata Y, Soma M, Morita M, Shimizu M. 1998. Field studies on imposex and organotin accumulation in the rock shell, *Thais clavigera*, from the Seto Inland Sea and the Sanriku region, Japan. *Sci Tot Environ* 214:65–70.

Horiguchi T, Shiraishi H, Shimizu M, Morita M. 1994. Imposex and organotin compounds in *Thais clavigera* and *T. bronni* in Japan. *J Mar Biol Ass UK* 74:651–669.

Horiguchi T, Shiraishi H, Shimizu M, Morita M. 1997a. Effects of triphenyltin chloride and five other organotin compounds on the development of imposex in the rock shell *Thais clavigera*. *Environ Pollut* 95:85–91.

Horiguchi T, Shiraishi H, Shimizu M, Morita M. 1997b. Imposex in sea snails, caused by organotin (tributyltin and triphenyltin) pollution in Japan: a survey. *Appl Organometal Chem* 11:451–455.

Horton D, Lewis TM. 1996. Effects of fenoxycarb on ovarian development, spring fecundity and longevity in winterform pear psylla. *Entomol Exper et Applicata* 81(2):181–187.

Huet M, Paulet YM, Glémarec M. 1996. Tributyltin (TBT) pollution in the coastal waters of west Brittany as indicated by imposex in *Nucella lapillus*. *Mar Environ Res* 41:57–167.

Hummon WD. 1974. Effects of DDT on longevity and reproductive rate in *Lepidodermella squammata* (Gastrotricha, Chaetonida). *Am Midl Nat* 92:327–339.

[IL DNR] Illinois Department of Natural Resources. 1999. Critical Trends Assessment Program. Web site: http://dnr.state.il.us/ctap/ctaphome.htm

James MO, Shiverick K. 1984. Cytochrome P450-dependent oxidation of progesterone, testosterone and ecdysone in the spiny lobster, *Panulirus argus*. *Arch Biochem Biophys* 233:1–9.

[JAMP] Joint Assessment and Monitoring Program. 1998. JAMP Guidelines for Contaminant-Specific Biological Effects Monitoring. London, UK: JAMP, Oslo and Paris Commission. 38 p.

Johnson RK, Weiderholm T, Rosenberg DM. 1993. Freshwater biomonitoring using individual organisms, populations, and species assemblages of benthic macroinvertebrates. In: Rosenberg DM, Resh VH, editors. Freshwater biomonitoring and benthic macroinvertebrates. New York: Chapman and Hall. p 40–158.

Johnston PA. 1987. Acute toxicity of inorganic selenium to *Daphnia magna* (Straus) and the effect of sub-acute exposure upon growth and reproduction. *Aquat Toxicol* 10:335–352.

Jones S, Lenz M. 1996. Fenoxycarb induced caste differentiation and mortality in *Coptotermes formosanus* (Isoptera:Rhinotermitidae). *J Econ Entomol* 89(4):906–914.

Joosse J. 1988. The hormones of mollusks. In: Laufer H, Downer RGH, editors. Endocrinology of selected invertebrate types. New York: Alan R. Liss. p 89–140.

Joosse J, Geraerts WPM. 1983. Endocrinology. In: Saleuddin ASM, Wilber KM, editors. The mollusca. New York: Academic Press. p 317–406.

Kaakeh W, Scharf ME, Bennett GW. 1997. Comparative contact activity and residual life of juvenile hormone analogs used for German cockroach (Dictyoptera: Blattellidae) control. *J Econ Entomol* 90(5):1247–1253.

Kahl MD, Makynen EA, Kosian PA, Ankley GT. 1997. Toxicity of 4-nonylphenol in a life-cycle test with the midge *Chironomus tentans*. *Ecotoxicol Environ Saf* 38:155–160.

Kao S-M, Su T-H. 1995. The effect of fenoxycarb and pyriproxifen on the egg production of the pharaoh ant *Monomorium pharaonis* (Hymenoptera: Formicidae). *Zhonghua Kunchong* 15(1):227–237.

Kendall R, Dickerson R, Giesy J, Suk W. 1998. editors. Principles and processes for evaluating ED in wildlife. Proceedings from a workshop; 6 March 1996; Kiawah Island SC. Pensacola FL: Society of Environmental Toxicology and Chemistry (SETAC). 515 p.

Kersting K. 1975. The use of microsystems for the evaluation of the effect of toxicants. *Hydrobiol Bull* 9:102–108.

Khan HK, Urs KCD. 1992. Developmental effects of two insect growth regulators (IGRs) on the prepupal and pupal stages of diamond back moth *Plutella xylostella* (Linn.). *J Insect Sci* 5(2):161–162.

Khristoforova NK, Gnezdilova SM, Vlasova GA. 1984. Effect of cadmium on gametogenesis and offspring of the sea urchin *Strongylocentrotus intermedius*. *Mar Ecol Progr Ser* 17:9–14.

King DW. 1964. Fine structure of the androgenic gland of the crab, *Pachygrapsus crassipes*. *Gen Comp Endocrinol* 4:533–544.

Kluytmans JH, Brands F, Zandee DI. 1988. Interactions of cadmium with the reproductive cycle of *Mytilus edulis* L. *Mar Environ Res* 24:189–192.

Kosalwat P, Knight AW. 1987. Chronic toxicity of copper to a partial life cycle test of the midge, *Chironomus decorus*. *Arch Environ Contam Toxicol* 16:283–290.

Lafont R. 1991. Reverse endocrinology, or "hormones" seeking functions. *Insect Biochem* 21:697–721.

Lapota D, Rosenberger DE, Platter-Rieger MF, Seligman PF. 1993. Growth and survival of *Mytilus edulis* larvae when exposed to low levels of dibutyltin and tributyltin. *Mar Biol* 115:413–419.

Laufer H, Ahl JSB, Sagi A. 1993. The role of juvenile hormones in crustacean reproduction. *Am Zool* 33:365–374.

Laughlin RB, Gustafson R, Pendoley P. 1988. Chronic embryo-larval toxicity of tributyltin (TBT) to the hard shell clam *Mercenaria mercenaria*. *Mar Ecol Progr Ser* 48:29–36.

Laughlin RB, Gustafson RG, Pendoley P. 1989. Acute toxicity of tributyltin (TBT) to early life history stages of the hard shell clam, *Mercenaria mercenaria*. *Bull Environ Contam Toxicol* 42:352–358.

Lee RF. 1985. Metabolism of tributyltin oxide by crabs, oysters and fish. *Mar Environ Res* 17:145–148.

Lee RF. 1991. Metabolism of tributyltin by marine animals and possible linkages to effect. *Mar Environ Res* 32:29–35.

Lee RF. 1993. Passage of xenobiotics and their metabolites from hepatopancreas into ovary and oocytes of blue crabs, *Callinectes sapidus*: possible implications for vitellogenesis. *Mar Env Res* 35:1–2.

Lee RF, Keeran WS, Pickwell GV. 1988. Marine invertebrate glutathione-S-transferases: purification, characterisation and induction. *Mar Env Res* 24:97–100.

Leonardi MG, Cappellozza S, Ianne P, Cappellozza L, Parenti P, Giordana B. 1996. Effects of the topical application of an insect growth regulator (fenoxycarb) on some physiological parameters in the fifth instar larvae of the silkworm *Bombyx mori*. *Comp Biochem Physiol* 113(2):361–365.

Lesh-Laurie GE. 1988. Coelenterate Endocrinology. In: Laufer H, Downer RGH, editors. Endocrinology of selected invertebrate types. New York: Alan R. Liss. p 3–29.

Leslie PH. 1945. On the use of matrices in certain population mathematics. *Biometrika* 35:213–245.

Lotka AJ. 1922. The stability of the normal age distribution. *Proc Natl Acad Sci U.S.A.* 8:339–345.

Lotka AJ. 1943. The place of the intrinsic rate of natural increase in population analysis. *Proc. 8th Am Sci Congr* 8:297–213.

Lyoussoufi A, Gadenne C, Rieux R, D'Arcier F. 1994. Effects of an insect growth regulator, fenoxycarb, on the diapause of the pear psylla *Cacopsylla pyri*. *Entomol Exper et Applicata* 72(3):239–244.

Macek KJ, Buxton KS, Sauter SS, Gnilka S, Dean JW. 1976. Chronic toxicity of atrazine to selected aquatic invertebrates and fishes. Duluth MN: Environmental Research Laboratory, U.S. Environmental Protection Agency. EPA 600/3-76-047.

Marletto F, Arzone A, Dolic M. 1997. The honeybee as a biological indicator of fenoxycarb pollution. *Apicolotore Moderno* 88(3):107–110.

Matthiessen P, Gibbs PE. 1998. Critical appraisal of the evidence for tributyltin-mediated endocrine disruption in mollusks. *Environ Toxicol Chem* 17:37–43.

Matthiessen P, Kilbride R, Mason C, Pendle M, Rees H, Waldock R. 1999. Monitoring the recovery of the benthic community in the River Crouch following TBT contamination. Report to the UK Department of the Environment, Transport and the Regions. Burnham-on-Crouch, UK: Centre for Environment, Fisheries and Aquaculture Science. EPG 1/5/92. 51 p.

Maugh TH. 1987. Chemicals: how many are there? *Science* 199:162.

McCartney MA. 1997. Sex allocation and male fitness gain in a colonial, hermaphroditic marine invertebrate. *Evolution* 51:127–140.

McKenney CL, Celestial DM. 1996. Modified survival, growth and reproduction in an estuarine mysid (*Mysidopsis bahia*) exposed to juvenile hormone analogue through a complete life cycle. *Aquat Toxicol* 35:11–20.

McKenney CL, Matthews E. 1990. Influence of an insect growth regulator on the larval development of an estuarine shrimp. *Environ Pollut* 64:169–178.

Mendel Z, Blumberg D, Ishaaya I. 1994. Effects of some insect growth regulators on natural enemies of scale insects (Hom.: Coccoidea). *Entomophaga* 39(2):199–209.

Mertz DB. 1971. Life history phenomena in increasing and decreasing populations. In: Patil GP, Pielou EC, Waters WE, editors. Statistical ecology, Volume 2. University Park PA: Pennsylvania State Univ. Press. p 361–399.

Milbrink G. 1983. Characteristic deformities in tubificid oligochaetes inhabiting polluted bays of Lake Vanern, Southern Sweden. *Hydrobiologia* 106:169–184.

Miller RA, Norris LA, Hawkes CL. 1973. Toxicity of 2,3,7,8-tetrachlorodibenzo-*p*-dioxin (TCCD) in aquatic organisms. *Environ Health Perspect* 5:177–186.

Minchin D, Bauer B, Oehlmann J, Schulte-Oehlmann U, Duggan CB. 1997. Biological indicators used to map organotin contamination from a fishing port, Killybegs, Ireland. *Mar Pollut Bull* 34:235–243.

Minchin D, Oehlmann J, Duggan CB, Stroben E, Keatinge M. 1995. Marine TBT antifouling contamination in Ireland, following legislation in 1987. *Mar Pollut Bull* 30:633–639.

Minchin D, Stroben E, Oehlmann J, Bauer B, Duggan CB, Keatinge M. 1996. Biological indicators used to map organotin contamination in Cork Harbour, Ireland. *Mar Pollut Bull* 32:188–195.

Moore CG, Stevenson JM. 1991. The occurrence of intersexuality in harpacticoid copepods and its relationship with pollution. *Mar Pollut Bull* 22:72–74.

Moore CG, Stevenson JM. 1994. Intersexuality in benthic harpacticoid copepods in the Firth of Forth, Scotland. *J Nat Hist* 28:1213–1230.

Nagabhushanam R, Kulkarni GK. 1981a. Effect of exogenous testosterone on the androgenic gland and testis of a marine penaeid prawn, *Parapenaeopsis hardwickii* (Miers) (Crustacea, Decapoda, Penaeidae). *Aquaculture* 23:19–27.

Nagabhushanam R, Kulkarni GK. 1981b. Freshwater palaemonid prawn, *Macrobrachium kistnensis* (Tewari)—effect of heavy metal pollutants. *Proc Indian Nat Sci Acad* 47B:380–386.

Nagabhushanam R, Reddy PS, Sarojini R. 1990. Tributyltin oxide induced alterations in exuvial weight and calcium content of the prawn, *Caridina rajadhari*. *Proc Indian Acad Sci (Animal Sciences)* 99:397–400.

[NRC] National Research Council. 1990. Managing troubled waters: the role of marine environmental monitoring. Washington DC: National Academy Press. p 3.

Neill WE. 1994. Spatial and temporal scaling and the organization of limnetic communities. In: Giller PS, Hildrew AG, Raffaelli DG. Aquatic ecology: scale, pattern, and process. 34th Symposium of the British Ecological Society. Oxford, UK: Blackwell Scientific Publishing. p 189-231.

[NOAA] National Oceanic and Atmospheric Administration. 1999. NOAA's National Status and Trends Program. Home Page. http://seaserver.nos.noaa.gov/projects/nsandt/nsandt.html

North Sea Quality Status Report. 1993. London, UK: Oslo and Paris Commissions/International Council for the Exploration of the Sea. 132 p.

Oberdörster E, Rittschof D, LeBlanc GA. 1998. Alteration of [^{14}C]testosterone metabolism after chronic exposure of *Daphnia magna* to tributyltin. *Arch Environ Contam Toxicol* 34:21–25.

Oberdörster E, Rittschof D, McClellan-Green P. 1998. Testosterone metabolism in imposex and normal *Ilyanassa obsoleta*: comparison of field and TBT Cl-induced imposex. *Mar Pollut Bull* 36:144–151.

Oehlmann J. 1994. Imposex bei Muriciden (Gastropoda, Prosobranchia), eine ökotoxikologische Untersuchung zu TBT-Effekten. Göttingen, Germany: CuvillierVerlag. 167 p.

Oehlmann J. 1998. Untersuchungen zum Einsatz von Pathomorphosen der ableitenden Geschlechtswege von Vorderkiemerschnecken (Gastropoda: Prosobranchia) für ein biologisches TBT-Effektmonitoring [Habilitation thesis]. Zittau, Germany: International Graduate School Zittau. 162 p.

Oehlmann J, Bettin C. 1996. TBT-induced imposex and the role of steroids in marine snails. *Malacol Rev* Suppl. 6 (Molluskan Reproduction):157–161.

Oehlmann J, Fioroni P, Stroben E, Markert B. 1996. Tributyltin (TBT) effects on *Ocinebrina aciculata* (Gastropoda: Muricidae): imposex development, sterilization, sex change and population decline. *Sci Total Environ* 188:205–223.

Oehlmann J, Schulte-Oehlmann U, Bauer B, Markert B. 1997. Einsatz TBT-induzierter Pathomorphosen bei Schnecken aus der Nord- und Ostsee zum biologischen Effektmonitoring. *German J Hydrogr* Suppl. 7:77–98.

Oehlmann J, Stroben E, Fioroni P. 1993. Fréquence et degré d'expression du pseudohermaphrodisme chez quelques Prosobranches Sténoglosses des côtes françaises (surtout de la baie de Morlaix et de la Manche). 2. Situation jusqu'au printemps de 1992. *Cah Biol Mar* 34:343–362.

Okoshi K, Mori K, Nomura T. 1987. Characteristics of shell chamber formation between the two local races in the Japanese oyster, *Crassostrea gigas*. *Aquaculture* 67:313–320.

Olszak RW, Pawlik B, Zajac RZ. 1994. The influence of some insect growth regulators on mortality and fecundity of the aphidophagous coccinellids *Adalia bipunctacta* L. and *Coccinella septempunctata* L. (Col. Coccinellidae). *J Appl Entomol* 117(1):58–63.

Oomen PA, De Ruijter A, Van Der Steen J. 1992. Method for honey bee brood feeding test with insect growth regulating insecticides. *EPPO Bull* 22: 613-616.

Ozretic B, Krajnovic-Ozretic M. 1985. Morphological and biochemical evidence of the toxic effect of pentachlorophenol on the developing embryos of the sea urchin. *Aquat Toxicol* 7:255–263.

Parks LG, LeBlanc GA. 1996. Reduction in steroid hormone biotransformation/elimination as a biomarker of pentachlorophenol chronic toxicity. *Aquat Toxicol* 34:291–303.

Peters RH. 1983. The ecological implications of body size. Cambridge, UK: Cambridge University Press.

Pinder LCV, Pottinger TG. 1998. Endocrine function in aquatic invertebrates and evidence for disruption by environmental pollutants. Draft report to the United Kingdom Environment Agency and the CEFIC Endocrine Modulators Steering Group. 178 p.

Plantevin G, Grenier S, Chavancy G. 1991. Effects of an insect growth regulator fenoxycarb on the post embryonic development of *Bombyx mori* Lepidoptera Bombycidae. *C R Acad Sci Ser III Sci Vie* 313(11):513–520.

Poloczanska ES, Ansell AD. 1999. Imposex in the whelks *Buccinum undatum* and *Neptunea antiqua* from the west coast of Scotland. *Mar Environ Res* 47:203–212.

Postma JF, Valk S de. 1996. Ecotoxicological monitoring at Loswal North and Loswal West using sea stars. Amsterdam, The Netherlands: AquaSense Laboratory. Technical Report no. 96.0409-2 (in Dutch).

Postma JF, Valk S de. 1997. Biomar-II project: biomarkers for reproductive toxicity in marine invertebrates. Amsterdam, The Netherlands: AquaSense Laboratory. Technical Report no. 97.0409 (in Dutch).

Prabhakaran S, Kamble ST. 1996. Effects of azadirachtin on different strains of German cockroach (Dictyoptera: Blattellidae). *Environ Entomol* 25(1):130–134.

Reddy PS, Nagabhushanam R, Sarojini R. 1992. Retardation of moulting in the prawn, *Caridini rajadhari*, exposed to tributyltin oxide (TBTO). *Proc Nat Acad Sci, India* 62B:353–356.

Reddy PS, Sarojini R, Nagabhushanam R. 1991. Impact of tributyltin oxide (TBTO) on limb regeneration of the prawn *Caridina rajadhari,* after exposure to different time intervals of amputation. *J Tissue Res* 1:35–39.

Rees HL, Waldock R, Matthiessen P, Pendle MA. 1999. Surveys of the epibenthos of the Crouch Estuary (UK) in relation to TBT contamination. *J Mar Biol Assoc UK* 79:209–223.

Reid BL, Brock VL, Bennett GW. 1994. Developmental, morphogenetic and reproductive effects of four polycyclic non-isoprenoid juvenoids in the German cockroach (Dictyoptera: Blattellidae). *J Entomol Sci* 29(1):31–42.

Reis-Henriques MA, Le Guellec D, Remy-Martin JP, Adessi GL. 1990. Studies of endogenous steroids from the marine mollusk *Mytilus edulis* L. by gas chromatography and mass spectrometry. *Comp Biochem Physiol* 95B:303–309.

Ronis MJJ, Mason AZ. 1996. The metabolism of testosterone by the periwinkle (*Littorina littorea*) in vitro and in vivo effects of tributyl tin. *Mar Environ Res* 42:161–166.

Root RB, Olson AM. 1969. Population increase of the cabbage aphid, *Brevicoryne brassicae*, on different host plants. *Can Entomol* 101:768–773.

Rosenberg DM, Resh VH. 1993. Introduction to freshwater biomonitoring and benthic macroinvertebrates. In: Rosenberg DM, Resh VH, editors. Freshwater biomonitoring and benthic macroinvertebrates. New York: Chapman and Hall. p 1–9.

Ruiz JM, Bryan GW, Gibbs PE 1995. Acute and chronic toxicity of tributyltin (TBT) to pediveliger larvae of the bivalve *Scrobicularia plana*. *Mar Biol* 124:119–126.

Ruiz JM, Bryan GW, Wigham GD, Gibbs PE. 1995. Effects of tributyltin (TBT) exposure on the reproduction and embryonic development of the bivalve *Scrobicularia plana*. *Mar Environ Res* 40:363–379.

Rumpf S, Frampton C, Dietrich DR. 1998. Effects of conventional insecticides and insect growth regulators on fecundity and other life table parameters of *Micromus tasmaniae* (Neuroptera: Hemerobiidae). *J Econ Entomol* 91(1):34–40.

Saavedra Alvarez MM, Ellis DV. 1990. Widespread neogastropod imposex in the northeast Pacific: implications for TBT contamination surveys. *Mar Pollut Bull* 21:244–247.

Salgado VL. 1992. The neurotoxic insecticidal mechanism of the nonsteroidal ecdysone agonist RH-5849 potassium channel block in nerve and muscle. *Pestic Biochem Physiol* 43(1):1–13.

Sangalang G, Jones G. 1997. Oocytes in testis and intersex in lobsters (*Homarus americanus*) from Nova Scotian sites: natural or site-related phenomenon? *Can Tech Rep Fish Aquat Sci* 2163:46.

Sarojini R, Nagabhushanam R, Fingerman M. 1994. A possible neurotransmitter-neuroendocrine mechanism in naphthalene-induced atresia of the ovary of the red swamp crayfish, *Procambrus clarkii*. *Comp Biochem Physiol* 108C:33–38.

Sarradin PM, Astruc A, Sabrier R, Astruc M. 1994. Survey of butyltin compounds in Arcachon Bay sediments. *Mar Pollut Bull* 28:621–628.

Sawby R, Klowden MJ, Sjogren RD. 1992. Sublethal effects of larval methoprene exposure on adult mosquito longevity. *J Am Mosq Control Assoc* 8(3):290–292.

Sayah F, Fayet C, Idaomar M, Karlinsky A. 1996. Effect of azadirachtin on vitellogenesis of *Labidura riparia* (Insecta: Dermaptera). *Tissue and Cell* 28(6):741–749.

Sayah F, Idaomar M, Soranzo L, Karlinsky A. 1998. Endocrine and neuroendocrine effects of azadirachtin in adult females of the earwig *Labidura riparia*. *Tissue and Cell* 30(1):86–94.

Schober U, Lampert W. 1977. Effects of sublethal concentrations of the herbicide Atrazin™ on growth and reproduction of *Daphnia pulex*. *Bull Environ Contam Toxicol* 17:269–277.

Schoenmakers HJN, van Bohemen ChG, Dieleman SJ. 1981. Effects of oestradiol-17b on the ovaries of the starfish *Asterias rubens*. *Develop Growth Differen* 23:125–135.

Schulte-Oehlmann U. 1997. Fortpflanzungsstörungen bei Süß- und Brackwasserschnecken - Einfluß der Umweltchemikalie Tributylzinn. Berlin, Germany: Wissenschaft und Technik Verlag. 208 p.

Schulte-Oehlmann U, Stroben E, Fioroni P, Oehlmann J. 1996. Beeinträchtigungen der Reproduktionsfähigkeit limnischer Vorderkiemerschnecken durch das Biozid Tributylzinn (TBT). In: Lozan JL, Kausch H, editors. Warnsignale aus Flüssen und Ästuaren: wissenschaftliche Fakten. Berlin, Germany: Parey. p 249–255.

Schultz TW, Freeman SR, Dumont JN. 1980. Uptake, depuration and distribution of selenium in *Daphnia* and its effects on survival and ultrastructure. *Arch Environ Contam Toxicol* 9:23–40.

Shurin JB, Dodson SI. 1997. Sublethal toxic effects of cyanobacteria and nonylphenol on environmental sex determination and development in *Daphnia*. *Environ Toxicol Chem* 16:1269–1276.

Smiley S. 1990. A review of echinoderm oogenesis. *J Electr Microsc Tech* 16:93–114.

Smith BS. 1980. The estuarine mud snail, *Nassarius obsoletus*: Abnormalities in the reproductive system. *J Mollusk Stud* 46:247–256.

Smith BS. 1981. Tributyltin compounds induce male characteristics on female mud snails *Nassarius obsoletus = Ilyanassa obsoleta*. *J Appl Toxicol* 1:141–144.

Smith PJ. 1996. Selective decline in imposex levels in the dogwhelk *Lepsiella scobina* following a ban on the use of TBT antifoulants in New Zealand. *Mar Pollut Bull* 32:362–365.

Smith PJ, McVeagh M. 1991. Widespread organotin pollution in New Zealand coastal waters as indicated by imposex in dogwhelks. *Mar Pollut Bull* 22:409–413.

Snyder MJ. 1998. Identification of a new cytochrome P450 family, CYP45, from the lobster, *Homarus americanus*, and expression following hormone and xenobiotic exposures. *Arch Biochem Biophys* 358:271–276.

Soltani N, Pitoizet N, Soltani-Mazouni N, Delachambre J, Delbecque JP. 1997. Activity of an anti-ecdysteroid compound (KK-42) on ovarian development and ecdysteroid secretion in mealworms. *Med Facul Landbouwkundige en Toepaste Biol Wetenschappen Universiteit Gent* 62(2B):531–537

Spence SK, Bryan GW, Gibbs PE, Masters D, Morris L, Hawkins SJ. 1990. Effects of TBT contamination on *Nucella* populations. *Func Ecol* 4:425–432.

Spooner N, Gibbs PE, Bryan GW, Goad LJ. 1991. The effects of tributyltin (TBT) upon steroid levels in the female dogwhelk, *Nucella lapillus*, leading to the development of imposex. *Mar Environ Res* 32:243–260.

Sprules WG, Casselman JM, Shuter BJ. 1983. Size distributions of pelagic particles in lakes. *Can J Fish Aquat Sci* 40:1761–1765.

Staub GC, Fingerman M. 1984a. Effect of naphthalene on color changes of the sand fiddler crab, *Uca pugilator*. *Comp Biochem Physiol* 77C:7–12.

Staub GC, Fingerman M. 1984b. A mechanism of action for the inhibition of black pigment dispersion in the fiddler crab, *Uca pugilator*, by naphthalene. *Comp Biochem Physiol* 79C:447–453.

Stewart C, de Mora SJ, Jones MRL, Miller MC. 1992. Imposex in New Zealand neogastropods. *Mar Pollut Bull* 24:204–209.

Stroben E, Oehlmann J, Schulte-Oehlmann U, Fioroni P. 1996. Seasonal variations in the genital ducts of normal and imposex-affected prosobranchs and its influence on biomonitoring indices. *Malacol Rev* Suppl. 6 (Molluskan Reproduction):173–184.

Subramanian MA, Varadaraj G. 1993. The effect of industrial effluents on moulting in *Macromia cingulata* (Rambur) (Anisoptera: Corduliidae). *Odonatologica* 22:229–232.

Swevers L, Lambert JGD, De Loof A. 1991. Synthesis and metabolism of vertebrate-type steroids by tissues of insects: a critical evaluation. *Experientia* 47:687–698.

Takagi M, Tsuda Y, Wada Y. 1995. Evaluation of the effective period of a juvenile hormone mimic pyriproxyfen against *Aedes albopictus*: preliminary experiments in the laboratory and field. *Trop Med* 37 (2):87–91.

Takeuchi N. 1987. Endocrine system of annelids. In: Matsumoto A, Ishii S, editors. Atlas of endocrine organs. Vertebrates and invertebrates. Berlin, Germany: Springer Verlag. p 255–269.

Tan KS. 1997. Imposex in three species of *Thais* from Singapore, with additional observations on *T. clavigera* (Kuster) from Japan. *Mar Pollut Bull* 34:577–581.

Tateishi K, Kiuchi M, Takeda S. 1993. New cuticle formation and molt inhibition by RH5849 in the common cutworm *Spodoptera litura* Lepidoptera Noctuidae. *Appl Entomol Zool* 28(2):177–184.

Tattersfield L, Matthiessen P, Campbell P, Grandy N, Länge R, editors. 1997. SETAC-Europe/OECD/EC Expert Workshop on Endocrine Modulators and Wildlife: Assessment and Testing (EMWAT); Veldhoven, The Netherlands; 10–13 April 1997. Brussels, Belgium: Society of Environmental Toxicology and Chemistry (SETAC)–Europe. 126 p.

Thain JE, Waldock MJ, Helm M. 1986. The effect of tributyltin on the reproduction of the oyster *Ostrea edulis*. *Int Council Explor Sea* CM 1986/E:14. 14 p.

Thorndyke MC, Georges D. 1986. Functional aspects of peptide neurohormones in protochordates. *Bull Soc Zool France* 111:57.

(SETAC)Trivedy K, Porcheron P, Lafont R, Magadum SB, Datta RK. 1996. Alterations of ecdysteroid titre and spinning by juvenile hormone analogue R394 in silkworm *Bombyx mori*. *Indian J Exp Biol* 34(6):543–546.

[USEPA] U.S. Environmental Protection Agency. 1998. Condition of the mid-Atlantic Estuaries. Washington DC: Office of Research and Development. Report Number: EPA 600-R-98-147. 50 p.

[USEPA] U.S. Environmental Protection Agency. 1999a. Near laboratory ecological research assessment (NLERA): The Little Miami River Watershed. USEPA web site: http://www.epa.gov/eerd/lmr.htm.

[USEPA] U.S. Environmental Protection Agency. 1999b. Biotic assemblages as indicators of stream condition. USEPA web site: http://www.epa.gov/eerd/biotass.htm.

[USEPA] U.S. Environmental Protection Agency. 1999c. Onboard the USEPA research vessel R/V Lake Guardian. USEPA web site: http://www.epa.gov/ginpo/ guard/ship.html.

[USGS] U.S. Geological Survey. 1999a. The biomonitoring of environmental status and trends program. USGS web page: http://www.best.usgs.gov/overview.html.

[USGS] U.S. Geological Survey. 1999b. Aquatic ecology: Upper Midwest Environmental Science Center. USGS web page: http://www.emtc.nbs.gov/aquatic ecology/aquateco.html.

[USGS] U.S. Geological Survey. 1999c. Aquatic ecology: threats to aquatic ecosystems: Upper Midwest Environmental Science Center. USGS web page: http://www.emtc.nbs.gov/aquatic ecology/threats/threats.html.

Valentine BJ, Gurr GM, Thwaite WG. 1996. Efficacy of the insect growth regulators tebufenozide and fenoxycarb for lepidopteran pest control in apples and their compatibility with biological control for integrated pest management. *Aust J Exp Agric* 36(4):501–506.

Van Brummelen TC, Van Gestel CAM, Verweij RA. 1996. Long-term toxicity of five polycyclic aromatic hydrocarbons for the terrestrial isopods *Oniscus asellus* and *Porcellio scaber*. *Environ Toxicol Chem* 15:1199–1210.

Velloso AHP, Rigitano RLD, Carvalho GAD. 1997. Effects of insect growth regulators on eggs and larvae of *Chrysoperla externa* (Hagen 1861) (Neuroptera: Chrysopidae). *Cienca e Agrotecnologia* 21(3):306–312.

Vethaak D, Jobling S, Waldock M, Bjerregård P, Dickerson R, Giesy J, Grothe D, Karbe L, Munkittrick K, Schlumpf M, Sumpter J. 1997. Approaches for the conduct of field surveys and toxicity identification and evaluation in identifying the hazards of endocrine modulating chemicals in wildlife. In: Tattersfield L, Matthiessen P, Campbell P, Grandy N, Länge R, editors. SETAC-Europe/OECD/EC Expert Workshop on Endocrine Modulators and Wildlife: Assessment and Testing; Veldhoven, The Netherlands; 10–13 April 1997. Brussels, Belgium: Society of Environmental Toxicology and Chemistry (SETAC)–Europe. 126 p.

Villalejo Fuerte M, CeballosVazquez BP, GarciaDominguez F. 1996. Reproductive cycle of *Laevicardium elatum* (Sowerby, 1833) (Bivalvia: Cardiidae) in Bahia Concepcion, Baja California Sur, Mexico. *J Shellfish Res* 15(3):741–745.

Volterra V. 1926. Fluctuations in the abundance of a species considered mathematically. *Nature* 118:558–560.

Voogt PA, Besten den PJ, Kusters GCM, Messing MWJ. 1987. Effects of cadmium and zinc on steroid metabolism and steroid level in the sea star *Asterias rubens* L. *Comp Biochem Physiol* 86C:83–89.

Voogt PA, Lambert JGD, Granneman JCM, Jansen M. 1992. Confirmation of the presence of estradiol-17 in sea star *Asterias rubens* by GC-MS. *Comp Biochem Physiol* 101B:13–16.

Waldock MJ, Thain JE. 1983. Shell thickening in *Crassostrea gigas*: Organotin antifouling or sediment induced? *Mar Pollut Bull* 14:411–415.

Waldock R, Rees HL, Matthiessen P, Pendle MA. 1999. Surveys of the benthic infauna of the Crouch Estuary (UK) in relation to TBT contamination. *J Mar Biol Assoc UK* 79:225–232.

Warwick WF. 1989. Morphological deformities in larvae of *Procladius skuse* (Diptera: Chironomidae) and their biomonitoring potential. *Can J Fish Aquat Sci* 46:1255–1271.

Warwick WF. 1991. Indexing deformities in ligulae and antennae of *Procladius* larvae (Diptera: Chironomidae): Application to contaminant-stressed environments. *Can J Fish Aquat Sci* 48:1151–1156.

Weis JS. 1976. The effect of mercury, cadmium and lead salts on limb regeneration in the fiddler crab, *Uca pugilator*. *Fishery Bull* 74:464–467.

Weis JS. 1978. Interactions of methylmercury, cadmium, and salinity on regeneration in fiddler crabs, *Uca pugilator, Uca pugnax*, and *Uca minax*. *Mar Biol* 49:119–124.

Weis JS. 1980. Effect of zinc on regeneration in the fiddler crab *Uca pugilator* and its interactions with methylmercury and cadmium. *Mar Environ Res* 3:249–255.

Weis JS, Cristini A, Rao KR. 1992. Effects of pollutants on molting and regeneration in crustacea. *Am Zool* 32:495–500.

Weis JS, Gottlieb J, Kwiatkowski J. 1987. Tributyltin retards regeneration and produces deformities in the limbs in the fiddler crab, *Uca pugilator*. *Arch Environ Contam Toxicol* 16:321–326.

Weis JS, Perlmutter J. 1987. Effects of tributyltin on activity and burrowing behavior of the fiddler crab, *Uca pugilator*. *Estuaries* 10:342–346.

West CW, Ankley GT, Nichols JW, Elonen GE, Nessa DE. 1997. Toxicity and bioaccumulation of 2,3,7,8-tetrachlorodibenzo-*p*-dioxin in long-term tests with the freshwater benthic invertebrates *Chironomus tentans* and *Lumbriculus variegatus*. *Environ Toxicol Chem* 16:1287–1294.

Williams DF, Vail KM. 1993. Pharoah ant Hymenoptera Formicidae fenoxycarb baits affect colony development. *J Econ Entomol* 86(4):1136–1143.

Wilson SP, Ahsanullah M, Thompson GB. 1993. Imposex in neogastropods: an indicator of tributyltin contamination in eastern Australia. *Mar Pollut Bull* 26:44–48.

Wing KD. 1988. A nonsteroidal ecdysone agonist: effects on a *Drosophila* cell line. *Science (Washington DC)* 241:467–469.

Woin P, Bronmark C. 1992. Effect of DDT and MCPA (4-chloro-2-methylphenoxyacetic acid) on reproduction of the pond snail, *Lymnaea stagnalis* L. *Bull Environ Contam Toxicol* 48:7–13.

Woolwine AA, Rodriguez L, Ostheimer E, Reagan T. 1997. Effects of insecticides on nontarget insects in sugarcane. *Arthrop Manag Tests* 22:135F.

Wootton AN, Herring CS, Spry JA, Wiseman A, Livingstone DR, Goldfarb PS. 1995. Evidence for the existence of cytochrome P450 gene families (CYP 1A, 3A, 4A, 11A) and modulation of gene expression (CYP 1A) in the mussel *Mytilus* spp. *Mar Env Res* 39:21–26.

Wright JF, Armitage PD, Furse MT, Moss D. 1989. Prediction of invertebrate communities using stream measurements. *Regulated Rivers: Research and Management* 4:147–155.

Yadwad VB. 1989. Effect of endosulfan on glutathione-S-transferase and glutathione content of the premoult field crab, *Paratelphusa hydrodromus*. *Bull Environ Contam Toxicol* 43:597–602.

Zou E, Fingerman M. 1997a. Synthetic estrogenic agents do not interfere with sex differentiation but do inhibit molting of the cladoceran *Daphnia magna*. *Bull Environ Contam Toxicol* 58:596–602.

Zou E, Fingerman M. 1997b. Effects of estrogenic xenobiotics on molting of the water flea, *Daphnia magna*. *Ecotox Environ Saf* 38:281–285.

Chapter 5

Conclusions and Recommendations

Peter L. deFur, Mark Crane, Lisa J. Tattersfield

Introduction

This final chapter presents the major conclusions from individual work groups and brings together conclusions and generic recommendations from the entire Endocrine Disruption in Invertebrates: Endocrinology, Testing, and Assessment Workshop. Detailed conclusions from Chapters 2, 3, and 4 are not repeated here, but are summarized into broad categories. The chapter concludes with recommendations for future research in several areas, including pure endocrinology, new laboratory testing, and fieldwork.

The broad audience envisaged for these recommendations includes policy-makers and regulators who are developing research efforts and implementing management programs on endocrine disruption (ED). This book should provide input for priority setting, which itself depends on understanding the extent of our collective knowledge of endocrinology and ED in invertebrates.

The introduction to the workshop (Chapter 1) summarized current regulatory and investigative efforts on endocrine disrupters, most of which focus on vertebrates. The U.S., Canada, Great Britain, Germany, the European Union, Japan, and other countries have initiated some regulatory activity to address ED. Although, to date, these efforts predominantly have emphasized vertebrates, research and regulatory programs also need to address the issue of ED in invertebrates.

The topic of invertebrate ED is critically important because of the role invertebrates play in all ecosystems. Invertebrates make up most of the diversity of animal life on earth, accounting for all but one of the phyla of multicellular animals. They are essential elements in every food web and are fundamental in the transfer of energy to higher trophic levels. Invertebrates also maintain other critical ecological functions such as plant pollination and detritus processing, as well as serving as food for humans.

The definition of endocrine disrupter (or endocrine disruption) was not seen to be a critical or difficult issue for this workshop. This was true probably because this

Endocrine Disruption in Invertebrates: Endocrinology, Testing, and Assessment. Peter L. deFur et al., editors.

workshop dealt with ED at an integrative level in scientific discussion, rather than in a policy or regulatory context. The workshop addressed questions on whether a compound affects an invertebrate hormone or system, how the effect occurs, the biological consequences at multiple levels, and how to detect such activity. Many of the semantic issues that have arisen in other committees (e.g., Endocrine Disrupter Screening and Testing Advisory Committee [EDSTAC] 1998) or volumes (Kendall et al. 1998) simply did not arise.

However, it was found that certain definitions can be problematic, and that misinterpretations can occur. This was evident when the Endocrinology Work Group integrated their initial outputs with those of the other groups, especially the Field Assessment Work Group. The Endocrinology Work Group noted that disruptive effects caused by a chemical may either 1) mimic effects caused by endocrine dysfunction or 2) alter a biological function that is normally under endocrine control. Both outcomes might be interpreted as ED but may not actually result from endocrine interference. A specific case among the invertebrates is that of the chitin synthetase inhibitors (CSIs) such as diflubenzuron (see Chapter 2). Chitin synthetase inhibitors prevent chitin synthesis and therefore normal cuticle production in insects and crustaceans. In the presence of diflubenzuron, the hormone ecdysone acts normally to initiate and coordinate the molting process, but the new cuticle made only of protein is unable to serve as the exoskeleton, and the animal cannot properly shed its old cuticle. The CSI thus interferes with the process of molting that is known to be under endocrine control by the hormone ecdysone, but there is no hormonal effect. A similar result of disruption of the molting is obtained with the nonsteroidal ecdysone mimic RH5992 (tebufenozide). This compound causes premature initiation of the molt in lepidopterans, followed by production of a new cuticle, but prevents shedding because of interference with the hormonal control of ecdysis behavior. Both of these compounds are called "insect growth regulators" (IGRs), but only RH5992 causes effects by disrupting the endocrine control of molting.

Chitin synthetase inhibitors should not be classified as endocrine disrupters, yet the effects of these compounds still need to be addressed in order to prevent unwanted impacts on nontarget species. The U.S. Department of Agriculture (USDA 1992) also reached that conclusion regarding the use of a CSI to control gypsy moths (*Lymantria dispar*) in forests of the eastern United States. Thus, the identification of a compound as an endocrine disrupter does not automatically limit or define the range of biological consequences or control measures to follow.

General Conclusions

The subject of ED in the invertebrate phyla is a daunting one. The invertebrates include an enormous number and diversity of animals, endocrine systems, and hormones. Additionally, our knowledge of the endocrinology of most invertebrate phyla is poor when compared with insects and crustaceans, and very poor when compared with vertebrates. However, the importance of invertebrates to ecological and human systems provides compelling reasons to discover more about both their basic endocrinology and the potential impact of endocrine-disrupting compounds (EDCs) on them.

Invertebrates occupy every habitat found on earth and exhibit a diverse array of life history patterns that require unique adaptations and physiological processes. As noted in Chapters 1 and 2, invertebrates use a variety of hormones and neurohormones in regulating physiology, growth, reproduction, development, and behavior. Because of this, one of the early conclusions of the workshop was that it was impossible to address all issues and questions related to ED in all invertebrates. The workshop could deal with general issues in the best known groups of invertebrates (insects and crustaceans) and point out areas where further research would most likely be fruitful.

Although invertebrates constitute the vast majority of animals on earth, possible ED in invertebrates has received less attention than vertebrates, despite the fact that all the fundamental elements of cell biology and endocrinology are found in the invertebrates. A major cause of this problem is the lack of information about the basic endocrinology of invertebrates, except in the case of insects and crustaceans. While research indicates that a great variety of biological phenomena among the invertebrates is controlled hormonally, the details of this control remain largely unexamined. The Endocrinology Work Group detailed this shortfall, and both the Laboratory Testing and Field Assessment Work Groups identified this limitation as key in their respective areas.

Invertebrates are a fruitful area for further investigation in all areas concerning ED, not only because of the need to protect invertebrates themselves but also because of their value as sentinels or indicators of effects on a wider range of species and ecosystems. Invertebrates are an imminently practical group for experimental work, screening and testing, and monitoring in the field. Several insect and crustacean species are currently used in this role, and protocols could be adapted relatively easily to allow the detection of endpoints related to ED. Other groups of invertebrates need further research before they can be used for application to ED testing and monitoring.

Although information on endocrinology is limited, there is a wealth of general ecotoxicological assay work with invertebrates, as outlined in Chapter 3. Laboratory testing and field assessment procedures are available for adaptation to evaluate ED in numerous phyla.

The identification and classification of ED in invertebrate phyla raises scientific questions that are not applicable to the vertebrates, and especially not to the mammals. The endocrine system of invertebrates is dominated by neurosecretion, rather than secretions from true ductless glands. Invertebrate neurosecretory systems are more closely integrated with nervous systems, and the stimulation of response elements is distributed more broadly among elements of the neuroendocrine system. Thus, in the invertebrates, separating strict endocrine pathways from other regulatory and response pathways can be difficult.

Many ambient conditions, as well as both natural and xenobiotic compounds, may affect the processes of development or reproduction in invertebrates through numerous alternative mechanisms. Altered reproduction or development, therefore, does not automatically imply that ED is involved (Chapter 3, "Invertebrate full life-cycle as the 'gold standard,'" p 111). An approach using information from a number of biological levels will be needed to elucidate mechanisms of effect. Development of suitable biomarkers and histopathological endpoints is required to demonstrate such mechanisms. Risk assessment, however, should be based on adverse effects irrespective of mechanistic cause.

Only a few known examples of ED in invertebrates have been directly investigated. Two quite different examples were examined in this workshop. One, the effects of tributyltin (TBT), comes from field observations and investigations; the other, insect hormone mimics developed as insecticides, derives from laboratory research. The effects of TBT on marine snails were discovered quite by accident when researchers observed the phenomenon of imposex in field populations of snails (Chapter 4). While it was clear from the outset that imposex was abnormal and likely a disruption of normal developmental processes, it only later became clear that hormonal processes were involved. Once it was acknowledged that this sort of condition could occur, additional occurrences of imposex were identified. In the second example, IGRs, of which insect hormone mimics and antagonists are examples, were intentionally synthesized to alter the growth, development, or metamorphosis of insects. Similarities among hormones or hormone systems in different invertebrate phyla then resulted in effects on some nontarget animals such as aquatic crustaceans and soil insects. Evolutionary conservatism among biological systems explains why biologically active compounds react in diverse biologically sensitive systems such as endocrine and neuroendocrine systems. These examples suggest that it is important to look for other examples of ED in invertebrates in complementary laboratory and field investigations. Field investigations need to focus on those locations and species where known exposures to EDCs have occurred for both terrestrial and aquatic habitats and species.

Endocrinology Work Group conclusions

Invertebrate endocrinology has not been well studied, particularly when compared with mammalian endocrine systems. Insect hormonal systems are the best known of all invertebrate hormone systems. This knowledge was gained because of the need to control insect pests and the subsequent development of target-specific insecticides. There is also a significant amount of information available on the endocrinology of some crustaceans, owing in large part to their economical importance as food. Information on the endocrinology of other invertebrate groups, however, is often limited and fragmentary.

For the majority of invertebrate phyla, hormonal function has been examined in only 1 or 2 species per phylum or group. In these cases, current knowledge of hormones and hormone function may be restricted to a single process or hormone system, e.g., reproduction. Endocrine processes that are relatively conserved among invertebrate groups provide the greatest opportunity to extrapolate study results to other invertebrate species. Currently, generalizations within phyla are difficult or impossible because of insufficient comparative research on different species.

As in the vertebrates, hormonal systems are closely integrated with nervous systems in all invertebrate phyla. Many of the control systems function through neurosecretion; these secretions include steroids, amines, and neuropeptides. More peptide hormones than steroid hormones are known in the invertebrates.

In vitro assays, such as receptor binding assays, must be developed if we are to better understand invertebrate endocrine processes. Such assays are an important part of comprehensive programs to screen chemicals for ED and to evaluate specific mechanisms of ED in research and monitoring efforts.

Vertebrate hormones are found in some of the invertebrate phyla, specifically in echinoderms and other deuterostomes. The steroid hormones in these taxa regulate the same types of functions, such as reproduction and sex differentiation, as they do in vertebrates. Although vertebrate hormones also are found in several additional phyla and are active on animals from still other phyla, it is not clear what functions, if any, the vertebrate hormones normally play in these invertebrates. Some processes are unique to invertebrates, such as limb regeneration, molting, metamorphosis of body form, and diapause. Hormonal systems controlling these processes may offer an opportunity to develop assay systems to screen or test chemicals for endocrine activity specific to invertebrates.

Many of the processes and functions (e.g., molting, metamorphosis, and regeneration) under hormonal control are unique to the invertebrates, and information on these processes could not be derived from evaluations of ED in vertebrates. A database should be developed in order to understand the susceptibility of uniquely invertebrate processes (i.e., those regulated by juvenile hormone [JH], ecdysteroids, or invertebrate neuropeptides) to environmental chemicals.

Individual invertebrate phyla are sufficiently unique in physiological function, development, and endocrine regulation that no single animal or group of animals provides sufficient information to serve as a screen and/or test for a range of phyla.

Testing Work Group conclusions

Toxicity testing alone, whether acute or chronic, cannot be used to identify the specific effects of EDCs. To make such identifications, it is necessary to measure changes in hormone levels, receptor binding, or some other marker or specific indicator of hormonal function. Existing ecotoxicological test methods can be used to assess appropriate biological processes, such as reproduction, development, and growth, in a wide range of animal phyla. As such, these toxicity tests are integrative of potential adverse effects irrespective of mechanisms. Adverse effects on whole-organism endpoints (e.g., development or reproduction) represent a greater level of potential ecological risk than does mechanistic evidence of endocrine activity per se. These tests may well serve as an effective starting point for ED testing, to which may be added more specific endpoints that may be indicative of ED.

Data limitations make it difficult to identify specific life stages or endpoints that should be incorporated into revised testing methods to evaluate potential effects of EDCs. Conducting full life-cycle tests with invertebrates was considered as an alternative, and the work group presented an approach for evaluating the utility of these tests.

Toxicity tests are available using reproduction, metamorphosis, growth, and mortality as endpoints in cnidarians, turbellarians, mollusks, annelids, mites, millipedes, centipedes, echinoderms, tunicates, crustaceans, and insects. Additional endpoints that are most likely to be useful in identifying ED are sex ratios, mating success, offspring viability, egg production (including the resting form) and hatching, molting, time-to-emergence, pigmentation, and morphological abnormalities.

The insects and crustaceans are the only 2 groups for which tests using surrogate species with well-understood endocrine systems have been developed and for which little test method development is needed. Comparative testing is needed to determine the degree to which one species or group may act as surrogates for a larger group. Assays using nematodes, rotifers, annelids, and some crustacean groups will require modest adaptation for use in screening and testing for ED. Moderate to extensive method development is likely needed for assays using echinoderms, cnidarians, mollusks, mites, and tunicates.

The limited knowledge of invertebrate endocrinology presents a barrier to mechanistically based tests for identifying and screening chemicals for ED. Tests on the reproduction, development, and survival of various species need specific elements of the endocrine mechanism integrated into these assays.

Screening and testing assays for ED need to be validated for invertebrates. This validation applies to assays that are either developed de novo for use in invertebrates or modified from vertebrate testing programs. Validation needs to include the use of reference compounds known to have endocrine-disrupting properties in the invertebrates under investigation. Potential reference compounds for this purpose are identified in Chapter 3. Based on results of validation studies, the most useful of these methods should be standardized for routine screening and assessment of EDCs.

Field Assessment Work Group conclusions

There are limited data that examine ED in field populations of invertebrates. The problem has largely been that no systematic investigations have focused on the question of whether EDCs are affecting invertebrate populations in the field.

Examples exist of terrestrial invertebrate field populations affected by insecticides specifically developed to act on endocrine systems. In the marine environment, only imposex in snails caused by the antifouling agent TBT provides a clear example of population-level effects from ED. This example has stood the test of time and scientific scrutiny. The fact that more than 150 species of marine mollusks have been affected by TBT in the field shows the powerful impact of seemingly subtle biochemical alterations.

Existing biological monitoring programs have yet to incorporate endpoints and mechanisms specific to ED into the suite that is measured regularly. Effects on relevant biological processes and functions (e.g., behavior, reproduction, and growth) may be observed in field surveys, but the mechanisms and processes are not known.

The Field Assessment Work Group concluded that the investigation of effects on invertebrate populations in the field is likely to yield further examples of ED, based on the widespread distribution of some key chemicals, commonalties in physiology and toxicology, and limited attention thus far given invertebrate ED. A number of areas were considered fruitful to investigate, such as the existing large-scale monitoring programs in the U.S. (e.g., the Environmental Monitoring and Assessment Program [EMAP] and Mussel Watch). In situ assays have also been used effectively to measure the toxicity of nonendocrine-disrupting compounds and to identify population-level effects. This assessment tool could be used to monitor ED as well.

The Field Assessment Work Group recognized the need to integrate output from both of the other work groups into the design of population-level and field surveys. Specifically, there is an urgent need for biological markers of exposure and of effects. Toxicity testing needs to incorporate clear, mechanistically linked endpoints that can be used in identifying endocrine-disrupting effects, as has been the case for imposex in snails and some of the other abnormalities in vertebrate populations (e.g., vitellogenin in male fish). The scope for development of biochemical and cellular biomarkers for use in monitoring programs will be limited until invertebrate endocrinology is better understood. However, current knowledge suggests that there is considerable opportunity for the development and use of morphological and reproductive indicators of ED, although these are rarely unequivocal about the causative mechanism, implying the need for follow-up diagnostic studies when such effects are observed.

In most cases, our knowledge is probably too incomplete at present to design programs for monitoring endocrine-disrupting exposure and effects in invertebrates. The Field Assessment Work Group therefore developed a series of research requirements that need to be addressed before such programs are launched. However, there is little doubt that carefully targeted monitoring programs are urgently needed because effects in invertebrates are probably widespread but generally undetected (Chapter 4).

The general strategy that should be used to design monitoring programs for ED effects in invertebrates can be divided into techniques for use in

1) general monitoring when impacts are unknown;
2) targeted monitoring to follow up on existing knowledge about possible ED effects;
3) active monitoring with, for example, caged sentinel species; and
4) bioassay-led fractionation procedures and other rapid screening techniques designed to throw light on causality.

Recommendations

- Basic research on invertebrate endocrinology is urgently needed to remedy our ignorance of mechanisms of action, physiological control, and hormone structure and function. This need is most apparent for the annelids, mollusks, nematodes, and echinoderms for which the interpretation of toxicity tests and field surveys is hampered by ignorance of endocrinology.
- Research is needed to test known EDCs with a variety of invertebrate bioassays. In particular, testing is required with specified reference compounds to determine potential effects especially of nonvertebrate-type endocrine disrupters in invertebrates.
- Field assessments and surveys need to be able to identify ED effects, and some specific resources need to be applied to the study of the occurrence and distribution of ED in invertebrates. These efforts will have to be informed by validated biomarkers and other indicators of ED exposure and/or effects in invertebrates.
- Spending is likely to be cost-effective if it supports research on those groups of invertebrates that are well known, abundant, ecologically important, or economically significant. This means that research efforts will add to and complement existing knowledge in the fields of ecology, toxicology, and endocrinology.
- Ecotoxicological approaches in current use should be adapted so that they can detect ED. Numerous field assessments, toxicology assays, and research programs, including model systems, are currently in use. These have the potential to be adapted rather quickly, with less cost, uncertainty, and delay than the development of new systems. They should be the first ones used in ED research and applications.

- The invertebrates potentially offer a wealth of knowledge in understanding comparative and ecological aspects of ED. For these reasons, invertebrate systems should be a high priority for research, screening and testing, and methods development.

References

[EDSTAC] Endocrine Disrupter Screening and Testing Advisory Committee. 1998. Report to the EPA. Washington DC: U.S. Environmental Protection Agency (USEPA) Office of Pollution Prevention and Toxic Substances.

Kendall RJ, Dickerson RL, Giesy JP, Suk WA. 1998. Principles and processes for evaluating endocrine disruption in wildlife. Pensacola FL: Society of Environmental Toxicology and Chemistry (SETAC). 515 p.

[USDA] U.S. Department of Agriculture. 1992. Ecological risk assessment for the national gypsy moth management program. Riverdale MD: USDA APHIS.

Abbreviations

1-MA	1-methyl adenine
20E	20-hydroxyecdysone
4NR	4 natural resources (Canadian working group of 4 governmental agencies)
AChE	acetylcholinesterase
AFNOR	Association Française de Normalization (France)
AhR	aryl hydrocarbon receptor
AJH	anti-juvenile hormone
AKH	adipokinetic hormone
ALP	alkali-labile phosphate
ANOVA	analysis of variance
APHA	American Public Health Association
APTS	6-(o-chlorophenyl) pyridyl-2-carboxyaldehyde pyrrolidinothiosemicarbazone
ASTM	American Society for Testing and Materials
ATP	adenosine triphosphatase
BDPH	black pigment-dispersing hormone
BEST	Biomonitoring of Environmental Status and Trends (USGS)
bHLH-PAS	basic helix-loop-helix Per-ARNT-Sim
Br	brain
BRC	broad complex
BSI	British Standards Institute
CA	corpora allata
cAMP	cyclic adenosine monophosphate
CC	corpora cardiaca
CCAP	crustacean cardioactive peptide

Endocrine Disruption in Invertebrates: Endocrinology, Testing, and Assessment. Peter L. deFur et al., editors.

CEFAS	Centre for Environment, Fisheries and Aquaculture Science (UK)
CEPA	Canadian Environmental Protection Act
cGMP	cyclic guanosine monophosphate
CHH	crustacean hyperglycemic hormone
CHH-B	crustacean hyperglycemic hormone B
CMA	Chemical Manufacturers Association
CNS	central nervous system
COMPREHEND	Community Program of Research on Environmental Hormones and Endocrine Disrupters
COREPA	common reactivity pattern
CSI	chitin synthesis inhibitor
CSIRO	Commonwealth Scientific and Industrial Research Organisation (Australia)
CSL	Central Science Laboratory (UK MAFF)
CTAP	Critical Trends Assessment Program (IL DNR)
DDE	dichlorodiphenyldichloroethylene
DDT	dichlorodiphenyltrichloroethane
DEP	diethyl phthalate
DES	diethylstilbestrol
DETR	Department of Environment, Transport and the Regions (UK)
DGGE	degradation gradient gel electrophoresis
DH	diuretic hormone
DIN	Deutsche Institut für Normung (Germany)
DRPH	distal retinal pigment light-adapting hormone
DTA	direct toxicity assessment
EC	effective concentration
EC DGXII	European Commission DGXII
ECETOC	European Centre for Ecotoxicology and Toxicology of Chemicals
ECH	erythrophore (red chromatophore) concentrating hormone
ECPA	European Crop Protection Association

EcR	ecdysone receptor
EcRE	ecdysone response element
ED	endocrine disruption
EDC	endocrine-disrupting compound
EDIETA	Endocrine Disruption in Invertebrates: Endocrinology, Testing, and Assessment
EDMAR	Endocrine Disruption in the Marine Environment (UK)
EDNH	egg development neurosecretory hormone
EDSTAC	Endocrine Disruptor Screening and Testing Advisory Committee
EDTA	endocrine disrupter testing and assessment
EEM	Environmental Effects Monitoring (Environment Canada)
EG	epitracheal gland
EH	eclosion hormone
ELH	egg-laying hormone
ELISA	enzyme-linked immunosorbent assay
EMAP	Environmental Monitoring and Assessment Program (USEPA)
EMD	ethyl 3-methyl 2-dodecenoate
EMSG	Endocrine Modulators Steering Group
EMWAT	Endocrine Modulators and Wildlife: Assessment and Testing
ESCORT	European Standard Characteristics of Beneficials Regulatory Testing
ETB	ethyl 4-[2-(tert-butylcarbonyloxy) butoxybenzoate
ETH	ecdysis-triggering hormone
Fmev	fluoromevalonolactone
FMRF	phenylalanine-methionine-anginine-phenylalanine
FXR	farnesol X receptor
GLW	glycine-leucine tryptophan
GSS	gonad-stimulating substance

Hoa-CHH-A	hyperglycemic hormone A
Hoa-CHH-B	hyperglycemic hormone B
Hoa-MIH	molt-inhibiting hormone
hsp	heat shock protein
ICPE	International Commission for the Protection of the Elbe
ICPM	International Commission for the Protection of the Meuse
IGR	insect growth regulator
IL DNR	Illinois Department of Natural Resources
IOBC/WPRS	International Organization for Biological and Integrated Control of Noxious Animals and Plants
IRC	International Rhine Commission
ISI	intersex index
ISO	International Organization for Standardization
IUCN	International Union for the Conservation of Nature
JAMP	Joint Assessment and Monitoring Program (OSPAR)
JEA	Japanese Environment Agency
JH	juvenile hormone
JHA	juvenile hormone analog
LAF_{an}	limb autotomy factor, anecdysial
LAF_{pro}	limb autotomy factor, proecdysial
LC	lethal concentration
LOES	National Investigation into Estrogenic Compounds in the Aquatic Environment (Netherlands)
MAFF	Ministry of Agriculture, Fisheries and Food (UK)
MDS	multidimensional scaling
MF	methyl farnesoate
MIH	molt-inhibiting hormone
MIP	molluscan insulin-like peptide

MOIH	mandibular organ-inhibiting hormone
MPF	maturation-promoting factor
MPhT	monophenyltin
NOAA	National Oceanic and Atmospheric Administration (U.S.)
NOEC	no-observed-effects concentration
NPDES	National Pollutant Discharge Elimination System (U.S.)
NRC	National Research Council (U.S.)
NSandT	National Status and Trends Program (NOAA)
OECD	Organization for Economic Cooperation and Development
OEH	ovarian ecdysteroidogenic hormone
OEL	oyster embryo larval
OEPP/EPPO	Organisation européenne et méditerranéenne pour la protection des plantes/European and Mediterranean Plant Protection Organization
ORD	Office of Research and Development (USEPA)
OSPAR	Oslo and Paris Commissions
PBAN	pheromone biosynthesis-activating neuropeptide
PCA	principal components analysis
PCB	polychlorinated biphenyl
PCP	pentachlorophenol
PD/LAH	pigment-dispersing/light-adapting hormone
PDH	pigment-dispersing hormone
PERC	Plymouth Environmental Research Centre (UK)
PN	proctodeal nerve
ppb	parts per billion
ppm	parts per million
PTG	prothoracic gland
PTTH	prothoracicotropic hormone
PVO	perivisceral organ

QUASIMEME	Quality Assurance in Marine Environmental Monitoring (OSPAR)
r	rate of population increase
RAR	retinoic acid receptor
RIKZ	Rijksinstituut voor Kust en Zee (National Institute for Coastal and Marine Management, Netherlands)
RIVPACS	River Invertebrate Prediction and Classification System (UK)
RIZA	Rijksinstituut voor Integraal Zoetwaterbeheer en Afvalwaterbehandeling (Institute for Inland Water Management and Waste Water Treatment, Netherlands)
RPCH	red pigment-concentrating hormone
RPLI	relative penis length index
RPSI	relative penis size index
RXR	retinoid X receptor
SAR	structure–activity relationship
SECOFASE	Sublethal Effects of Chemicals on Fauna Soil Ecosystem (European Union)
SNIFFER	Scotland and Northern Ireland Forum for Environmental Research
STP	sewage treatment plant
TBT	tributyltin
TCDD	2,3,7,8-tetrachlorodibenzo-p-dioxin
TCHT	tricyclohexyltin
TIE	toxicity identification evaluation
TPhT	triphenyltin
TPrT	tripropyltin
TR	thyroid hormone receptor
TSRI	Toxic Substances Research Initiative (Canada)
UK DETR	United Kingdom Department of Environment, Transport and the Regions
UK EA	UK Environment Agency

USEPA	United States Environmental Protection Agency
USFS	U.S. Forest Service
USFWS	U.S. Fish and Wildlife Service
USGS	U.S. Geological Survey
USP	ultraspiracle
VDSI	vas deferens sequence index
VIH	vitellogenesis-inhibiting hormone
VROM	Directorate General of Public Works and Water Management (The Netherlands)
WET	whole effluent toxicity

Index

Endocrine Disruption in Invertebrates: Endocrinology, Testing, and Assessment. Peter L. deFur et al., editors.

H

I

M

N

O

P

Q

R

S

T